Advanced Textbooks in Control and Signal Processing

Series Editors

Professor Michael J. Grimble, Professor of Industrial Systems and Director
Professor Michael A. Johnson, Professor of Control Systems and Deputy Director
Industrial Control Centre, Department of Electronic and Electrical Engineering,
University of Strathclyde, Graham Hills Building, 50 George Street, Glasgow G1 1QE, U.K.

E.F. Camacho and C. Bordons

Model Predictive control

Second Edition

 Springer

Professor E.F. Camacho
Associate Professor C. Bordons

Escuela Superior de Ingenieros, Universidad de Sevilla,
Camino de los Descubrimientos S/N, 41092 Sevilla, Spain

British Library Cataloguing in Publication Data
Camacho, E. F.
    Model predictive control. – 2nd ed. – (Advanced textbooks
    in control and signal processing)
    1.Predictive control
    I.Title II.Bordons, C. (Carlos)
    629.8
    ISBN 1852336943

Library of Congress Cataloging-in-Publication Data
Camacho, E.F.
    Model predictive control / E.F. Camacho and C. Bordons.
        p. cm. – (Advanced textbooks in control and signal processing)
    Includes bibliographical references and index.
    ISBN 978-1-85233-694-3   ISBN 978-0-85729-398-5 (eBook)
    DOI 10.1007/978-0-85729-398-5
    1. Predictive control. I. Bordons, C. (Carlos), 1962-   II. Title. III. Series.
TJ217.6.C34 2003
629.8--dc22   2003070755

Advanced Textbooks in Control and Signal Processing ISSN 1439-2232

ISBN 978-1-85233-694-3   Printed on acid-free paper

© Springer-Verlag London 2007
Originally published by Springer-Verlag London Limited in 2007

MATLAB® and Simulink® are registered trademarks of The MathWorks, Inc., 3 Apple Hill Drive, Natick, MA 01760-2098, U.S.A. http://www.mathworks.com

9 8 7 6 5 4 3 2

springer.com

To Janet
E.F.C.

To Carlos and Marta
C.B.

Series Editors' Foreword

The topics of control engineering and signal processing continue to flourish and develop. In common with general scientific investigation, new ideas, concepts and interpretations emerge quite spontaneously and these are then discussed, used, discarded or subsumed into the prevailing subject paradigm. Sometimes these innovative concepts coalesce into a new sub-discipline within the broad subject tapestry of control and signal processing. This preliminary battle between old and new usually takes place at conferences, through the Internet and in the journals of the discipline. After a little more maturity has been acquired by the new concepts then archival publication as a scientific or engineering monograph may occur.

A new concept in control and signal processing is known to have arrived when sufficient material has evolved for the topic to be taught as a specialised tutorial workshop or as a course to undergraduate, graduate or industrial engineers. *Advanced Textbooks in Control and Signal Processing* are designed as a vehicle for the systematic presentation of course material for both popular and innovative topics in the discipline. It is hoped that prospective authors will welcome the opportunity to publish a structured and systematic presentation of some of the newer emerging control and signal processing technologies in the textbook series.

The books of E.F. Camacho and C. Bordons on model predictive control provide a valuable archive of the development of this particular control technology and theoretical paradigm. In 1995 Professors Camacho and Bordons published their monograph *Model Predictive Control in the Process Industries* (ISBN 3-540-19924-1) in the Springer-Verlag London *Advances in Industrial Control* series. As the title demonstrates, this monograph emphasized the widespread use of the model predictive control technique in the process industries. It was the use of simple models and the ability of the method easily to accommodate system constraints that gave the method its advantage over classical control. Another feature was the optimisation framework of the method where minimising energy and resource usage are widely used concepts in the process industries.

The *Advances in Industrial Control* monograph on model predictive control was a very successful book. Somehow the mix of introductions to Model Predictive Control theory and the empirical practical guidelines developed by the authors was readily absorbed by industrial engineers and academic researchers alike. So that just three years later in 1998, the monograph was revised and reincarnated as a volume in the *Advanced Textbooks in Control and Signal Processing* series simply titled *Model Predictive Control* (ISBN 3-540-76241-8).

Now a further five years has passed and the subject of model predictive control continues to grow along with the stature and experience of the distinguished authors, Professors Camacho and Bordons. This second edition has three new chapters and an up-graded applications chapter. The mix of theory and empirical practical insight remains the same but the new chapters are on nonlinear model predictive control, applications to hybrid systems and on fast implementation methods. The new applications included are for an olive oil mill and a robot problem. Thus the second edition archives recent theoretical developments to nonlinear and hybrid systems whilst the robot application broadens the applications archive to areas other than the process industries.

We welcome this second edition of Professors Camacho and Bordons' *Model Predictive Control*. Engineers and control researchers new to the predictive control methods will find the early chapters of the book provide an excellent historical and tutorial introduction to the techniques. Seasoned researchers will be interested to add to their knowledge an assessment of the potential of predictive control methods for nonlinear and hybrid systems. In five years' time we may even be looking forward to a further update of this very successful control engineering method in a third edition of a fine *Advanced Textbooks in Control and Signal Processing* volume!

M.J. Grimble and M.A. Johnson
Industrial Control Centre
Glasgow, Scotland, U.K.
October 2003

Preface

Model Predictive Control (MPC) has developed considerably over the last two decades, both within the research control community and in industry. This success can be attributed to the fact that Model Predictive Control is, perhaps, the most general way of posing the process control problem in the time domain. Model Predictive Control formulation integrates optimal control, stochastic control, control of processes with dead time, multivariable control and future references when available. Another advantage of Model Predictive Control is that because of the finite control horizon used, constraints and, in general nonlinear processes which are frequently found in industry, can be handled. Although Model Predictive Control has been found to be quite a robust type of control in most reported applications, stability and robustness proofs have been difficult to obtain because of the finite horizon used. This has been a drawback for a wider dissemination of Model Predictive Control in the control research community. Some new and very promising results in this context allow one to think that this control technique will experience greater expansion within this community in the near future. On the other hand, although a number of applications have been reported in both industry and research institutions, Model Predictive Control has not yet reached in industry the popularity that its potential would suggest. One reason for this is that its implementation requires some mathematical complexities which are not a problem in general for the research control community, where mathematical packages are normally fully available, but which represent a drawback for the use of the technique by control engineers in practice.

One of the goals of this text is to contribute to filling the gap between the empirical way in which practitioners tend to use control algorithms and the powerful but sometimes abstractly formulated techniques developed by control researchers. The book focuses on implementation issues for Model Predictive Controllers and intends to present easy ways of implementing them in industry. The book also aims to serve as a guide to implement Model Predictive Control and as a motivation for doing so by showing that using such a powerful control technique does not require complex control algorithms.

The book is aimed mainly at practitioners, although it can be followed by a wide range of readers, as only basic knowledge of control theory and sample data systems is required. A general survey of the field, and guidance in the choice of appropriate implementation techniques, as well as many illustrative examples, are given for practicing engineers and senior undergraduate and graduate students. The book covers most Model Predictive Control algorithms with a special emphasis on Generalized Predictive Control. This control method uses a transfer function model of the process in terms of gains, time constants and dead times which are well understood in industry. This method is middle of the road between industry and academy, where state space-based methods are more attractive because they allow easy analysis of stability and robustness.

We have not tried to give a full description of all MPC algorithms and their properties, although the main ones and their main properties are described. Neither do we claim this technique to be the best choice for the control of every process, although we feel that it has many advantages. Therefore we have not tried to make a comparative study of different Model Predictive Control algorithms amongst themselves and versus other control strategies.

The text gathers recent results and developments that have appeared in the active field of Model Predictive Control since the first edition was published in 1999. The text is composed of material collected from lectures given to senior undergraduate students and articles written by the authors, and is also based on a previous book (*Model Predictive Control in the Process Industry*, Springer, 1995), written by the authors.

This second edition is not just an updated version of the previous book; it also includes exercises and companion software. This MATLAB®-based software package can be freely downloaded from the book's companion web site (http://www.esi.us.es/MPCBOOK) and allows the examples that appear in the book to be reproduced.

E. F. Camacho and C. Bordons
Seville, March 2004

Acknowledgements

The authors would like to thank a number of people who in various ways have made this book possible. Firstly we thank Janet Buckley, who translated part of the book from our native language to English and corrected and polished the style of the rest. Our thanks also to Manuel Berenguel, who implemented and tested the controllers on the solar power plant, and to Juan Gómez-Ortega who implemented the controller on the mobile robots. The contributions of Daniel Limón to the stability analysis in nonlinear MPC, and of Daniel R. Ramírez and Teodoro Alamo for their help on robust MPC and revising the manuscript are deeply appreciated. We also want to thank Winston García-Gabín for the discussion about transmission zeros; José R. Cueli, Fernando Dorado, Sandra Piñón, Miguel Peña, David Muñoz and Alfonso Cepeda who helped us with some of the examples given; Carmen Fernández and Rafael Payseo for their help in developing the software and Julio Normey who helped with the analysis of the effects of predictions on robustness.

Our thanks to our colleagues, especially to Francisco R. Rubio and Javier Aracil, and to many other colleagues and friends from the department. Part of the material included in the book is the result of research work funded by MCYT and EC. We gratefully acknowledge these institutions for their support.

Finally, both authors thank their families for their support, patience and understanding of family time lost during the writing of the book.

Glossary

Notation

$\mathbf{A}(\cdot)$ boldface upper case letters denote polynomial matrices.

$A(\cdot)$ italic and upper case letters denote polynomials.

M italic upper case letters denote real matrices.

$\mathbf{b}$ boldface lower letters indicate real vectors composed of elements at different time instants.

$\mathbf{M}$ boldface upper case letters denote real matrices composed of other matrices or vectors.

Symbols

s complex variable used in Laplace transform

z^{-1} backward shift operator

z forward shift operator and complex variable used in $z-$ transform

$(M)_{ij}$ element ij of matrix M

$(v)_i$ $i^{th}-$ element of vector v

$(\cdot)^T$ transpose of $(\cdot)$

$diag(x_1, \cdots, x_n)$ diagonal matrix with diagonal elements equal to $x_1, \cdots, x_n$

$|(\cdot)|$ absolute value of $(\cdot)$

$\|\mathbf{v}\|_Q^2$ $\mathbf{v}^T Q \mathbf{v}$

$\|\mathbf{v}\|_l$ $l-$ norm of $\mathbf{v}$

$\|\mathbf{v}\|_\infty$ infinity norm of $\mathbf{v}$

$I_{n \times n}$ $(n \times n)$ identity matrix

I identity matrix of appropriate dimensions

$\mathbf{0}_{p \times q}$ $(p \times q)$ matrix with all entries equal to zero

$\mathbf{0}$ matrix of appropriate dimensions with all entries equal to zero

$\mathbf{1}_n$ column vector of dimension n with all entries equal to one

$\mathbf{1}$ column vector with all entries equal to one

$< x, z >$ dot product of vectors x and z

$E[\cdot]$ expectation operator

$\hat{\cdot}$ expected value

$\hat{x}(t + j|t)$ expected value of $x(t + j)$ with available information at instant t

$\delta(P(\cdot))$ degree of polynomial $P(\cdot)$

$\triangle$ $1 - z^{-1}$. increment operator

$det(M)$ determinant of matrix M

$\min\limits_{x \in \mathbf{X}} J(x)$ the minimum value of $J(x)$ for all values of $x \in \mathbf{X}$

Model parameters and variables

m number of input variables

n number of output variables

$u(t)$ input variables at instant t

$y(t)$ output variables at instant t

$x(t)$ state variables at instant t

$e(t)$ discrete white noise with zero mean

d dead time of the process expressed in sampling time units

$\mathbf{A}(z^{-1})$ process left polynomial matrix for the LMFD

$\mathbf{B}(z^{-1})$ process right polynomial matrix for the LMFD

$\mathbf{C}(z^{-1})$ colouring polynomial matrix

Controller parameters and variables

N_1 lower value of prediction horizon

N_2 higher value of prediction horizon

N number of points of prediction horizon ($N = N_2 - N_1$)

N_3 control horizon (N_u)

λ weighting factor for control increments

δ weighting factor for predicted error

u vector of future control increments for the control horizon

y vector of predicted outputs for prediction horizon

f vector of predicted free response

w vector of future references

$\overline{U}$ vector of maximum allowed values of manipulated variables

$\underline{U}$ vector of minimum allowed values of manipulated variables

$\overline{u}$ vector of maximum allowed values of manipulated variable slew rates

$\underline{u}$ vector of minimum allowed values of manipulated variable slew rates

$\overline{y}$ vector of maximum allowed values of output variables

$\underline{y}$ vector of minimum allowed values of output variables

$\tilde{\mathbf{A}}(z^{-1})$ polynomial $\mathbf{A}(z^{-1})$ multiplied by $\triangle$

Acronyms

ANN Artificial Neural Network

CARIMA Controlled Autoregressive Integrated Moving Average

CARMA Controlled Autoregressive Moving Average

CRHPC Constrained Receding Horizon Predictive Control

DMC Dynamic Matrix Control

EHAC Extended Horizon Adaptive Control

EPSAC Extended Prediction Self-Adaptive Control

FIR Finite Impulse Response

FLOP Floating Point Operation

GMV Generalized Minimum Variance

GPC Generalized Predictive Control

HIECON Hierarchical Constraint Control

IDCOM Identification and Command

KKT Karush-Kuhn-Tucker

LCP Linear Complementary Problem

LMFD Left Matrix Fraction Description

LMI Linear Matrix Inequalities

LP Linear Programming

LQ Linear Quadratic

LQG Linear Quadratic Gaussian

LRPC Long Range Predictive Control

LTR Loop Transfer Recovery

MAC Model Algorithmic Control

MILP Mixed Integer Linear Programming

MIMO Multi-Input Multi-Output

MIP Mixed Integer Programming

MIQP Mixed Integer Quadratic Programming

MLD Mixed Logical Dynamical

MPC Model Predictive Control
MPHC Model Predictive Heuristic Control
MUSMAR Multi-Step Multivariable Adaptive Control
MURHAC Multipredictor Receding Horizon Adaptive Control
NLP Nonlinear Programming
OPC Optimum Predictive Control
OUD Outside Unit Disk
PCT Predictive Control Technology
PFC Predictive Functional Control
PID Proportional Integral Derivative
PWA Piecewise Affine
QP Quadratic Programming
RMPCT Robust Model Predictive Control Technology
SCADA Supervisory Control and Data Acquisition
SCAP Adaptive Predictive Control System
SGPC Stable Generalized Predictive Control
SISO Single-Input Single-Output
SMCA Setpoint Multivariable Control Architecture
SQP Sequential Quadratic Programming
UPC Unified Predictive Control

Contents

1

Introduction to Model Predictive Control

Model Predictive Control (MPC) originated in the late seventies and has developed considerably since then. The term Model Predictive Control does not designate a specific control strategy but rather an ample range of control methods which make explicit use of a model of the process to obtain the control signal by minimizing an objective function. These design methods lead to controllers which have practically the same structure and present adequate degrees of freedom. The ideas, appearing in greater or lesser degree in the predictive control family, are basically:

- explicit use of a model to predict the process output at future time instants (horizon);
- calculation of a control sequence minimizing an objective function; and
- receding strategy, so that at each instant the horizon is displaced towards the future, which involves the application of the first control signal of the sequence calculated at each step.

The various MPC algorithms (also called receding horizon Predictive Control or RHPC) only differ amongst themselves in the model used to represent the process and the noises and cost function to be minimized. This type of control is of an open nature, within which many works have been developed and are widely received by the academic world and industry. There are many applications of predictive control successfully in use at the current time, not only in the process industry but also applications to the control of other processes ranging from robots [78] to clinical anaesthesia [127]. Applications in the cement industry, drying towers, and robot arms are described in [54], whilst developments for distillation columns, PVC plants, steam generators, or servos are presented in [179] and [182]. The good performance of these applications shows the capacity of the MPC to achieve highly efficient control systems able to operate during long periods of time with hardly any intervention.

MPC presents a series of advantages over other methods, amongst which the following stand out:

- It is particularly attractive to staff with only a limited knowledge of control because the concepts are very intuitive and at the same time the tuning is relatively easy.
- It can be used to control a great variety of processes, from those with relatively simple dynamics to more complex ones, including systems with long delay times or nonminimum phase or unstable ones.
- The multivariable case can easily be dealt with.
- It intrinsically has compensation for dead times.
- It introduces feed forward control in a natural way to compensate for measurable disturbances.
- The resulting controller is an easy-to-implement control law.
- Its extension to the treatment of constraints is conceptually simple, and these can be systematically included during the design process.
- It is very useful when future references (robotics or batch processes) are known.
- It is a totally open methodology based on certain basic principles which allows for future extensions.

As is logical, however, it also has its drawbacks. One of these is that although the resulting control law is easy to implement and requires little computation, its derivation is more complex than that of the classical PID controllers. If the process dynamic does not change, the derivation of the controller can be done beforehand, but in the adaptive control case all the computation has to be carried out at every sampling time. When constraints are considered, the amount of computation required is even higher. Although this, with the computing power available today, is not an essential problem, one should bear in mind that many industrial process control computers are not at their best regarding their computing power and, above all, that most of the available time at the process computer normally has to be used for purposes other than the control algorithm itself (communication, dialogues with the operators, alarms, recording, etc.). Even so, the greatest drawback is the need for an appropriate model of the process to be available. The design algorithm is based on prior knowledge of the model and is independent of it, but it is obvious that the benefits obtained will be affected by the discrepancies existing between the real process and the model used.

In practice, MPC has proved to be a reasonable strategy for industrial control, in spite of the original lack of theoretical results at some crucial points such as stability and robustness.

1.1 MPC Strategy

The methodology of all the controllers belonging to the MPC family is characterized by the following strategy, represented in Figure 1.1:

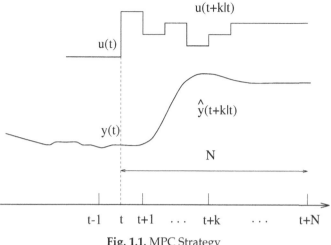

Fig. 1.1. MPC Strategy

1. The future outputs for a determined horizon N, called the prediction horizon, are predicted at each instant t using the process model. These predicted outputs $y(t + k \mid t)$ [1] for $k = 1 \ldots N$ depend on the known values up to instant t (past inputs and outputs) and on the future control signals $u(t+k \mid t), k = 0 \ldots N-1$, which are those to be sent to the system and calculated.

2. The set of future control signals is calculated by optimizing a determined criterion to keep the process as close as possible to the reference trajectory $w(t + k)$ (which can be the setpoint itself or a close approximation of it). This criterion usually takes the form of a quadratic function of the errors between the predicted output signal and the predicted reference trajectory. The control effort is included in the objective function in most cases. An explicit solution can be obtained if the criterion is quadratic, the model is linear, and there are no constraints; otherwise an iterative optimization method has to be used. Some assumptions about the structure of the future control law are also made in some cases, such as that it will be constant from a given instant.

3. The control signal $u(t \mid t)$ is sent to the process whilst the next control signals calculated are rejected, because at the next sampling instant $y(t + 1)$ is already known and step 1 is repeated with this new value and all the sequences are brought up to date. Thus the $u(t + 1 \mid t + 1)$ is calculated (which in principle will be different from the $u(t + 1 \mid t)$ because of the new information available) using the receding horizon concept.

---

[1] The notation indicates the value of the variable at the instant $t + k$ calculated at instant t.

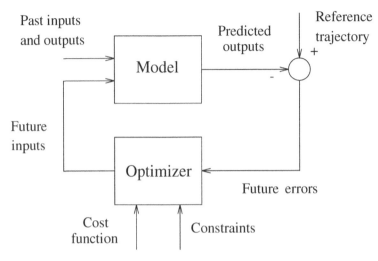

Fig. 1.2. Basic structure of MPC

In order to implement this strategy, the basic structure shown in Figure 1.2 is used. A model is used to predict the future plant outputs, based on past and current values and on the proposed optimal future control actions. These actions are calculated by the optimizer taking into account the cost function (where the future tracking error is considered) as well as the constraints.

The process model plays, in consequence, a decisive role in the controller. The chosen model must be able to capture the process dynamics to precisely predict the future outputs and be simple to implement and understand. As MPC is not a unique technique but rather a set of different methodologies, there are many types of models used in various formulations.

One of the most popular in industry is the Truncated Impulse Response Model, which is very simple to obtain as it only needs the measurement of the output when the process is excited with an impulse input. It is widely accepted in industrial practice because it is very intuitive and can also be used for multivariable processes, although its main drawbacks are the large number of parameters needed and that only open-loop stable processes can be described this way. Closely related to this kind of model is the Step Response Model, obtained when the input is a step.

The State Space Model is, perhaps, more widespread in the academic research community as the derivation of the controller is very simple even for the multivariable case. The state space description allows for an easier expression of stability and robustness criteria. The Transfer Function Model is also used in the academic research community and although the derivation of the controller is more difficult, it requires fewer parameters. Dead time, so frequent in industry, can be handled easier than with other descriptions. This type of model is better understood in industry than state space models, as some of the concepts used in the transfer function context such as dead time, gains, and time constants are usually employed in industry. This description is somewhat middle of the road between academy and industry, and that is why it has been chosen in this text as the main model description.

The optimizer is another fundamental part of the strategy as it provides the control actions. If the cost function is quadratic, its minimum can be obtained as an explicit function (linear) of past inputs and outputs and the future reference trajectory. In the presence of inequality constraints the solution has to be obtained by more computationally taxing numerical algorithms. The size of the optimization problems depends on the number of variables and the prediction horizons used and usually turns out to be a relatively modest optimization problem which does not require solving sophisticated computer codes. However, the amount of time needed for the constrained and robust cases can be various orders of magnitude higher than that needed for the unconstrained case and the bandwidth of the process to which constrained MPC can be applied is considerably reduced.

Notice that the MPC strategy is very similar to the control strategy used in driving a car. The driver knows the desired reference trajectory for a finite control horizon and by taking into account the car characteristics (mental model of the car) decides which control actions (accelerator, brakes, steering) to take to follow the desired trajectory. Only the first control actions are taken at each instant, and the procedure is repeated for the next control decision in a receding horizon fashion. Notice that when using classical control schemes, such as PIDs, the control actions are taken based on past errors. If the car driving analogy is extended, as has been done by one of the commercial MPC vendors (SCAP) [134] in its publicity, the PID way of driving a car would be equivalent to driving the car just using the mirror as shown in Figure 1.3. This analogy is not totally fair with PIDs, because more information (the reference trajectory) is used by MPC. Notice that if a future point in the desired reference trajectory is used as the setpoint for the PID, the differences between both control strategies would not seem so abysmal.

1.2 Historical Perspective

From the end of the 1970s various articles appeared showing an incipient interest in MPC in the industry, principally the Richalet *et al.* publications [181][182] presenting Model Predictive Heuristic Control (MPHC) (later known as Model Algorithmic Control (MAC)) and those of Cutler and Ramakter [62] with Dynamic Matrix Control (DMC). A dynamic process model is explicitly used in both algorithms (impulse response in the first and step response in the second) to predict the effect of the future control actions at

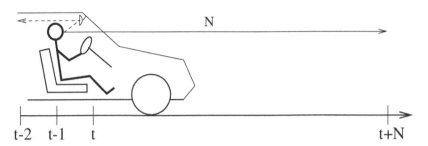

Fig. 1.3. MPC analogy

the output; these are determined by minimizing the predicted error subject to operational restrictions. The optimization is repeated at each sampling period with up-to-date information about the process. These formulations were heuristic and algorithmic and took advantage of the increasing potential of digital computers at the time.

These controllers were closely related to the minimum time optimal control problem and to linear programming [211]. The receding horizon principle, one of the central ideas of MPC, was proposed by Propoi as long ago as 1963 [169], within the frame of "open-loop optimal feedback", which was extensively dealt with in the seventies.

MPC quickly became popular, particularly in chemical process industries, due to the simplicity of the algorithm and to the use of the impulse or step response model which, although possessing many more parameters than the formulations in the state space or input-output domain, is usually preferred as being more intuitive and requiring less *a priori* information for its identification. A complete report of its application in the petrochemical sector during the eighties can be found in [73]. The majority of these applications were carried out on multivariable systems including constraints. In spite of this success, these formulations lacked formal theories providing stability and robustness results; in fact, the finite-horizon case seemed too difficult to analyze except in very specific cases.

Another line of work arose independently around adaptive control ideas, developing strategies essentially for monovariable processes formulated with input-output models. Peterka's Predictor-Based Self-Tuning Control [165] can be included here. It was designed to minimize, for the most recent predicted values, the expected value of a quadratic criterion on a given control horizon (finite or asymptotically infinite) or Ydstie's Extended Horizon Adaptive Control (EHAC) [207]. This method tries to keep the future output (calculated by a Diophantine equation) close to the reference at a period of time after the process delay and permits different strategies. Extended Prediction Self Adaptive Control (EPSAC) by De Keyser and Van Cuawenberghe

[107] proposes a constant control signal starting from the present moment[2] while using a suboptimal predictor instead of solving a Diophantine equation. Generalized Predictive Control (GPC) developed by Clarke *et al.* in 1987 [58] also appears within this context. This uses ideas from Generalized Minimum Variance (GMV) [56] and is perhaps one of the most popular methods at the moment and will be the object of detailed study in the following chapters. There are numerous predictive controller formulations based on the same common ideas, amongst which can be included: Multistep Multivariable Adaptive Control (MUSMAR) [83], Multipredictor Receding Horizon Adaptive Control (MURHAC) [121], Predictive Functional Control (PFC) [180], or Unified Predictive Control (UPC) [194].

MPC has also been formulated in the state space context [140]. This not only allows for the use of well known theorems of the state space theory, but also facilitates their generalization to more complex cases such as multivariable processes, nonlinear processes and systems with stochastic disturbances and noise in the measured variables. By extending the step response model and using known state estimation techniques, processes with integrators can also be treated. The state estimation techniques arising from stochastic optimal control can be used for predictions without adding complications [116]. This perspective leads to simple tuning rules for stability and robustness: the MPC controller can be interpreted as being a compensator based on a state observer and its stability, performance and robustness are determined by the poles of the observer (which can be directly fixed by adjustable parameters) and the poles of the regulator (determined by the horizons, weights, etc.). An analysis of the inherent characteristics of all the MPC algorithms (especially of the GPC) from the point of view of the gaussian quadratic optimal linear theory can be found in the book by Bitmead *et al.* [27].

Although the first works on GPC proved some specific stability theorems using state space relationships and studied the influence of filter polynomials on robustness improvement, the original lack of general stability results for finite horizon receding controllers was recognized as a drawback. Because of this, a fresh line of work on new predictive control methods with guaranteed stability appeared in the nineties. Two methods, CRHPC (Clarke and Scattolini [61]) and SIORHC (Mosca *et al.* [143]), were independently developed and proven to be stable by imposing endpoint equality constraints on the output after a finite horizon. Bearing in mind the same objective, Kouvaritakis *et al.* [112] presented *stable* GPC, a formulation which guarantees closed-loop stability by stabilizing the process prior to the minimization of the objective function.

Very impressive results were obtained for what seemed to be a problem too difficult to tackle, that of the stability of constrained receding horizon controllers [177],[185], [213]. Even when the optimization algorithm finds a solution, this does not guarantee closed-loop stability, that is, optimality does

---

[2] Note that due to the receding horizon the real signal need not be kept constant.

not imply stability. The use of terminal penalties and/or constraints, Lyapunov functions, or invariant sets has given rise to a family of techniques that guarantee the stability of the system. This problem has been tackled from different points of view and several contributions have appeared in recent years, always analyzing the regulator problem (drive the state to zero) in a state space framework. The main proposed formulations with guaranteed stability are summarized in [137] where general sufficient conditions to design a stabilizing constrained MPC are presented.

Results have also been obtained using robust control design approaches in the MPC context. The key idea is to take into account uncertainties about the process in an explicit manner and to design MPC to optimize the objective function for the worst situation of the uncertainties. These challenging results allow one to think that MPC will experience an even greater dissemination in both the academic world and the control practitioner community. In this context, one of the leading manufacturers of distributed control equipment, Honeywell, incorporated Robust Multivariable Predictive Control (RMPC^{TM}) into its TDC 3000 control system and announced it contains several breakthroughs in technology.

Model Predictive Control is considered to be a mature technique for linear and rather slow systems like the ones usually encountered in the process industry. More complex systems, such as nonlinear, hybrid, or very fast processes, were considered beyond the realm of MPC. During the last few years some impressive results have been produced in these fields. Bemporad *et al.* [23] have shown that a constrained MPC results in a piecewise affine controller that can be implemented with little computational burden. Applications of MPC to nonlinear and to hybrid processes have also appeared in the literature.

1.3 Industrial Technology

This section is focused on those predictive control technologies that have great impact on the industrial world and are commercially available, dealing with several topics such as a short application summary and the limitations of the existing technology.

Although there are companies that make use of technology they have developed inhouse that is not offered externally, the following can be considered representative of the current state of the art of Model Predictive Control technology. Their product names and acronyms are:

- AspenTech: Dynamic Matrix Control (DMC)
- Adersa: Identification and Command (IDCOM) , Hierarchical Constraint Control (HIECON), and Predictive Functional Control (PFC)
- Honeywell Profimatics: Robust Model Predictive Control Technology (RMPCT) and Predictive Control Technology (PCT)

- Setpoint Inc.: Setpoint Multivariable Control Architecture (SMCA) and IDCOM-M (multivariable)
- Treiber Controls: Optimum Predictive Control (OPC)
- ABB: 3dMPC
- Pavillion Technologies Inc.: Process Perfecter
- Simulation Sciences: Connoisseur

Some of these algorithms will be treated in more detail in following chapters. Notice that each product is not the algorithm alone, but is accompanied by additional packages, usually identification or plant test packages.

There are thousands of applications of MPC in industry. The majority of applications (see surveys by Qin and Badgwell [170] [171]) are in the area of refining, one of the original application fields of MPC, where it has a solid background. An important number of applications can be found in petrochemicals and chemicals. Significant growth areas include pulp and paper, food processing, aerospace, and automotive industries. Other areas such as gas, utility, furnaces, or mining and metallurgy also appear in the report. Some applications in the cement industry or pulp factories can be found in [134]. Although MPC technology has not yet penetrated deeply into areas where process nonlinearities are strong and frequent changes in operation conditions occur, the number of nonlinear MPC applications is clearly increasing.

The existing industrial MPC technology has several limitations, as pointed out by Muske and Rawlings [147]. The most outstanding ones are:

- Overparameterized models: most commercial products use the step or impulse response model of the plant, that are known to be overparameterized. For instance, a first-order process can be described by a transfer function model using only three parameters (gain, time constant, and dead time), whilst a step response model will require more than 30 coefficients to describe the same dynamics. Besides, these models are not valid for unstable processes. These problems can be overcome by using an auto-regressive parametric model.
- Tuning: the tuning procedure is not clearly defined since the trade-off between tuning parameters and closed loop behaviour is generally not very clear. Tuning in the presence of constraints may be even more difficult, and even for the nominal case, is not easy to guarantee closed-loop stability; that is why so much effort must be spent on prior simulations. The feasibility of the problem is one of most challenging topics of MPC nowadays, and it will be treated in detail in Chapter 7.
- Suboptimality of the dynamic optimization: several packages provide suboptimal solutions to the minimization of the cost function in order to speed up the solution time. It can be accepted in high-speed applications (tracking systems) where solving the problem at every sampling time may not be feasible, but it is difficult to justify for process control ap-

plications unless it can be shown that the suboptimal solution is always very nearly optimal.

- Model uncertainty: although model identification packages provide estimates of model uncertainty, only one product (RMPCT) uses this information in the control design. All other controllers can be detuned to improve robustness, although the relation between performance and robustness is not very clear.
- Constant disturbance assumption: although perhaps the most reasonable assumption is to consider that the output disturbance will remain constant in the future, better feedback would be possible if the distribution of the disturbance could be characterized more carefully.
- Analysis: a systematic analysis of stability and robustness properties of MPC is not possible in its original finite horizon formulation. The control law is in general time-varying and cannot be represented in the standard closed-loop form, especially in the constrained case. Furthermore, the results obtained by the academic research community about stability and robustness can only be applied to very small (in terms of state space and control horizons) processes.

The technology is continually evolving and the next generation will have to face new challenges in open topics such as model identification, unmeasured disturbance estimation and prediction, systematic treatment of modelling error, and uncertainty or the open field of nonlinear model predictive control.

1.4 Outline of the Chapters

The book aims to study the most important issues of MPC regarding its application to process control. To achieve this objective, it is organized as follows.

Chapter 2 describes the main elements that appear in any MPC formulation and reviews the best-known methods. A brief review of the most outstanding methods is made. Chapter 3 focuses on commercial Model Predictive controllers. Because of its popularity, Generalized Predictive Control (GPC) is treated in greater detail in Chapter 4. Two related methods which have shown good stability properties (CRHPC and SGPC) are also described.

Chapter 5 shows how GPC can easily be applied to a wide variety of plants in the process industry using some Ziegler-Nichols types of tuning rules. By using these, the implementation of GPC is considerably simplified, and the computational burden and time that the implementation of GPC may bear, especially for the adaptive case, are avoided. The rules have been obtained for plants that can be modelled by the reaction curve method and plants having an integrating term, that is, most of the plants in the process industry. The robustness of the method is studied. In order to do this, both structured and unstructured uncertainties are considered. The closed loop is studied, defining the uncertainty limits that preserve stability of the real process when it is being controlled by a GPC designed for the nominal model.

The way of implementing MPC on multivariable processes, which can often be found in industry, is treated in Chapter 6. Some examples dealing with implementation topics such as dead times are provided.

Although in practice all processes are subject to constraints, most of the existing controllers do not consider them explicitly. One of the advantages of MPC is that constraints can be dealt with explicitly. Chapter 7 is dedicated to showing how MPC can be implemented on processes subject to constraints. Although constraints play an important role in industrial practice, they are not considered in many formulations. The minimization of the objective function can no longer be done explicitly and a numerical solution is necessary. Existing numerical techniques are revised and some examples and algorithms are included.

Chapter 8 deals with the robust implementation of MPC. Although a robustness analysis is performed in Chapter 5 for GPC of processes that can be described by the reaction curve method, this chapter indicates how MPC can be implemented by explicitly taking into account model inaccuracies or uncertainties. The controller is designed to minimize the objective function for the worst situation.

This new edition includes three new chapters addressing the latest developments in the field. Chapter 9 is dedicated to describing some techniques that can be used to apply MPC to nonlinear processes. The way in which MPC can be applied to control hybrid processes (i.e., processes containing continuous and discrete parts) is described in Chapter 10. Chapter 11 describes strategies to implement fast MPC of constrained processes and uncertain processes which can be used even in the case of processes requiring small sampling times.

This book could not end without presenting some practical implementations of MPC. Chapter 12 presents results obtained in different plants. First, a real application of GPC to a solar power plant is presented. This process presents changing perturbations that make it suitable for an adaptive control policy. The same controller is developed on a commercial distributed control system and applied to a pilot plant. Operating results show that a technique that has sometimes been rejected by practitioners because of its complexity can easily be programmed in any standard control system, obtaining better results and being as easy to use as traditional PID controllers. The application of a GPC to a diffusion process of a sugar factory is presented as well as a predictive strategy in the process of olive oil extraction in a mill. Finally a nonlinear controller is implemented for path tracking of a mobile robot.

2

Model Predictive Controllers

This chapter describes the elements that are common to all Model Predictive controllers, showing the various alternatives used in the different implementations. Some of the most popular methods will later be reviewed to demonstrate their most outstanding characteristics.

2.1 MPC Elements

All the MPC algorithms possess common elements, and different options can be chosen for each element giving rise to different algorithms. These elements are:

- prediction model,
- objective function and
- obtaining the control law.

2.1.1 Prediction Model

The model is the cornerstone of MPC; a complete design should include the necessary mechanisms for obtaining the best possible model, which should be complete enough to fully capture the process dynamics and allow the predictions to be calculated, and at the same time to be intuitive and permit theoretic analysis. The use of the process model is determined by the necessity to calculate the predicted output at future instants $\hat{y}(t + k \mid t)$. The different strategies of MPC can use various models to represent the relationship between the outputs and the measurable inputs, some of which are manipulated variables and others can be considered to be measurable disturbances which can be compensated for by feedforward action. A disturbance model can also be taken into account to describe the behaviour not reflected by the process model, including the effect of nonmeasurable inputs, noise and model errors. The model can be separated into two parts: the actual process model and the disturbances model. Both parts are needed for the prediction.

Process Model

Practically every possible form of modelling a process appears in a given MPC formulation, the following being the most commonly used:

- Impulse response. Also known as weighting sequence or convolution model, it appears in MAC and as a special case in GPC and EPSAC. The output is related to the input by the equation

$$y(t) = \sum_{i=1}^{\infty} h_i u(t - i)$$

where h_i is the sampled output when the process is excited by a unitary impulse (see Figure 2.1a). This sum is truncated and only N values are considered (thus only stable processes without integrators can be represented), having

$$y(t) = \sum_{i=1}^{N} h_i u(t - i) = H(z^{-1})u(t) \tag{2.1}$$

where $H(z^{-1}) = h_1 z^{-1} + h_2 z^{-2} + \cdots + h_N z^{-N}$, where z^{-1} is the backward shift operator. Another inconvenience of this method is the large number of parameters necessary, as N is usually a high value (on the order of 40 to 50). The prediction will be given by:

$$\hat{y}(t + k \mid t) = \sum_{i=1}^{N} h_i u(t + k - i \mid t) = H(z^{-1})u(t + k \mid t)$$

This method is widely accepted in industrial practice because it is very intuitive and clearly reflects the influence of each manipulated variable on a determined output. Note that if the process is multivariable, the different outputs will reflect the effect of the m inputs in the following way:

$$y_j(t) = \sum_{k=1}^{m} \sum_{i=1}^{N} h_i^{kj} u^k(t - i)$$

One great advantage of this method is that no prior information about the process is needed, so that the identification process is simplified, and at the same time it allows complex dynamics such as nonminimum phase or delays to be described easily.

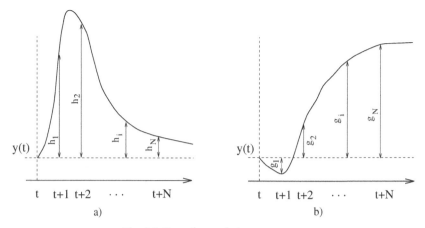

Fig. 2.1. Impulse and step response

- Step response. Used by DMC and its variants, this is very similar to impulse response except that the input signal is a step. For stable systems, the truncated response is given by:

$$y(t) = y_0 + \sum_{i=1}^{N} g_i \triangle u(t-i) = y_0 + G(z^{-1})(1-z^{-1})u(t) \qquad (2.2)$$

where g_i are the sampled output values for the step input and $\triangle u(t) = u(t) - u(t-1)$ as shown in Figure 2.1b. The value of y_0 can be taken to be 0 without loss of generality, so that the predictor will be:

$$\hat{y}(t+k \mid t) = \sum_{i=1}^{N} g_i \triangle u(t+k-i \mid t)$$

As an impulse can be considered as the difference between two steps with a lag of one sampling period, it can be written for a linear system that:

$$h_i = g_i - g_{i-1} \qquad\qquad g_i = \sum_{j=1}^{i} h_j$$

This method has the same advantages and disadvantages as the impulse response method.

- Transfer function. Used by GPC, UPC, EPSAC, EHAC, MUSMAR or MURHAC (amongst others), this uses the concept of transfer function $G = B/A$ so that the output is given by:

$$A(z^{-1})y(t) = B(z^{-1})u(t)$$

with

$$A(z^{-1}) = 1 + a_1 z^{-1} + a_2 z^{-2} + \cdots + a_{na} z^{-na}$$
$$B(z^{-1}) = b_1 z^{-1} + b_2 z^{-2} + \cdots + b_{nb} z^{-nb}$$

Thus the prediction is given by

$$\hat{y}(t + k \mid t) = \frac{B(z^{-1})}{A(z^{-1})} u(t + k \mid t)$$

This representation is also valid for unstable processes and has the advantage that it only needs a few parameters, although *a priori* knowledge of the process is fundamental, especially that of the order of the A and B polynomials.

- State space. Used in PFC, for example, it has the following representation:

$$x(t) = Ax(t - 1) + Bu(t - 1)$$
$$y(t) = Cx(t)$$

where x is the state and A, B and C are the matrices of the system, input and output respectively. The prediction for this model is given by [11]

$$\hat{y}(t + k \mid t) = C\hat{x}(t + k \mid t) = C[A^k x(t) + \sum_{i=1}^{k} A^{i-1} Bu(t + k - i \mid t)]$$

It has the advantage that it can be used for multivariable processes in a straightforward manner. The control law is simply the feedback of a linear combination of the state vector, although sometimes the state basis chosen has no physical meaning. The calculations may be complicated with the additional necessity of including an observer if the states are not accessible.

- Others. Nonlinear models can also be used to represent the process, but they cause the optimization problem to be more complicated. Neural nets [198] and fuzzy logic [192] are other forms of representation used in some applications.

Disturbances Model

The choice of the model used to represent the disturbances is as important as the choice of the process model. A model widely used is the Controlled Auto-Regressive and Integrated Moving Average (CARIMA) in which the disturbances, that is, the differences between the measured output and the output calculated by the model, are given by

$$n(t) = \frac{C(z^{-1})e(t)}{D(z^{-1})}$$

where the polynomial $D(z^{-1})$ explicitly includes the integrator $\triangle = 1 - z^{-1}$, $e(t)$ is a white noise of zero mean and the polynomial C is normally considered to equal one. This model is considered appropriate for two types of disturbances, random changes occurring at random instants (for example, changes in the quality of the material) and "Brownian motion" and it is used directly in GPC, EPSAC, EHAC UPC and with slight variations in other methods. Note that by including an integrator an offset-free steady-state control is achieved.

Using the Diophantine equation

$$1 = E_k(z^{-1})D(z^{-1}) + z^{-k}F_k(z^{-1}) \tag{2.3}$$

one has

$$n(t) = E_k(z^{-1})e(t) + z^{-k}\frac{F_k(z^{-1})}{D(z^{-1})}e(t) \qquad n(t+k) = E_k(z^{-1})e(t+k) + F_k(z^{-1})n(t)$$

and the prediction will be

$$\hat{n}(t + k \mid t) = F_k(z^{-1})n(t) \tag{2.4}$$

If equation (2.4) is combined with a transfer function model (like the one used in GPC), making $D(z^{-1}) = A(z^{-1})(1 - z^{-1})$, the output prediction can be obtained:

$$\hat{y}(t + k \mid t) = \frac{B(z^{-1})}{A(z^{-1})}u(t + k \mid t) + F_k(z^{-1})(y(t) - \frac{B(z^{-1})}{A(z^{-1})}u(t))$$

$$\hat{y}(t + k \mid t) = F_k(z^{-1})y(t) + \frac{B(z^{-1})}{A(z^{-1})}(1 - z^{-k}F_k(z^{-1}))u(t + k \mid t)$$

and using (2.3) the following expression is obtained for the k-step ahead predictor

$$\hat{y}(t + k \mid t) = F_k(z^{-1})y(t) + E_k(z^{-1})B(z^{-1}) \triangle u(t + k \mid t)$$

In the particular case of ARIMA the constant disturbance

$$n(t) = \frac{e(t)}{1 - z^{-1}}$$

can be included whose best predictions will be $\hat{n}(t + k \mid t) = n(t)$. This disturbance model, together with the step response model is the one used on DMC and related methods.

An extension of this is the drift disturbance used in PFC

$$n(t) = \frac{e(t)}{(1 - z^{-1})^2} \qquad .$$

with $\hat{n}(t + k \mid t) = n(t) + (n(t) - n(t - 1))k$ being the optimum prediction. Other polynomial models of high order can likewise be used.

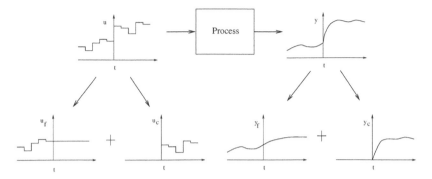

Fig. 2.2. Free and forced responses

Free and Forced Response

A typical characteristic of most MPC is the use of *free* and *forced* response concepts. The idea is to express the control sequence as the addition of the two signals:

$$u(t) = u_f(t) + u_c(t)$$

The signal $u_f(t)$ corresponds to the past inputs and is kept constant and equal to the last value of the manipulated variable in future time instants. That is,

$$u_f(t - j) = u(t - j) \text{ for } j = 1, 2, \cdots$$
$$u_f(t + j) = u(t - 1) \text{ for } j = 0, 1, 2, \cdots$$

The signal $u_c(t)$ is made equal to zero in the past and equal to the next control moves in the future. That is,

$$u_c(t - j) = 0 \quad \text{for } j = 1, 2, \cdots$$
$$u_c(t + j) = u(t + j) - u(t - 1) \quad \text{for } j = 0, 1, 2, \cdots$$

The prediction of the output sequence is separated into two parts, as can be seen in Figure 2.2. One of them $(y_f(t))$, the *free* response, corresponds to the prediction of the output when the process manipulated variable is made equal to $u_f(t)$, and the other, the *forced* response $(y_c(t))$, corresponds to the prediction of the process output when the control sequence is made equal to $u_c(t)$. The *free* response corresponds to the evolution of the process due to its present state, while the forced response is due to future control moves.

2.1.2 Objective Function

The various MPC algorithms propose different cost functions for obtaining the control law. The general aim is that the future output (y) on the considered horizon should follow a determined reference signal (w) and, at the same time, the control effort ($\triangle u$) necessary for doing so should be penalized. The general expression for such an objective function will be:

$$J(N_1, N_2, N_u) = \sum_{j=N_1}^{N_2} \delta(j)[\hat{y}(t + j \mid t) - w(t + j)]^2 + \sum_{j=1}^{N_u} \lambda(j)[\triangle u(t + j - 1)]^2$$

$$(2.5)$$

In some methods the second term, which considers the control effort, is not taken into account, whilst in others (UPC) the values of the control signal (not its increments) also appear directly. In the cost function it is possible to consider:

- parameters: N_1 and N_2 are the minimum and maximum prediction horizons and N_u is the control horizon, which does not necessarily have to coincide with the maximum horizon, as will be seen later. The meaning of N_1 and N_2 is rather intuitive. They mark the limits of the instants in which it is desirable for the output to follow the reference. Thus, if a high value of N_1 is taken, it is because it is of no importance if there are errors in the first instants. This will originate a smooth response of the process. Note that in processes with dead time d there is no reason for N_1 to be less than d because the output will not begin to evolve until instant $t + d$. Also if the process is nonminimum phase, this parameter will allow the first instants of inverse response to be eliminated from the objective function. The coefficients $\delta(j)$ and $\lambda(j)$ are sequences that consider the future behaviour; usually constant values or exponential sequences are considered. For example, it is possible to obtain an exponential weight of $\delta(j)$ along the horizon by using:

$$\delta(j) = \alpha^{N_2 - j}$$

If α is given a value between 0 and 1, the errors farthest from instant t are penalized more than those nearer to it, giving rise to smoother control with less effort. If, on the other hand, $\alpha > 1$, the first errors are more penalized, provoking tighter control. In PFC the error is only counted at certain points (coincidence points); this is easily achieved in the objective function giving value one to the elements of sequence $\delta(j)$ at said points and zero at the others. All these values can be used as tuning parameters to cover an ample scope of options, from standard control to a made-to-measure design strategy for a particular process.

- reference trajectory: One of the advantages of predictive control is that if the future evolution of the reference is known a priori, the system can react before the change has effectively been made, thus avoiding the effects of delay in the process response. The future evolution of reference $r(t + k)$ is known beforehand in many applications, such as robotics, servos or batch processes; in other applications a noticeable improvement in performance can be obtained even though the reference is constant by

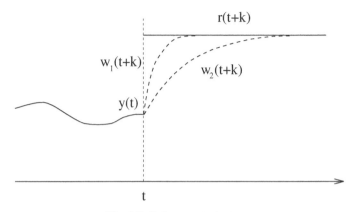

Fig. 2.3. Reference trajectory

simply knowing the instant when the value changes and getting ahead of this circumstance. In minimization (2.5), the majority of methods usually use a reference trajectory $w(t + k)$ which does not necessarily have to coincide with the real reference. It is normally a smooth approximation from the current value of the output $y(t)$ towards the known reference by means of the first-order system:

$$w(t) = y(t) \quad w(t+k) = \alpha w(t+k-1)+(1-\alpha)r(t+k) \quad k = 1 \ldots N \quad (2.6)$$

α is a parameter between 0 and 1 (the closer to 1 the smoother the approximation) that constitutes an adjustable value that will influence the dynamic response of the system. In Figure 2.3 the form of trajectory is shown from when the reference $r(t + k)$ is constant and for two different values of α; small values of this parameter provide fast tracking (w_1), if it is increased then the reference trajectory becomes w_2, giving rise to a smoother response.

Another strategy is the one used in PFC, which is useful for variable set-points:

$$w(t + k) = r(t + k) - \alpha^k(y(t) - r(t))$$

The reference trajectory can be used to specify closed-loop behaviour; this idea is used in GPC or EPSAC defining an auxiliary output

$$\psi(t) = P(z^{-1})y(t)$$

where the error in the objective function is given by $\psi(t + k) - w(t + k)$. The filter $P(z^{-1})$ has unit static gain and the generation of a reference trajectory with dynamics defined by $1/P(z^{-1})$ and an initial value of that of the measured output is achieved. In [57] it is demonstrated that if a *deadbeat* control in $\psi(t)$ is achieved so that

$$\psi(t) = B(z^{-1})w(t)$$

$B(z^{-1})$ being a determined polynomial with unit gain, the closed-loop response of the process will clearly be:

$$y(t) = \frac{B(z^{-1})}{P(z^{-1})} w(t)$$

In short, that is equivalent to placing the closed-loop poles at the zeros of design polynomial $P(z^{-1})$.

- constraints: In practice all processes are subject to constraints. The actuators have a limited field of action and a determined slew rate, as is the case of the valves, limited by the positions of totally open or closed and by the response rate. Constructive reasons, safety or environmental ones, or even the sensor scopes themselves can cause limits in the process variables such as levels in tanks, flows in piping, or maximum temperatures and pressures; moreover, the operational conditions are normally defined by the intersection of certain constraints for basically economic reasons, so that the control system will operate close to the boundaries. All of this makes the introduction of constraints in the function to be minimized necessary. Many predictive algorithms intrinsically take into account constraints (MAC, DMC) and have therefore been very successful in industry, whilst others can incorporate them *a posteriori* (GPC)[38]. Normally, bounds in the amplitude and in the slew rate of the control signal and limits in the output will be considered

$$u_{min} \leq \quad u(t) \quad \leq u_{max} \qquad \forall t$$
$$du_{min} \leq u(t) - u(t-1) \leq du_{max} \qquad \forall t$$
$$y_{min} \leq \quad y(t) \quad \leq y_{max} \qquad \forall t$$

By adding these constraints to the objective function, the minimization becomes more complex, so that the solution cannot be obtained explicitly as in the unconstrained case.

2.1.3 Obtaining the Control Law

In order to obtain values $u(t+k \mid t)$ it is necessary to minimize the functional J of Equation (2.5). To do this the values of the predicted outputs $\hat{y}(t+k \mid t)$ are calculated as a function of past values of inputs and outputs and future control signals, making use of the model chosen and substituted in the cost function, obtaining an expression whose minimization leads to the looked-for values. An analytical solution can be obtained for the quadratic criterion if the model is linear and there are not constraints, otherwise an iterative method of optimization should be used. Whatever the method, obtaining the solution is not easy because there will be $N_2 - N_1 + 1$ independent variables, a value which can be high (on the order of 10 to 30). In order to reduce this degree of freedom a certain structure may be imposed on the control law.

Furthermore, it has been found [105] that this structuralizing of the control law produces an improvement in robustness and in the general behaviour of the system, basically due to the fact that allowing the free evolution of the manipulated variables (without being structured) may lead to undesirable high-frequency control signals and at the worst to instability. This control law structure is sometimes imposed by the use of the control horizon concept (N_u) used in DMC, GPC, EPSAC and EHAC, that consists of considering that after a certain interval $N_u < N_2$ there is no variation in the proposed control signals, that is:

$$\triangle u(t + j - 1) = 0 \qquad j > N_u$$

which is equivalent to giving infinite weights to the changes in the control from a certain instant. The extreme case would be to consider N_u equal to 1 with which all future actions would be equal to $u(t)$[1]. Another way of structuring the control law is by using base functions, a procedure used in PFC which consists of representing the control signal as a linear combination of certain predetermined base functions:

$$u(t + k) = \sum_{i=1}^{n} \mu_i(t) B_i(k) \tag{2.7}$$

The B_i are chosen according to the nature of the process and the reference, they are normally polynomial type

$$B_0 = 1 \quad B_1 = k \quad B_2 = k^2 \ldots$$

As has been indicated previously, an explicit solution does not exist in the presence of constraints, so that quadratic programming methods have to be used (these methods will be studied in Chapter 7). However, an explicit solution does exist for certain types of constraints, for example, when the condition that the output attains the reference value at a determined instant is imposed, this method is used in Constrained Receding Horizon Predictive Control (CRHPC) [61], which is very similar to GPC and which guarantees stability results.

2.2 Review of Some MPC Algorithms

Some of the most popular methods will now be reviewed in order to demonstrate their most outstanding characteristics. Comparative studies can be found in [73],[108], [113] and [170]. The methods considered to be most representative, DMC, MAC, GPC, PFC, EPSAC and EHAC will briefly be dealt with. Some of them will be studied in greater detail in following chapters. Chapter 3 is devoted to DMC, MAC and PFC while GPC and its derivations are treated in Chapter 4.

---

[1] Remember that due to the receding horizon, the control signal is recalculated in the following sample.

Dynamic Matrix Control

Dynamic Matrix Control uses the step response (2.2) to model the process, only taking into account the first N terms, therefore assuming the process to be stable and without integrators. As regards the disturbances, their value will be considered to be the same as at instant t all along the horizon, that is, to be equal to the measured value of the output (y_m) minus the one estimated by the model $(\hat{y}(t \mid t))$.

$$\hat{n}(t + k \mid t) = \hat{n}(t \mid t) = y_m(t) - \hat{y}(t \mid t)$$

and therefore the predicted value of the output will be

$$\hat{y}(t + k \mid t) = \sum_{i=1}^{k} g_i \triangle u(t + k - i) + \sum_{i=k+1}^{N} g_i \triangle u(t + k - i) + \hat{n}(t + k \mid t)$$

where the first term contains the future control actions to be calculated, the second contains past values of the control actions and is therefore known, and the last represents the disturbances. The cost function may consider future errors only, or it can include the control effort, in which case it presents the generic form (2.5). One of the characteristics of this method that makes it very popular in the industry is the addition of constraints, in such a way that equations of the form

$$\sum_{i=1}^{N} C_{yi}^{j} \hat{y}(t + k \mid t) + C_{ui}^{j} u(t + k - i) + c^{j} \leq 0 \qquad j = 1 \ldots N_c$$

must be added to the minimization. Optimization (numerical because of the presence of constraints) is carried out at each sampling instant and the value of $u(t)$ is sent to the process as is normally done in all MPC methods. The inconveniences of this method are, on one hand, the size of the process model required and, on the other hand, the inability to work with unstable processes.

Model Algorithmic Control

Also known as Model Predictive Heuristic Control, Model Algorithmic Control is marketed under the name of IDCOM (Identification-Command). It is very similar to the previous method with a few differences. Firstly, it uses an impulse response model (2.1) valid only for stable processes, in which the value of $u(t)$ appears instead of $\triangle u(t)$. Furthermore, it makes no use of the control horizon concept so that in the calculations as many control signals as future outputs appear. It introduces a reference trajectory as a first-order system which evolves from the actual output to the setpoint according to a determined time constant, following Expression (2.6). The variance of the error between this trajectory and the output is what one aims at minimizing in the objective function. The disturbances can be treated as in DMC or their estimations can be carried out by the following recursive expression:

$$\hat{n}(t + k \mid t) = \alpha \hat{n}(t + k - 1 \mid t) + (1 - \alpha)(y_m(t) - \hat{y}(t \mid t))$$

with $\hat{n}(t \mid t) = 0$. α is an adjustable parameter $(0 \leq \alpha < 1)$ closely related to the response time, the bandwidth and the robustness of the closed-loop system [73]. It also takes into account constraints in the actuators and in the internal variables or secondary outputs. Various algorithms can be used for optimizing in the presence of constraints, from the ones presented initially by Richalet *et al.* that can also be used for identifying the impulse response, to others that are shown in [187].

Predictive Functional Control

This controller was developed by Richalet [178] for the case of fast processes. It uses a state space model of the process and allows for nonlinear and unstable linear internal models. Nonlinear dynamics can be entered in the form of a nonlinear state space model. PFC has two distinctive characteristics: the use of *coincidence points* and *basis functions*.

The concept of coincidence points is used to simplify the calculation by considering only a subset of points in the prediction horizon. The desired and the predicted future outputs are required to coincide at these points, not in the whole prediction horizon.

The controller parameterizes the control signal using a set of polynomial basis functions, as given by Equation (2.7). This allows a relatively complex input profile to be specified over a large horizon using a small number of parameters. Choosing the family of basis functions establishes many of the features of the computed input profile. These functions can be selected to follow a polynomial setpoint with no lag, an important feature for mechanical servo control applications.

The cost function to be minimized is:

$$J = \sum_{j=1}^{n_H} [\hat{y}(t + h_j) - w(t + h_j)]^2$$

where $w(t + j)$ is usually a first-order approach to the known reference.

The PFC algorithm can also accommodate maximum and minimum input acceleration constraints which are useful in mechanical servo control applications.

Extended Prediction Self Adaptive Control

The implementation of EPSAC is different to the previous methods. For predicting, the process is modelled by the transfer function

$$A(z^{-1})y(t) = B(z^{-1})u(t - d) + v(t)$$

where d is the delay and $v(t)$ the disturbance. The model can be extended by a term $D(z^{-1})d(t)$, with $d(t)$ being a measurable disturbance in order to include *feedforward* effect. Using this method the prediction is obtained as shown in [108]. One characteristic of the method is that the control law structure is very simple, as it is reduced to considering that the control signal is going to stay constant from instant t, that is, $\triangle u(t+k) = 0$ for $k > 0$. In short, the control horizon is reduced to 1 and therefore the calculation is reduced to one single value: $u(t)$. To obtain this value a cost function is used of the form:

$$\sum_{k=d}^{N} \gamma(k)[w(t+k) - P(z^{-1})\hat{y}(t+k \mid t)]^2$$

where $P(z^{-1})$ is a design polynomial with unit static gain and factor $\gamma(k)$ being a weighting sequence, similar to those appearing in (2.5). The control signal can be calculated analytically (which is an advantage over the previous methods) in the form:

$$u(t) = \frac{\sum\limits_{k=d}^{N} h_k \gamma(k)[w(t+k) - P(z^{-1})\hat{y}(t+k \mid t)]}{\sum\limits_{k=d}^{N} \gamma(k)h_k^2}$$

where h_k is the discrete impulse response of the system.

Extended Horizon Adaptive Control

This formulation considers the process modelled by its transfer function without taking a model of the disturbances into account:

$$A(z^{-1})y(t) = B(z^{-1})u(t - d)$$

It aims at minimizing the discrepancy between the model and the reference at instant $t + N$: $\hat{y}(t + N \mid t) - w(t + N)$, with $N \geq d$. The solution to this problem is not unique (unless $N = d$)[207]; a possible strategy is to consider that the control horizon is 1, that is,

$$\triangle u(t + k - 1) = 0 \qquad 1 < k \leq N - d$$

or to minimize the control effort:

$$J = \sum_{k=0}^{N-d} u^2(t + k)$$

There is an incremental version of EHAC that allows the disturbances in the load to be dealt with easily; it consists of considering

$$J = \sum_{k=0}^{N-d} \triangle u^2(t+k)$$

In this formulation a predictor of N steps is used as follows

$$\hat{y}(t+N \mid t) = y(t) + F(z^{-1}) \triangle y(t) + where E(z^{-1})B(z^{-1}) \triangle u(t+N-d)$$

$E(z^{-1})$ and $F(z^{-1})$ are polynomials satisfying the equation

$$(1 - z^{-1}) = A(z^{-1})E(z^{-1})(1 - z^{-1}) + z^{-N}F(z^{-1})(1 - z^{-1})$$

with the degree of E being equal to $N - 1$. One advantage of this method is that a simple explicit solution can easily be obtained, resulting in

$$u(t) = u(t-1) + \frac{\alpha_0(w(t+N) - \hat{y}(t+N \mid t))}{\displaystyle\sum_{k=0}^{N-d} \alpha_i^2}$$

where α_k is the coefficient corresponding to $\triangle u(t+k)$ in the prediction equation. Thus the control law only depends on the process parameters and can therefore easily be made self-tuning if it has an online identifier. As can be seen the only parameter of adjustment is the horizon of prediction N, which simplifies its use but provides little freedom for the design. One sees that the reference trajectory cannot be used because the error is only considered at one instant $(t+N)$, neither is it possible to ponder the control efforts at each point, so that certain frequencies in the performance cannot be eliminated.

Generalized Predictive Control

The output predictions of the Generalized Predictive Controller are based on using a CARIMA model:

$$A(z^{-1})y(t) = B(z^{-1})z^{-d} u(t-1) + C(z^{-1})\frac{e(t)}{\triangle}$$

where the unmeasurable disturbance is given by a white noise coloured by $C(z^{-1})$. As its true value is difficult to know, this polynomial can be used for optimal disturbance rejection, although its role in robustness enhancement is more convincing.

The derivation of the optimal prediction is done by solving a Diophantine equation whose solution can be found by an efficient recursive algorithm.

This algorithm, as with all algorithms using transfer function models, can easily be implemented in an adaptive mode using an online estimation algorithm such as recursive least squares.

GPC uses a quadratic cost function of the form:

$$J(N_1, N_2, N_u) = \sum_{j=N_1}^{N_2} \delta(j)[\hat{y}(t+j \mid t) - w(t+j)]^2 + \sum_{j=1}^{N_u} \lambda(j)[\triangle u(t+j-1)]^2$$

where the weighting sequences $\delta(j)$ and $\lambda(j)$ are usually chosen constant or exponentially increasing and the reference trajectory $w(t+j)$ can be generated by a simple recursion which starts at the current output and tends exponentially to the setpoint.

The theoretical basis of the GPC algorithm has been widely studied, and it has been shown [57] that, for limiting cases of parameter choices, this algorithm is stable and also that well-known controllers such as mean level and deadbeat control are inherent in the GPC structure.

2.3 State Space Formulation

State space models can be used to formulate the predictive control problem. The main theoretical results of MPC related to stability come from a state space formulation, which can be used for both monovariable and multivariable processes and can easily be extended to nonlinear processes. The following equations are used in the linear case to capture process dynamics:

$$\begin{aligned} x(t+1) &= Ax(t) + Bu(t) \\ y(t) &= Cx(t) \end{aligned} \tag{2.8}$$

In the single-input single-output (SISO) case, $y(t)$ and $u(t)$ are scalars and $x(t)$ is the state vector. A multiple-input multiple-output (MIMO) process has the same description but with input vectors u of dimension m and y of dimension n. In this section, for notation simplicity, only the SISO case is considered. The MIMO case is addressed in Chapter 6 together with other formulations of MPC for MIMO processes.

An incremental state space model can also be used if the model input is the control increment $\triangle u(t)$ instead of the control signal $u(t)$. This model can be written in the general state space form taking into account that $\triangle u(t) = u(t) - u(t-1)$. The following representation is obtained combining this expression with (2.8):

$$\begin{bmatrix} x(t+1) \\ u(t) \end{bmatrix} = \begin{bmatrix} A & B \\ 0 & I \end{bmatrix} \begin{bmatrix} x(t) \\ u(t-1) \end{bmatrix} + \begin{bmatrix} B \\ I \end{bmatrix} \triangle u(t)$$

$$y(t) = \begin{bmatrix} C & 0 \end{bmatrix} \begin{bmatrix} x(t) \\ u(t-1) \end{bmatrix}$$

Defining a new state vector as $\overline{x}(t) = [x(t) \quad u(t-1)]^T$, the incremental model takes the general form (2.8):

$$\bar{x}(t+1) = M\bar{x}(t) + N \triangle u(t)$$
$$y(t) = Q\bar{x}(t) \qquad (2.9)$$

where the relationship between (M, N, Q) and the nonincremental form matrices (A, B, C) can easily be obtained by comparing (2.8) and (2.9).

In order to minimize the objective function (2.5), output predictions over the horizon must be computed. If the incremental model is used, predictions can be obtained using (2.9) recursively, resulting in:

$$\hat{y}(t+j) = QM^j\hat{x}(t) + \sum_{i=0}^{j-1} QM^{j-i-1}N \triangle u(t+i)$$

Notice that the prediction needs an unbiased estimation of the state vector $x(t)$. If the state vector is not accessible an observer must be included, which calculates the estimation by means of

$$\hat{x}(t \mid t) = \hat{x}(t \mid t-1) + K(y_m(t) - y(t \mid t-1))$$

where $y_m(t)$ is the measured output. If the plant is subject to white noise disturbances affecting the process and the output with known covariance matrices, the observer becomes a Kalman filter [11] and the gain K is calculated solving a Riccati equation.

Now, the predictions along the horizon are given by

$$\mathbf{y} = \begin{bmatrix} \hat{y}(t+1|t) \\ \hat{y}(t+2|t) \\ \vdots \\ \hat{y}(t+N_2|t) \end{bmatrix} = \begin{bmatrix} QM\hat{x}(t) + QN \triangle u(t) \\ QM^2\hat{x}(t) + \sum_{i=0}^{1} QM^{1-i}N \triangle u(t+i) \\ \vdots \\ QM^{N_2}\hat{x}(t) + \sum_{i=0}^{N_2-1} QM^{N_2-1-i}N \triangle u(t+i) \end{bmatrix}$$

which can be expressed in vector form as

$$\mathbf{y} = \mathbf{F}\hat{x}(t) + \mathbf{H}\mathbf{u} \qquad (2.10)$$

where $\mathbf{u} = [\triangle u(t) \ \triangle u(t+1) \ \ldots \ \triangle u(t+N_u-1)]^T$ is the vector of future control increments, $\mathbf{H}$ is a block lower triangular matrix with its nonnull elements defined by $\mathbf{H}_{ij} = QM^{i-j}N$ and matrix $\mathbf{F}$ is defined as

$$\mathbf{F} = \begin{bmatrix} QM \\ QM^2 \\ \vdots \\ QM^{N_2} \end{bmatrix}$$

Notice that (2.10) is composed of two terms: the first depends on the current state and therefore is known at instant t, while the second depends on the vector of future control actions, which is the decision variable that must be calculated. The control sequence **u** is calculated minimizing the objective function (2.5), that (in the case of $\delta(j) = 1$ and $\lambda(j) = \lambda$) can be written as:

$$J = (\mathbf{Hu} + \mathbf{F}\hat{x}(t) - \mathbf{w})^T (\mathbf{Hu} + \mathbf{F}\hat{x}(t) - \mathbf{w}) + \lambda \mathbf{u}^T \mathbf{u}$$

If the are no constraints, an analytical solution exists that provides the optimum as:

$$\mathbf{u} = (\mathbf{H}^T \mathbf{H} + \lambda \mathbf{I})^{-1} \mathbf{H}^T (\mathbf{w} - \mathbf{F}\hat{x}(t))$$

As a receding horizon strategy is used, only the first element of the control sequence, $\triangle u(t)$, is sent to the plant and all the computation is repeated at the next sampling time. Notice that a state observer is needed, since the control law depends on $\hat{x}(t)$.

It must be noted that when the control and the maximum prediction horizons approach infinity and there are no constraints, the predictive controller becomes the well-known linear quadratic regulator (LQR) problem (see Appendix B). The optimal control sequence is generated by a static state feedback law where the feedback gain matrix is computed via the solution of an algebraic Riccati equation. This equivalence allows the theoretical study of MPC problems based on results coming from the optimal control field, as in the case of closed-loop stability (see Section 9.5).

If the state space model of Equation (2.8) is used, the predictions are computed in a slightly different manner, as shown by Maciejowski [131]. Now

$$\mathbf{y} = \begin{bmatrix} CA \\ CA^2 \\ \vdots \\ CA^{N_2} \end{bmatrix} \hat{\mathbf{x}}(t) + \begin{bmatrix} CB \\ CA^2 B \\ \vdots \\ \sum\limits_{i=0}^{N_2-1} CA^i B \end{bmatrix} u(t-1)$$

$$+ \begin{bmatrix} B & \cdots & 0 \\ C(AB + B) & \cdots & 0 \\ \vdots & \ddots & \vdots \\ \sum\limits_{i=0}^{N_2-1} CA^i B & \cdots & \sum\limits_{i=0}^{N_2-N_u} CA^i B \end{bmatrix} \mathbf{u}$$

which can be expressed in vector form as

$$\mathbf{y} = \mathbf{\Psi}\hat{x}(t) + \mathbf{\Upsilon} u(t-1) + \mathbf{\Theta} \mathbf{u}$$

Notice that a new term has appeared that depends on $u(t-1)$ and does not affect the optimization since it does not depend on the decision variable **u**. Therefore, the control action is calculated as

$$\mathbf{u} = (\mathbf{\Theta}^T \mathbf{\Theta} + \lambda I)^{-1} \mathbf{\Theta}^T (\mathbf{w} - \mathbf{\Psi}\hat{x}(t) - \mathbf{\Upsilon} u(t-1))$$

Whatever kind of model is used, the control law is a static state feedback law that needs a state observer. In the case where constraints must be taken into account, the solution must be obtained by a Quadratic Programming (QP) algorithm, as will be studied in Chapter 7.

3

Commercial Model Predictive Control Schemes

As has been shown in previous chapters, there is a wide family of predictive controllers, each member of which is defined by the choice of the common elements such as the prediction model, the objective function and obtaining the control law.

This chapter is dedicated to an overview of some MPC algorithms widely used in industry. The first two belong to a major category of predictive control approaches, those employing convolutional models, also called non-parametric methods. These approaches are based on step response or impulse response models; the most representative formulations are Dynamic Matrix Control (DMC) and Model Algorithmic Control (MAC). The third MPC algorithm presented in this chapter is Predictive Functional Control (PFC), which uses a set of basis functions to form the future control sequence.

It should be clear that the descriptions given here are necessarily incomplete, since only the general characteristics of each method are presented and each controller has proprietary features which are not known.

3.1 Dynamic Matrix Control

DMC was developed at the end of the seventies by Cutler and Ramaker [62] of Shell Oil Co. and has been widely accepted in the industrial world, mainly by petrochemical industries [170].

Nowadays DMC is something more than an algorithm, and part of its success is due to the fact that the commercial product covers topics such as model identification and global plant optimization. In this section only the *standard* algorithm is analyzed, without addressing technical details such as software and hardware compatibilities, user interface requirements, personnel training or configuration and maintenance issues. The great success of DMC in industry comes from its ability to deal with multivariable processes. In this chapter, only the Single-Input Single-Output (SISO) case is addressed, leaving the Multiple-Input Multiple-Output (MIMO) case for Section 6.5. Fundamentals of this controller can be more easily understood in the SISO; the extension of the method to MIMO plants is basically a matter of notation.

3.1.1 Prediction

The process model employed in this formulation is the step response of the plant, while the disturbance is considered to be constant along the horizon. The procedure to obtain the predictions is as follows.

As a step response model is employed:

$$y(t) = \sum_{i=1}^{\infty} g_i \,\triangle\, u(t - i)$$

the predicted values along the horizon will be:

$$\hat{y}(t + k \mid t) = \sum_{i=1}^{\infty} g_i \,\triangle\, u(t + k - i) + \hat{n}(t + k \mid t) =$$

$$\sum_{i=1}^{k} g_i \,\triangle\, u(t + k - i) + \sum_{i=k+1}^{\infty} g_i \,\triangle\, u(t + k - i) + \hat{n}(t + k \mid t)$$

Disturbances are considered to be constant, that is, $\hat{n}(t + k \mid t) = \hat{n}(t \mid t) = y_m(t) - \hat{y}(t \mid t)$. Then it can be written that:

$$\hat{y}(t + k \mid t) = \sum_{i=1}^{k} g_i \,\triangle\, u(t + k - i) + \sum_{i=k+1}^{\infty} g_i \,\triangle\, u(t + k - i) + y_m(t)$$

$$- \sum_{i=1}^{\infty} g_i \,\triangle\, u(t - i) = \sum_{i=1}^{k} g_i \,\triangle\, u(t + k - i) + f(t + k)$$

where $f(t + k)$ is the free response of the system, that is, the part of the response that does not depend on the future control actions, and is given by:

$$f(t + k) = y_m(t) + \sum_{i=1}^{\infty} (g_{k+i} - g_i) \,\triangle\, u(t - i) \tag{3.1}$$

If the process is asymptotically stable, the coefficients g_i of the step response tend to a constant value after N sampling periods, so it can be considered that

$$g_{k+i} - g_i \approx 0, \qquad i > N$$

and therefore the free response can be computed as:

$$f(t+k) = y_m(t) + \sum_{i=1}^{N} (g_{k+i} - g_i) \triangle u(t-i)$$

Notice that if the process is not asymptotically stable, then N does not exist and $f(t+k)$ cannot be computed (although a generalization exists in the case of the instability being produced by pure integrators).

Now the predictions can be computed along the prediction horizon ($k = 1, \ldots, p$), considering m control actions:

$$\hat{y}(t+1 \mid t) = g_1 \triangle u(t) + f(t+1)$$
$$\hat{y}(t+2 \mid t) = g_2 \triangle u(t) + g_1 \triangle u(t+1) + f(t+2)$$
$$\vdots$$
$$\hat{y}(t+p \mid t) = \sum_{i=p-m+1}^{p} g_i \triangle u(t+p-i) + f(t+p)$$

Defining the system's *dynamic matrix* $\mathbf{G}$ as

$$\mathbf{G} = \begin{bmatrix} g_1 & 0 & \cdots & 0 \\ g_2 & g_1 & \cdots & 0 \\ \vdots & \vdots & \ddots & \vdots \\ g_m & g_{m-1} & \cdots & g_1 \\ \vdots & \vdots & \ddots & \vdots \\ g_p & g_{p-1} & \cdots & g_{p-m+1} \end{bmatrix}$$

it can be written that

$$\hat{\mathbf{y}} = \mathbf{G}\mathbf{u} + \mathbf{f} \tag{3.2}$$

Observe that $\mathbf{G}$ is made up of m (the control horizon) columns of the system's step response appropriately shifted down in order. $\hat{\mathbf{y}}$ is a p-dimensional vector containing the system predictions along the horizon, $\mathbf{u}$ represents the m-dimensional vector of control increments and $\mathbf{f}$ is the free response vector.

This is the expression that relates the future outputs with the control increments, so it will be used to calculate the necessary action to achieve a specific system behaviour.

Notice that $\mathbf{f}$ depends on the state vector $x(t)$, which in this case is given by $x(t)^T = [y_m(t)\ u(t-1)\ u(t-2) \ldots u(t-N-1)]$ and can be expressed as $\mathbf{f} = \mathbf{F}x(t)$, and consequently the prediction can be written as:

$$\hat{\mathbf{y}} = \mathbf{G}\mathbf{u} + \mathbf{F}x(t)$$

3.1.2 Measurable Disturbances

Measurable disturbances can easily be added to the prediction equations, since they can be treated as system inputs. Expression (3.2) can be used to calculate the predicted disturbances

$$\hat{\mathbf{y}}_d = \mathbf{D}\mathbf{d} + \mathbf{f}_d$$

where $\hat{\mathbf{y}}_d$ is the contribution of the measurable disturbance to the system output, $\mathbf{D}$ is a matrix similar to $\mathbf{G}$ containing the coefficients of the system response to a step in the disturbance, $\mathbf{d}$ is the vector of disturbance increment and $\mathbf{f}_d$ is the part of the response that does not depend on the disturbance.

In the most general case of measurable and nonmeasurable disturbances, the complete free response of the system (the fraction of the output that does not depend on the manipulated variable) can be considered as the sum of four effects: the response to the input $u(t)$, to the measurable disturbance $d(t)$, to the nonmeasurable disturbance and to the actual process state:

$$\mathbf{f} = \mathbf{f}_u + \mathbf{D}\mathbf{d} + \mathbf{f}_d + \mathbf{f}_n$$

Therefore the prediction can be computed by the general known expression:

$$\hat{\mathbf{y}} = \mathbf{G}\mathbf{u} + \mathbf{f}$$

3.1.3 Control Algorithm

The industrial success of DMC has mainly come from its application to high-dimension multivariable systems with the use of constraints. This section describes the control algorithm starting from the simpler case of a monovariable system without constraints; later it is extended to the general multivariable and constrained cases.

The objective of a DMC controller is to drive the output as close to the setpoint as possible in a least-squares sense with the possibility of the inclusion of a penalty term on the input moves. Therefore, the manipulated variables are selected to minimize a quadratic objective that can consider the minimization of future errors alone

$$J = \sum_{j=1}^{p} [\hat{y}(t+j \mid t) - w(t+j)]^2$$

or it can include the control effort, in which case it presents the generic form

$$J = \sum_{j=1}^{p} [\hat{y}(t+j \mid t) - w(t+j)]^2 + \sum_{j=1}^{m} \lambda [\triangle u(t+j-1)]^2$$

If there are no constraints, the solution to the minimization of the cost function $J = \mathbf{e}\mathbf{e}^T + \lambda \mathbf{u}\mathbf{u}^T$, where $\mathbf{e}$ is the vector of future errors along the

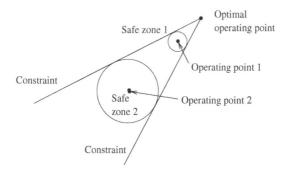

Fig. 3.1. Economic operating point of a typical process

prediction horizon and **u** is the vector composed of the future control increments $\triangle u(t), \ldots, \triangle u(t + m)$, can be obtained analytically by computing the derivative of J and making it equal to 0, which provides the general result:

$$\mathbf{u} = (\mathbf{G}^T \mathbf{G} + \lambda I)^{-1} \mathbf{G}^T (\mathbf{w} - \mathbf{f})$$

Remember that, as in all predictive strategies, only the first element of vector **u** ($\triangle u(t)$) is really sent to the plant. It is not advisable to implement the entire sequence over the next m intervals in automatic succession. This is because is impossible to perfectly estimate the disturbance vector, and therefore it is impossible to anticipate precisely the unavoidable disturbances that cause the actual output to differ from the predictions used to compute the sequence of control actions. Furthermore, the setpoint can also change over the next m intervals.

The Constrained Problem

Though computationally more involved than simpler algorithms, the flexible constraint-handling capabilities of the method (and MPC in general) are very attractive for practical applications, since the economic operating point of a typical process unit often lies at the intersection of constraints [168], as shown in Figure 3.1. It can be seen that, due to safety reasons, it is necessary to keep a safe zone around the operating point, since the effect of perturbations can make the process violate constraints. This zone can be reduced, and therefore the economic profit improved, if the controller is able to handle constraints (operating point 1).

Constraints in both inputs and outputs can be posed in such a way that equations of the form

$$\sum_{i=1}^{N} C_{yi}^j \hat{y}(t + k \mid t) + C_{ui}^j u(t + k - i) + c^j \leq 0 \qquad j = 1 \ldots N_c$$

must be added to the minimization. As future projected outputs can be re-lated directly back to the control increment vector through the dynamic ma-trix, all input and output constraints can be collected into a matrix inequality involving the input vector, $\mathbf{R}\mathbf{u} \leq \mathbf{c}$ (for further details see Chapter 7). There-fore the problem takes the form of a standard Quadratic Programming (QP) formulation. The optimization is now numerical because of the presence of constraints and is carried out by means of standard commercial optimiza-tion QP code at each sampling instant, and then the value of $u(t)$ is sent to the process, as is normally done in all MPC methods. In this case the method is known as QDMC, due to the Quadratic Programming algorithm employed.

3.2 Model Algorithmic Control

Maybe the simplest and most intuitive formulation of Predictive Control is the one based on the key ideas of Richalet *et al.* [182], known as MAC and Model Predictive Heuristic Control (MPHC), whose software is called IDCOM (Identification-Command). This method is very similar to DMC with a few differences. It makes use of a truncated step response of the process and pro-vides a simple explicit solution in the absence of constraints. This method has clearly been accepted by practitioners and is extensively used in many con-trol applications [70] where most of its success is due to the process model used. It is known that transfer function models can give results with large errors when there is a mismatch in the model order. On the other hand, the impulse response representation is a good choice, since the identification of impulse responses is relatively simple.

3.2.1 Process Model and Prediction

The system output at instant t is related to the inputs by the coefficients of the truncated impulse response as follows:

$$y(t) = \sum_{j=1}^{N} h_j u(t-j) = H(z^{-1})u(t)$$

This model predicts that the output at a given time depends on a linear com-bination of past input values; the weights h_i are the impulse response coeffi-cients. As the response is truncated to N elements, the system is assumed to be stable and causal. Using this internal model, a k-step ahead predictor can be written as

$$\hat{y}(t+k \mid t) = \sum_{j=1}^{N} h_j u(t+k-j) + \hat{n}(t+k \mid t)$$

where the sum can be divided into two terms

$$f_r(t+k) = \sum_{j=k+1}^{N} h_j u(t+k-j) \quad f_o(t+k) = \sum_{j=1}^{k} h_j u(t+k-j)$$

such that f_r represents the free response, being the expected value of $y(t+j)$ assuming zero future control actions, and f_o is the forced response, that is, the additional component of output response due to the proposed set of future control actions. It is now assumed that the disturbances will remain constant in the future with the same value as at instant t, that is, $\hat{n}(t+k \mid t) = \hat{n}(t \mid t)$, which is the measured output minus the output predicted by the nominal model:

$$\hat{n}(t+k \mid t) = \hat{n}(t \mid t) = y(t) - \sum_{j=1}^{N} h_j u(t-j)$$

Then the prediction is given by:

$$\hat{y}(t+k \mid t) = f_r + f_o + \hat{n}(t \mid t)$$

If M is the horizon and $\mathbf{u}_+$ the vector of proposed control actions (not increments), $\mathbf{u}_-$ of past control actions, $\mathbf{y}$ the predicted outputs, $\mathbf{n}$ the disturbances, and the reference vector $\mathbf{w}$ is a smooth approach to the current setpoint

$$\mathbf{u}_+ = \begin{bmatrix} u(t) \\ u(t+1) \\ \vdots \\ u(t+M-1) \end{bmatrix} \quad \mathbf{u}_- = \begin{bmatrix} u(t-N+1) \\ u(t-N+2) \\ \vdots \\ u(t-1) \end{bmatrix} \quad \mathbf{y} = \begin{bmatrix} \hat{y}(t+1) \\ \hat{y}(t+2) \\ \vdots \\ \hat{y}(t+M) \end{bmatrix}$$

$$\mathbf{n} = \begin{bmatrix} \hat{n}(t+1) \\ \hat{n}(t+2) \\ \vdots \\ \hat{n}(t+M) \end{bmatrix} \quad \mathbf{w} = \begin{bmatrix} w(t+1) \\ w(t+2) \\ \vdots \\ w(t+M) \end{bmatrix}$$

and defining the matrices

$$\mathbf{H}_1 = \begin{bmatrix} h_1 & 0 & \cdots & 0 \\ h_2 & h_1 & \cdots & 0 \\ \cdots & \cdots & \ddots & \cdots \\ h_M & h_{M-1} & \cdots & h_1 \end{bmatrix} \quad \mathbf{H}_2 = \begin{bmatrix} h_N & \cdots & h_i & \cdots & h_2 \\ 0 & \cdots & h_j & \cdots & h_3 \\ \cdots & \ddots & \cdots & \cdots & \cdots \\ 0 & \cdots & h_N & \cdots & h_{M+1} \end{bmatrix}$$

the predictor can be written as

$$\mathbf{y} = \mathbf{H}_1 \, \mathbf{u}_+ + \mathbf{H}_2 \, \mathbf{u}_- + \mathbf{n}$$

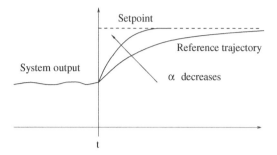

Fig. 3.2. Influence of α on the reference tracking

3.2.2 Control Law

The primary objective of the controller is to determine the sequence of control moves that will minimize the sum of the squared deviations of the predicted output from the reference trajectory.

The reference trajectory used in MAC is normally a smooth approximation from the current value of the system output towards the known reference by means of a first-order system of the form:

$$w(t+k) = \alpha w(t+k-1) + (1-\alpha)r(t+k) \quad k = 1\ldots N, \quad \text{with } w(t) = y(t)$$

It is important to note that the shape of the reference trajectory (which depends on the choice of α) determines the desired speed of approach to the setpoint. This is of great interest in practice because it provides a natural way to control the aggressiveness of the algorithm: increasing the time constant leads to a slower but more robust controller (see Figure 3.2). Therefore this is an important tuning parameter for this controller, and its choice is very closely linked to the robustness of the closed-loop system [159]. Parameter α is a more direct and more intuitive tuning parameter than factors such as weighting sequences or horizon lengths employed by other formulations.

The objective function minimizes the error as well as the control effort. If future errors are expressed as

$$\mathbf{e} = \mathbf{w} - \mathbf{y} = \mathbf{w} - \mathbf{H_2}\mathbf{u_-} - \mathbf{n} - \mathbf{H_1}\mathbf{u_+} = \mathbf{w} - \mathbf{f} - \mathbf{H_1}\mathbf{u_+}$$

where vector $\mathbf{f}$ contains the terms depending on known values (past inputs, current output and references). Then the cost function can be written as

$$J = \mathbf{e}^T\mathbf{e} + \lambda \mathbf{u_+}^T \mathbf{u_+}$$

where λ is the penalization factor for the input variable variations. If no constraints are considered, the solution can be obtained explicitly, giving:

$$\mathbf{u_+} = (\mathbf{H_1^T}\mathbf{H_1} + \lambda\mathbf{I})^{-1}\mathbf{H_1}^T(\mathbf{w} - \mathbf{f}) \tag{3.3}$$

As it is a receding horizon strategy, only the first element of this vector $u(t)$ is used, rejecting the rest and repeating the calculations at the next sampling time.

The calculation of the control law (3.3) is relatively simple compared to other formulations, although it requires the inversion of an $M \times M$ matrix. Notice that if the number of future inputs to be calculated is chosen as a value $P < M$, then this matrix is of dimension $P \times P$, since $\mathbf{H}_1$ is of dimension $M \times P$, hence reducing the necessary calculations.

The simplicity of the algorithm, as well as the possibility of including constraints, has converted this formulation into one of the most frequently used in industry nowadays.

3.3 Predictive Functional Control

The Predictive Functional Controller PFC was proposed by Richalet [178] for fast processes and is characterized by two distinctive features: the control signal is structured as a linear combination of certain predetermined *basis functions* and the concept of *coincidence points* to evaluate the cost function along the horizon.

3.3.1 Formulation

Consider the following state space model

$$x(t) = Mx(t-1) + Nu(t-1)$$
$$y(t) = Qx(t)$$

as representing the process behaviour. The prediction is obtained adding an autocompensation term calculated as a function of the observed differences between the model and the past outputs:

$$\hat{y}(t+k \mid t) = y(t+k) + \hat{e}(t+k \mid t)$$

The predicted plant-model error is assumed to have the form

$$\hat{e}(t+k \mid t) = y_m(t) - \hat{y}(t \mid t-1) + \sum_{j=1}^{r} e_j k^j$$

where $y_m(t)$ is the output measured value and coefficients e_j are obtained by a least-squares fit to previous errors.

The future control signal is structured as a linear combination of the basis functions B_i, which are chosen according to the nature of the process and the reference:

$$u(t+k) = \sum_{i=1}^{n_B} \mu_i(t) B_i(k) \tag{3.4}$$

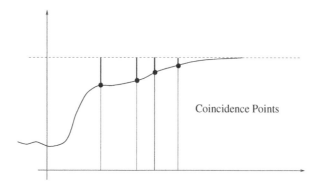

Fig. 3.3. Coincidence points

Normally these functions are polynomial type: steps $(B_1(k) = 1)$, ramps $(B_2(k) = k)$, or parabolas $(B_3(k) = k^2)$, as the majority of references can be expressed as combinations of these functions. With this strategy a complex input profile can be specified using a small number of unknown parameters.

The choice of basis functions defines the input profile and can assure a predetermined behaviour (smooth signal, for instance). This can result in an advantage when controlling nonlinear systems. An important feature for mechanical servo applications is that if a polynomial basis is chosen, then the order can be selected to follow a polynomial setpoint.

The cost function to be minimized is:

$$J = \sum_{j=1}^{n_H} [\hat{y}(t + h_j) - w(t + h_j)]^2$$

where $w(t + j)$ is usually a first-order approach to the known reference, as in (2.6) or

$$w(t + k) = r(t + k) - \alpha^k (r(t) - y(t))$$

In order to smooth the control signal, a quadratic factor of the form $\lambda[\triangle u(k)]^2$ can be added to the cost function.

The predicted error is not considered all along the horizon, only in certain instants h_j, $j = 1, \ldots, n_H$ called *coincidence points* (see figure 3.3). These points can be considered tuning parameters and must be chosen taking into account their influence on the stability and robustness of the control system. Their number must be at least equal to the selected number of basis functions.

Calculation of the Control Law

The calculation of the control law implies computing the values of $\mu_i(t)$ of Equation (3.4). Notice that these coefficients are chosen to be optimal at each instant t and therefore they are different at each step.

In the case of SISO processes without constraints, the control law can be obtained as follows. First, the output is decomposed into free and forced outputs, and the structure of the control signal is employed to give

$$y(t+k) = QM^k x(t) + \sum_{i=1}^{n_B} y_{B_i}(k)\mu_i(t)$$

where y_{B_i} is the system response to the basis function B_i.

Now the cost function can be written as

$$J = \sum_{j=1}^{n_H} [\hat{y}(t+h_j) - w(t+h_j)]^2 = \sum_{j=1}^{n_H} [\mathbf{y}_B(h_j)\mu - d(t+h_j)]^2$$

where

$$\mu = [\mu_1(t)\ldots\mu_{n_B}(t)]^T$$
$$\mathbf{y}_B(h_j) = [y_{B_1}(h_j)\ldots y_{B_{n_B}}(h_j)]$$
$$d(t+h_j) = w(t+h_j) - QM^j x(t) - e(t+h_j)$$

The cost function can be written in vector form defining $\mathbf{d} = [d(t+h_1)\ldots d(t+h_{n_H})]^T$ as the term which contains values that are known at time t and where $\mathbf{Y}_B$ is a matrix whose rows are the vectors $\mathbf{y}_B$ at the coincidence points h_j, $j = 1,\ldots n_h$, giving:

$$J = (\mathbf{Y}_B\mu - \mathbf{d})^T (\mathbf{Y}_B\mu - \mathbf{d})$$

Minimizing J with respect to the coefficients μ:

$$\frac{\partial J}{\partial \mu} = 0 \Rightarrow \mathbf{Y}_B^T \mathbf{Y}_B \mu - \mathbf{Y}_B^T \mathbf{d} = 0$$

and therefore the vector of the weights of the basis functions is given by the solution of

$$\mathbf{Y}_B \mu = \mathbf{d}$$

Then the control signal, taking into account the receding horizon strategy, is given by:

$$u(t) = \sum_{i=1}^{n_B} \mu_i(t) B_i(0)$$

This algorithm can only be used for stable models, since the pole cancellations can lead to stability problems when unstable or high-oscillatory modes appear. In this case, a procedure that decomposes the model into two stable ones can be employed [178]. The method can be used for nonlinear processes using nonlinear state space models.

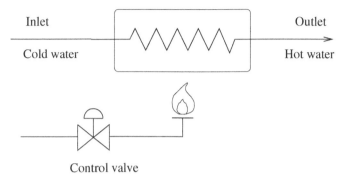

Inlet   Outlet

Cold water   Hot water

Control valve

Fig. 3.4. Water heater

3.4 Case Study: A Water Heater

This example shows the design of a DMC to control the outlet temperature of a water heater. Notice that a MAC can be designed following the same steps.

Consider a water heater where the cold water is heated by means of a gas burner. The outlet temperature depends on the energy added to the water through the gas burner (see Figure 3.4). Therefore this temperature can be controlled by the valve which manipulates the gas flow to the heater.

The step response model of this process must be obtained to design the controller. The step response is obtained by applying a step in the control valve. Coefficients g_i can be obtained directly from the response shown in figure 3.5. It can be seen that the output stabilizes after 30 periods, so the model is given by

$$y(t) = \sum_{i=1}^{30} g_i \,\Delta\, u(t-i),$$

where the coefficients g_i are shown in the following table:

g_1	g_2	g_3	g_4	g_5	g_6	g_7	g_8	g_9	g_{10}
0	0	0.271	0.498	0.687	0.845	0.977	1.087	1.179	1.256
g_{11}	g_{12}	g_{13}	g_{14}	g_{15}	g_{16}	g_{17}	g_{18}	g_{19}	g_{20}
1.320	1.374	1.419	1.456	1.487	1.513	1.535	1.553	1.565	1.581
g_{21}	g_{22}	g_{23}	g_{24}	g_{25}	g_{26}	g_{27}	g_{28}	g_{29}	g_{30}
1.592	1.600	1.608	1.614	1.619	1.623	1.627	1.630	1.633	1.635

The response shown in Figure 3.5 corresponds to a system with a transfer function given by:

$$G(z) = \frac{0.2713 z^{-3}}{1 - 0.8351 z^{-1}}$$

Notice that although the g_i coefficients are obtained in practice from real plant tests, for this example, where the response has been generated with a simple model, the step response coefficients can easily be obtained from the transfer function by the expression

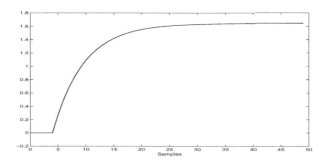

Fig. 3.5. Step response

$$g_j = -\sum_{i=1}^{j} a_i g_{j-i} + \sum_{i=0}^{j-1} b_i \qquad g_k = 0 \quad \text{for} \quad k \leq 0 \qquad (3.5)$$

where a_i and b_i are the coefficients of the denominator and numerator of the discrete transfer function respectively (starting from $i = 0$).

In this example the first two coefficients of the step model are zero since the system has a dead time of two sampling periods.

Considering a prediction horizon of 10 and a control horizon of 5, the dynamic matrix is obtained from the coefficients of the step response and is given by:

$$\mathbf{G} = \begin{bmatrix} 0 & 0 & 0 & 0 & 0 \\ 0 & 0 & 0 & 0 & 0 \\ 0.271 & 0 & 0 & 0 & 0 \\ 0.498 & 0.271 & 0 & 0 & 0 \\ 0.687 & 0.498 & 0.271 & 0 & 0 \\ 0.845 & 0.687 & 0.498 & 0.271 & 0 \\ 0.977 & 0.845 & 0.687 & 0.498 & 0.271 \\ 1.087 & 0.977 & 0.845 & 0.687 & 0.498 \\ 1.179 & 1.087 & 0.977 & 0.845 & 0.687 \\ 1.256 & 1.179 & 1.087 & 0.977 & 0.845 \end{bmatrix}$$

Taking $\lambda = 1$, matrix $(\mathbf{G}^T\mathbf{G} + \lambda I)^{-1}\mathbf{G}^T$ is calculated and therefore the control law is given by the product of the first row of this matrix ($\mathbf{K}$) times the vector that contains the difference between the reference trajectory and the free response

$$\triangle u(t) = \mathbf{K}(\mathbf{w} - \mathbf{f})$$

with

$$\mathbf{K} = [0 \ \ 0 \ \ 0.1465 \ \ 0.1836 \ \ 0.1640 \ \ 0.1224 \ \ 0.0780 \ \ 0.0410 \ \ 0.0101 \ \ -0.0157]$$

where the free response is easily calculated using equation (3.1):

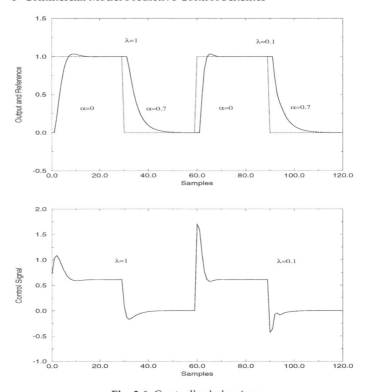

Fig. 3.6. Controller behaviour

$$f(t + k) = y_m(t) + \sum_{i=1}^{30}(g_{k+i} - g_i) \triangle u(t - i)$$

Figure 3.6 shows the system response to a change in the outlet temperature setpoint for different shapes of the control weighting factor and the reference trajectory. The first setpoint change is made with a value of $\lambda = 1$ and $\alpha = 0$. In the second change α is changed to 0.7 and later the control weighting factor is changed to 0.1 for the same values of α. It can be seen that a small value of α makes the system response faster with a slight oscillation, while a small value of λ gives bigger control actions. The combination $\lambda = 0.1$, $\alpha = 0$ provides the fastest response, but the control effort seems to be too vigorous.

The inlet temperature can become a disturbance, since any change in its value will disturb the process from its steady-state operating point. This temperature can be measured and the controller can take into account its value in order to reject its effect before it appears in the system output. That is, it can be treated by DMC as a measurable disturbance and explicitly incorporated into the formulation. To do this, a model of the effect of the inlet temperature changes on the outlet temperature can easily be obtained by a step test.

In this example the disturbance is modelled by

$$y(t) = \sum_{i=1}^{30} d_i \triangle u(t-i)$$

with

d_1	d_2	d_3	d_4	d_5	d_6	d_7	d_8	d_9	d_{10}
0	0	0.050	0.095	0.135	0.172	0.205	0.234	0.261	0.285
d_{11}	d_{12}	d_{13}	d_{14}	d_{15}	d_{16}	d_{17}	d_{18}	d_{19}	d_{20}
0.306	0.326	0.343	0.359	0.373	0.386	0.397	0.407	0.417	0.425
d_{21}	d_{22}	d_{23}	d_{24}	d_{25}	d_{26}	d_{27}	d_{28}	d_{29}	d_{30}
0.433	0.439	0.445	0.451	0.456	0.460	0.464	0.468	0.471	0.474

which corresponds to the transfer function:

$$G(z) = \frac{0.05z^{-3}}{1 - 0.9z^{-1}}$$

Notice that the first ten d_i coefficients are used to build matrix **D** in the same way as matrix **G**.

Figure 3.7 shows a simulation where a disturbance occurs from $t = 20$ to $t = 60$. In case the controller explicitly considers the measurable disturbances, it is able to reject them, since the controller starts acting when the disturbance appears, not when its effect appears in the outlet temperature. On the other hand, if the controller does not take into account the measurable disturbance, it reacts later, when the effect on the output is considerable.

3.5 Exercises

3.1. If the cost function of a predictive controller is $J = \mathbf{ee}^T + \lambda\mathbf{uu}^T$, with $\mathbf{e} = \mathbf{Gu} + \mathbf{f} - \mathbf{w}$, demonstrate that the minimum is given by $\mathbf{u} = (\mathbf{G}^T\mathbf{G} + \lambda I)^{-1}\mathbf{G}^T(\mathbf{w} - \mathbf{f})$ in the unconstrained case.

3.2. For the water heater of Section 3.4:

1. Obtain the impulse response model.
2. Create matrices $\mathbf{H}_1$ and $\mathbf{H}_2$ for the MAC controller with $P = 3$ and $M = 5$.
3. Calculate the control law and simulate the same experiments as in the example.

3.3. Given a process described by

$$x(t+1) = \begin{bmatrix} 0.8 & 0.1 \\ 0.1 & 0.9 \end{bmatrix} x(t) + \begin{bmatrix} 0.1 \\ 0 \end{bmatrix} u(t), \quad y(t) = \begin{bmatrix} 0.1 & 0.9 \end{bmatrix} x(t)$$

simulate a PFC to move the process output from 0 to 1, with $N = 30$. Compare the effect of using one or three basis functions.

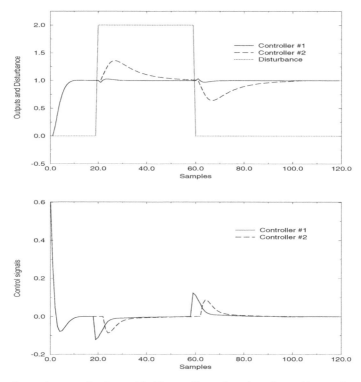

Fig. 3.7. Disturbance rejection with (Controller #1) and without (Controller #2) considering measurable disturbances

3.4. Change the number of model elements of the water heater from $N = 30$ to $N = 10$ and simulate the effect on the closed-loop behaviour.

3.5. For the same example, change the values of m, p and λ and see the results.

3.6. Obtain a state space model for the water heater, design a PFC controller and compare the results with those obtained with DMC.

3.7. Design a DMC for the process $G(s) = \frac{225(s-1)}{(s+1)(s^2+30s+225)}$. In order to do this:

1. Obtain the discrete equivalent when sampling at $T = 0.2$ second.
2. Compute the step response model by performing a unitary step at the input.
3. Calculate the control law with $N = 40$, $p = 15$, $m = 2$ and $\lambda = 1$.
4. Simulate the response to a setpoint change of 1.

4

Generalized Predictive Control

This chapter describes one of the most popular predictive control algorithms: Generalized Predictive Control (GPC). The method is developed in detail, showing the general procedure to obtain the control law and its most outstanding characteristics. The original algorithm is extended to include the cases of measurable disturbances and change in the predictor. Close derivations of this controller such as CRHPC and Stable GPC are also treated here, illustrating the way they can be implemented.

4.1 Introduction

The GPC method was proposed by Clarke *et al.* [58] and has become one of the most popular MPC methods in both industry and academia. It has been successfully implemented in many industrial applications [54], showing good performance and a certain degree of robustness. It can handle many different control problems for a wide range of plants with a reasonable number of design variables, which have to be specified by the user depending upon prior knowledge of the plant and control objectives.

The basic idea of GPC is to calculate a sequence of future control signals in such a way that it minimizes a multistage cost function defined over a prediction horizon. The index to be optimized is the expectation of a quadratic function measuring the distance between the predicted system output and some predicted reference sequence over the horizon plus a quadratic function measuring the control effort. This approach was used in [118] and [119] to obtain a generalized pole placement controller which is an extension of the well-known pole placement controllers [4], [205] and belongs to the class of extended horizon controllers.

Generalized Predictive Control has many ideas in common with the predictive controllers previously mentioned since it is based upon the same concepts but it has some differences. As will be seen, it provides an analytical solution (in the absence of constraints), it can deal with unstable and non-minimum phase plants and it incorporates the concept of control horizon as well as the consideration of weighting control increments in the cost function. The general set of choices available for GPC leads to a greater variety of control objectives compared to other approaches, some of which can be considered as subsets or limiting cases of GPC.

4.2 Formulation of Generalized Predictive Control

Most single-input single-output (SISO) plants, when considering operation around a particular setpoint and after linearization, can be described by

$$A(z^{-1})y(t) = z^{-d}B(z^{-1})u(t-1) + C(z^{-1})e(t)$$

where $u(t)$ and $y(t)$ are the control and output sequences of the plant and $e(t)$ is a zero mean white noise. A, B and C are the following polynomials in the backward shift operator z^{-1}:

$$A(z^{-1}) = 1 + a_1 z^{-1} + a_2 z^{-2} + \ldots + a_{na} z^{-na}$$
$$B(z^{-1}) = b_0 + b_1 z^{-1} + b_2 z^{-2} + \ldots + b_{nb} z^{-nb}$$
$$C(z^{-1}) = 1 + c_1 z^{-1} + c_2 z^{-2} + \ldots + c_{nc} z^{-nc}$$

where d is the dead time of the system. This model is known as a Controller Auto-Regressive Moving-Average (CARMA) model. It has been argued [58] that for many industrial applications in which disturbances are non-stationary an integrated CARMA (CARIMA) model is more appropriate. A CARIMA model is given by

$$A(z^{-1})y(t) = B(z^{-1})z^{-d}\,u(t-1) + C(z^{-1})\frac{e(t)}{\triangle} \tag{4.1}$$

with

$$\triangle = 1 - z^{-1}$$

For simplicity in the following, the C polynomial is chosen to be 1. Notice that if C^{-1} can be truncated it can be absorbed into A and B. The general case of a coloured noise will be treated later.

The Generalized Predictive Control (GPC) algorithm consists of applying a control sequence that minimizes a multistage cost function of the form

$$J(N_1, N_2, N_u) = \sum_{j=N_1}^{N_2} \delta(j)[\hat{y}(t+j\mid t) - w(t+j)]^2 + \sum_{j=1}^{N_u} \lambda(j)[\triangle u(t+j-1)]^2 \tag{4.2}$$

where $\hat{y}(t+j \mid t)$ is an optimum j step ahead prediction of the system output on data up to time t, N_1 and N_2 are the minimum and maximum costing horizons, N_u is the control horizon, $\delta(j)$ and $\lambda(j)$ are weighting sequences and $w(t+j)$ is the future reference trajectory, which can be calculated as shown in (2.6).

The objective of predictive control is to compute the future control sequence $u(t), u(t+1), \ldots$ in such a way that the future plant output $y(t+j)$ is driven close to $w(t+j)$. This is accomplished by minimizing $J(N_1, N_2, N_u)$.

In order to optimize the cost function the optimal prediction of $y(t+j)$ for $j \geq N_1$ and $j \leq N_2$ will be obtained. Consider the following Diophantine equation:

$$1 = E_j(z^{-1})\tilde{A}(z^{-1}) + z^{-j}F_j(z^{-1}) \text{ with } \tilde{A}(z^{-1}) = \triangle A(z^{-1}) \qquad (4.3)$$

The polynomials E_j and F_j are uniquely defined with degrees $j-1$ and na, respectively. They can be obtained by dividing 1 by $\tilde{A}(z^{-1})$ until the remainder can be factorized as $z^{-j}F_j(z^{-1})$. The quotient of the division is the polynomial $E_j(z^{-1})$.

If Equation (4.1) is multiplied by $\triangle E_j(z^{-1})z^j$,

$$\begin{aligned}\tilde{A}(z^{-1})E_j(z^{-1})y(t+j) &= E_j(z^{-1})B(z^{-1})\triangle u(t+j-d-1) \\ &+ E_j(z^{-1})e(t+j)\end{aligned} \qquad (4.4)$$

Considering (4.3), Equation (4.4) can be written as

$$(1-z^{-j}F_j(z^{-1}))y(t+j) = E_j(z^{-1})B(z^{-1})\triangle u(t+j-d-1)+E_j(z^{-1})e(t+j)$$

which can be rewritten as:

$$y(t+j) = F_j(z^{-1})y(t)+E_j(z^{-1})B(z^{-1})\triangle u(t+j-d-1)+E_j(z^{-1})e(t+j) \quad (4.5)$$

As the degree of polynomial $E_j(z^{-1}) = j-1$, the noise terms in Equation (4.5) are all in the future. The best prediction of $y(t+j)$ is therefore

$$\hat{y}(t+j \mid t) = G_j(z^{-1})\triangle u(t+j-d-1) + F_j(z^{-1})y(t)$$

where $G_j(z^{-1}) = E_j(z^{-1})B(z^{-1})$.

It is very simple to show that the polynomials E_j and F_j can be obtained recursively. The recursion of the Diophantine equation has been demonstrated in [58]. A simpler demonstration is given in the following. There are other formulations of GPC not based on the recursion of the Diophantine equation [2].

Consider that polynomials E_j and F_j have been obtained by dividing 1 by $\tilde{A}(z^{-1})$ until the remainder of the division can be factorized as $z^{-j}F_j(z^{-1})$. These polynomials can be expressed as:

$$F_j(z^{-1}) = f_{j,0} + f_{j,1}z^{-1} + \ldots + f_{j,na}z^{-na}$$
$$E_j(z^{-1}) = e_{j,0} + e_{j,1}z^{-1} + \ldots + e_{j,j-1}z^{-(j-1)}$$

Suppose that the same procedure is used to obtain E_{j+1} and F_{j+1}, that is, dividing 1 by $\tilde{A}(z^{-1})$ until the remainder of the division can be factorized as $z^{-(j+1)}F_{j+1}(z^{-1})$ with

$$F_{j+1}(z^{-1}) = f_{j+1,0} + f_{j+1,1}z^{-1} + \ldots + f_{j+1,na}z^{-na}$$

It is clear that only another step of the division performed to obtain the polynomials E_j and F_j has to be taken in order to obtain the polynomials E_{j+1} and F_{j+1}. The polynomial E_{j+1} will be given by

$$E_{j+1}(z^{-1}) = E_j(z^{-1}) + e_{j+1,j}z^{-j}$$

with $e_{j+1,j} = f_{j,0}$

The coefficients of polynomial F_{j+1} can then be expressed as:

$$f_{j+1,i} = f_{j,i+1} - f_{j,0}\,\tilde{a}_{i+1} \quad i = 0\ldots na - 1$$

The polynomial G_{j+1} can be obtained recursively as follows:

$$G_{j+1} = E_{j+1}B = (E_j + f_{j,0}z^{-j})B$$
$$G_{j+1} = G_j + f_{j,0}z^{-j}B$$

That is, the first j coefficient of G_{j+1} will be identical to those of G_j and the remaining coefficients will be given by:

$$g_{j+1,j+i} = g_{j,j+i} + f_{j,0}\,b_i \qquad i = 0\ldots nb$$

To solve the GPC problem the set of control signals $u(t), u(t+1), \ldots, u(t+N)$ has to be obtained in order to optimize Expression (4.2). As the system considered has a dead time of d sampling periods, the output of the system will be influenced by signal $u(t)$ after sampling period $d + 1$. The values N_1, N_2 and N_u defining the horizon can be defined by $N_1 = d + 1$, $N_2 = d + N$ and $N_u = N$. Notice that there is no point in making $N_1 < d + 1$ as terms added to expression (4.2) will only depend on the past control signals. On the other hand, if $N_1 > d + 1$ the first points in the reference sequence, being the ones guessed with most certainty, will not be taken into account.

Now consider the following set of j ahead optimal predictions:

$$\hat{y}(t + d + 1 \mid t) = G_{d+1}\,\triangle\,u(t) + F_{d+1}y(t)$$
$$\hat{y}(t + d + 2 \mid t) = G_{d+2}\,\triangle\,u(t + 1) + F_{d+2}y(t)$$

$$\vdots$$

$$\hat{y}(t + d + N \mid t) = G_{d+N}\,\triangle\,u(t + N - 1) + F_{d+N}y(t)$$

which can be written as:

$$\mathbf{y} = \mathbf{Gu} + \mathbf{F}(z^{-1})y(t) + \mathbf{G'}(z^{-1})\,\triangle\,u(t - 1) \qquad (4.6)$$

where

$$\mathbf{y} = \begin{bmatrix} \hat{y}(t+d+1 \mid t) \\ \hat{y}(t+d+2 \mid t) \\ \vdots \\ \hat{y}(t+d+N \mid t) \end{bmatrix} \qquad \mathbf{u} = \begin{bmatrix} \triangle u(t) \\ \triangle u(t+1) \\ \vdots \\ \triangle u(t+N-1) \end{bmatrix}$$

$$\mathbf{G} = \begin{bmatrix} g_0 & 0 & \dots & 0 \\ g_1 & g_0 & \dots & 0 \\ \vdots & \vdots & \vdots & \vdots \\ g_{N-1} & g_{N-2} & \dots & g_0 \end{bmatrix}$$

$$\mathbf{G}'(z^{-1}) = \begin{bmatrix} (G_{d+1}(z^{-1}) - g_0)z \\ (G_{d+2}(z^{-1}) - g_0 - g_1 z^{-1})z^2 \\ \vdots \\ (G_{d+N}(z^{-1}) - g_0 - g_1 z^{-1} - \dots - g_{N-1} z^{-(N-1)})z^N \end{bmatrix}$$

$$\mathbf{F}(z^{-1}) = \begin{bmatrix} F_{d+1}(z^{-1}) \\ F_{d+2}(z^{-1}) \\ \vdots \\ F_{d+N}(z^{-1}) \end{bmatrix}$$

Notice that the last two terms in Equation (4.6) only depend on the past and can be grouped into $\mathbf{f}$ leading to:

$$\mathbf{y} = \mathbf{Gu} + \mathbf{f}$$

Notice that if all initial conditions are zero, the free response $\mathbf{f}$ is also zero. If a unit step is applied to the input at time t; that is,

$$\triangle u(t) = 1, \triangle u(t+1) = 0, \dots, \triangle u(t+N-1) = 0$$

the expected output sequence $[\hat{y}(t+1), \hat{y}(t+2), \dots, \hat{y}(t+N)]^T$ is equal to the first column of matrix $\mathbf{G}$. That is, the first column of matrix $\mathbf{G}$ can be calculated as the step response of the plant when a unit step is applied to the manipulated variable. The free response term can be calculated recursively by

$$\mathbf{f}_{j+1} = z(1 - \tilde{A}(z^{-1}))\mathbf{f}_j + B(z^{-1}) \triangle u(t-d+j)$$

with $\mathbf{f}_0 = y(t)$ and $\triangle u(t+j) = 0$ for $j \geq 0$.

Expression (4.2) can be written as

$$J = (\mathbf{Gu} + \mathbf{f} - \mathbf{w})^T (\mathbf{Gu} + \mathbf{f} - \mathbf{w}) + \lambda \mathbf{u}^T \mathbf{u} \qquad (4.7)$$

where

$$\mathbf{w} = \begin{bmatrix} w(t+d+1) & w(t+d+2) & \cdots & w(t+d+N) \end{bmatrix}^T$$

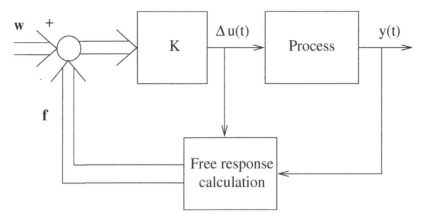

Fig. 4.1. GPC control law

Equation (4.7) can be written as

$$J = \frac{1}{2}\mathbf{u}^T\mathbf{H}\mathbf{u} + \mathbf{b}^T\mathbf{u} + f_0 \tag{4.8}$$

where

$$\mathbf{H} = 2(\mathbf{G}^T\mathbf{G} + \lambda\mathbf{I})$$
$$\mathbf{b}^T = 2(\mathbf{f} - \mathbf{w})^T\mathbf{G}$$
$$f_0 = (\mathbf{f} - \mathbf{w})^T(\mathbf{f} - \mathbf{w})$$

The minimum of J, assuming there are no constraints on the control signals, can be found by making the gradient of J equal to zero, which leads to:

$$\mathbf{u} = -\mathbf{H}^{-1}\mathbf{b} = (\mathbf{G}^T\mathbf{G} + \lambda\mathbf{I})^{-1}\mathbf{G}^T(\mathbf{w} - \mathbf{f}) \tag{4.9}$$

Notice that the control signal that is actually sent to the process is the first element of vector $\mathbf{u}$, given by:

$$\triangle u(t) = \mathbf{K}(\mathbf{w} - \mathbf{f}) \tag{4.10}$$

where $\mathbf{K}$ is the first row of matrix $(\mathbf{G}^T\mathbf{G}+\lambda\mathbf{I})^{-1}\mathbf{G}^T$. This has a clear meaning that can easily be derived from Figure 4.1: if there are no future predicted errors, that is, if $\mathbf{w} - \mathbf{f} = \mathbf{0}$, then there is no control move, since the objective will be fulfilled with the free evolution of the process. However, in the other case, there will be an increment in the control action proportional (with a factor $\mathbf{K}$) to that future error. Notice that the action is taken with respect to *future* errors, not *past* errors, as is the case in conventional feedback controllers.

Notice that only the first element of $\mathbf{u}$ is applied and the procedure is repeated at the next sampling time. The solution to the GPC given involves the inversion (or triangularization) of an $N \times N$ matrix which requires a substantial amount of computation. In [58] the concept of control horizon is used to reduce the amount of computation needed, assuming that the projected control signals are going to be constant after $N_u < N$. This leads to the inversion of an $N_u \times N_u$ matrix which reduces the amount of computation (in particular, if $N_u = 1$ it is reduced to a scalar computation, as in EPSAC), but restricts the optimality of the GPC. A fast algorithm to implement self-tuning GPC for processes that can be modelled by the reaction curve method is presented in the next chapter. The use of Hopfield neural networks has also been proposed [172] to obtain fast GPCs.

4.3 The Coloured Noise Case

When the noise polynomial $C(z^{-1})$ of Equation (4.1) is not equal to 1 the prediction changes slightly. In order to calculate the predictor in this situation, the following Diophantine equation is solved:

$$C(z^{-1}) = E_j(z^{-1})\tilde{A}(z^{-1}) + z^{-j}F_j(z^{-1}) \tag{4.11}$$

with $\delta(E_j(z^{-1})) = j - 1$ and $\delta(F_j(z^{-1})) = \delta(\tilde{A}(z^{-1})) - 1$.

Multiplying equation (4.1) by $\triangle E_j(z^{-1})z^j$ and using (4.11)

$$C(z^{-1})(y(t+j) - E_j(z^{-1})e(t+j)) = E_j(z^{-1})B(z^{-1})\triangle u(t+j-1) + F_j(z^{-1})y(t)$$

As the noise terms are all in the future, the expected value of the left-hand side of this equation is:

$$E[C(z^{-1})(y(t+j) - E_j(z^{-1})e(t+j))] = C(z^{-1})\hat{y}(t+j|t)$$

The expected value of the output can be generated by the equation:

$$C(z^{-1})\hat{y}(t+j|t) = E_j(z^{-1})B(z^{-1}) \triangle u(t+j-1) + F_j(z^{-1})y(t) \tag{4.12}$$

Notice that this prediction equation could be used to generate the predictions in a recursive way. An explicit expression for the optimal j step ahead prediction can be obtained by solving the Diophantine equation

$$1 = C(z^{-1})M_j(z^{-1}) + z^{-k}N_j(z^{-1}) \tag{4.13}$$

with $\delta(M_j(z^{-1})) = j - 1$ and $\delta(N_j(z^{-1})) = \delta(C(z^{-1})) - 1$.

Multiplying Equation (4.12) by $M_j(z^{-1})$ and using (4.13),

$$\hat{y}(t+j|t) = M_j E_j(z^{-1})B(z^{-1})\triangle u(t+j-1) + M_j(z^{-1})F_j(z^{-1})y(t) + N_j(z^{-1})y(t)$$

which can be expressed as

$$\hat{y}(t+j|t) = G(z^{-1}) \, \triangle \, u(t+j-1) + G_p(z^{-1}) \, \triangle \, u(t+j-1)$$
$$+ (M_j(z^{-1})F_j(z^{-1}) + N_j(z^{-1}))y(t)$$

with $\delta(G(z^{-1})) < j$. These predictions can be used in the cost function which can be minimized as in the white noise case.

Another way of computing the prediction is by considering the filtered signals from the plant input/output data

$$y^f(t) = \frac{1}{C(z^{-1})}y(t) \qquad\qquad u^f(t) = \frac{1}{C(z^{-1})}u(t)$$

so that the resulting overall model becomes

$$A(z^{-1})y^f(t) = B(z^{-1})u^f(t) + \frac{e(t)}{\triangle}$$

and the white noise procedure for computing the prediction can be used. The predicted signal $\hat{y}^f(t+j|t)$ obtained this way has to be filtered by $C(z^{-1})$ in order to get $\hat{y}(t+j|t)$.

4.4 An Example

In order to show how a Generalized Predictive Controller can be implemented, a simple example is presented. The controller will be designed for a first-order system for the sake of clarity.

The following discrete equivalence can be obtained when a first-order continuous plant is discretized

$$(1 + az^{-1})y(t) = (b_0 + b_1 z^{-1})u(t-1) + \frac{e(t)}{\triangle}$$

In this example the delay d is equal to 0 and the noise polynomial $C(z^{-1})$ is considered to be equal to 1.

The algorithm to obtain the control law described in the previous section will be used on the preceding system, obtaining numerical results for the parameter values $a = -0.8$, $b_0 = 0.4$ and $b_1 = 0.6$, the horizons being $N_1 = 1$ and $N_2 = N_u = 3$. As has been shown, predicted values of the process output over the horizon are first calculated and rewritten in the form of Equation (4.6), and then the control law is computed using Expression (4.9).

Predictor polynomials $E_j(z^{-1})$, $F_j(z^{-1})$ from $j = 1$ to $j = 3$ will be calculated solving the Diophantine Equation (4.3), with

$$\tilde{A}(z^{-1}) = A(z^{-1})(1 - z^{-1}) = 1 - 1.8z^{-1} + 0.8z^{-2}$$

In this simple case where the horizon is not too long, the polynomials can be directly obtained by dividing 1 by $\tilde{A}(z^{-1})$ with simple calculations. As has been explained earlier, they can also be computed recursively, starting with the values obtained at the first step of the division, that is:

$$E_1(z^{-1}) = 1 \qquad F_1(z^{-1}) = 1.8 - 0.8z^{-1}$$

Whatever the procedure employed, the values obtained are:

$$E_2 = 1 + 1.8z^{-1} \qquad F_2 = 2.44 - 1.44z^{-1}$$
$$E_3 = 1 + 1.8z^{-1} + 2.44z^{-2} \quad F_3 = 2.952 - 1.952z^{-1}$$

With these values and the polynomial $B(z^{-1}) = 0.4 + 0.6z^{-1}$, the values of $G_i(z^{-1})$ are

$$G_1 = 0.4 + 0.6z^{-1}$$
$$G_2 = 0.4 + 1.32z^{-1} + 1.08z^{-2}$$
$$G_3 = 0.4 + 1.32z^{-1} + 2.056z^{-2} + 1.464z^{-3}$$

and so the predicted outputs can be written as:

$$
\begin{bmatrix} \hat{y}(t+1 \mid t) \\ \hat{y}(t+2 \mid t) \\ \hat{y}(t+3 \mid t) \end{bmatrix}
=
\begin{bmatrix} 0.4 & 0 & 0 \\ 1.32 & 0.4 & 0 \\ 2.056 & 1.32 & 0.4 \end{bmatrix}
\begin{bmatrix} \triangle u(t) \\ \triangle u(t+1) \\ \triangle u(t+2) \end{bmatrix}
$$
$$
+ \underbrace{\begin{bmatrix} 0.6 \triangle u(t-1) + 1.8y(t) - 0.8y(t-1) \\ 1.08 \triangle u(t-1) + 2.44y(t) - 1.44y(t-1) \\ 1.464 \triangle u(t-1) + 2.952y(t) - 1.952y(t-1) \end{bmatrix}}_{\mathbf{f}}
$$

The following step is to calculate $\mathbf{H}^{-1}\mathbf{b}$. If λ is taken as equal to 0.8

$$
(\mathbf{G}^T\mathbf{G} + \lambda\mathbf{I})^{-1}\mathbf{G}^T =
\begin{bmatrix} 0.133 & 0.286 & 0.147 \\ -0.154 & -0.165 & 0.286 \\ -0.029 & -0.154 & 0.1334 \end{bmatrix}
$$

As only $\triangle u(t)$ is needed for the calculations, only the first row of the matrix is used, obtaining the following expression for the control law:

$$\triangle u(t) = -0.604 \triangle u(t-1) - 1.371y(t) + 0.805y(t-1)$$
$$+ 0.133w(t+1) + 0.286w(t+2) + 0.147w(t+3)$$

where $w(t+i)$ is the reference trajectory which can be considered constant and equal to the current setpoint or a first-order approach to the desired value. Then the control signal is a function of this desired reference and of past inputs and outputs and is given by:

$$u(t) = 0.396u(t-1) + 0.604u(t-2) - 1.371y(t) + 0.805y(t-1)$$
$$+ 0.133w(t+1) + 0.286w(t+2) + 0.147w(t+3) \qquad (4.14)$$

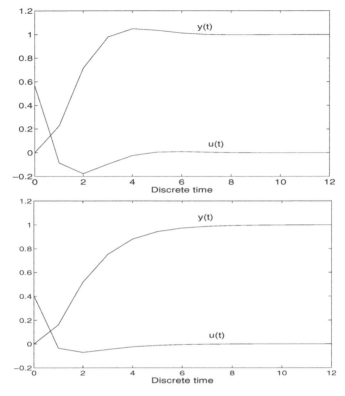

Fig. 4.2. System response

Simulation results show the behaviour of the closed-loop system. In the first graph of Figure 4.2 the reference is constant and equal to 1, and in the second one there is a smooth approach to the same value, obtaining a slightly different response, slower but without overshoot.

The GPC control law can also be calculated without the use of the Diophantine equation.

To obtain the control law it is necessary to know matrix $\mathbf{G}$ and the free response $\mathbf{f}$, to compute $\mathbf{u} = (\mathbf{G}^T\mathbf{G} + \lambda\mathbf{I})^{-1}\mathbf{G}^T(\mathbf{w} - \mathbf{f})$. Matrix $\mathbf{G}$ is composed of the plant step response coefficients, so that the elements of the first column of this matrix are the first N coefficients, that can be computed as

$$g_j = -\sum_{i=1}^{j} a_i g_{j-i} + \sum_{i=0}^{j-1} b_i \text{ with } g_k = 0 \quad \forall k < 0$$

where b_i and a_i are the parameters of the numerator and denominator of the transfer function.

Therefore, as the prediction horizon is 3, $A = 1 - 0.8z^{-1}$ and $B = 0.4 + 0.6z^{-1}$

$$g_0 = b_0 = 0.4$$
$$g_1 = -a_1 g_0 + b_0 + b_1 = 1.32$$
$$g_2 = -a_1 g_1 - a_2 g_0 - a_3 g_0 + b_0 + b_1 = 2.056$$

and the matrix is given by

$$\mathbf{G} = \begin{bmatrix} 0.4 & 0 & 0 \\ 1.32 & 0.4 & 0 \\ 2.056 & 1.32 & 0.4 \end{bmatrix}$$

which logically coincides with the one obtained by the previous method.

The free response can also be calculated without the use of the Diophantine equation, just noting that it is the response of the plant assuming that future controls equal the previous control $u(t-1)$ and that the disturbance is constant. Thus, using the transfer function

$$y(t) = 0.8y(t-1) + 0.4u(t-1) + 0.6u(t-2)$$
$$y(t+1) = 0.8y(t) + 0.4u(t) + 0.6u(t-1)$$

If both equations are added and $y(t+1)$ is extracted

$$y(t+1) = 1.8y(t) - 0.8y(t-1) + 0.4 \triangle u(t) + 0.6 \triangle u(t-1)$$

Now, considering that in the free response only the control increments before instant t appear:

$$f(t+1) = 1.8y(t) - 0.8y(t-1) + 0.6 \triangle u(t-1)$$
$$f(t+2) = 1.8f(t+1) - 0.8y(t) = 2.44y(t) - 1.44y(t-1) + 1.08 \triangle u(t-1)$$
$$f(t+3) = 1.8f(t+2) - 0.8f(t+1)$$
$$= 2.952y(t) - 1.952y(t-1) + 1.464 \triangle u(t-1)$$

Vector $\mathbf{f}$ obtained this way is the same as the one previously obtained, so the control law is the one given by Equation (4.14).

4.5 Closed-Loop Relationships

Closed-loop relations can be obtained for the unconstrained GPC. The closed-loop system can be posed in the classical pole-placement structure of Figure 4.3

The control law can be stated as

$$R(z^{-1}) \triangle u(t) = T(z^{-1})w(t) - S(z^{-1})y(t) \qquad (4.15)$$

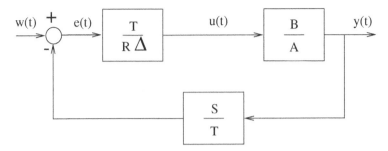

Fig. 4.3. Classical pole-placement structure

where R, S and T are polynomials in the backward shift operator. This control law can be considered as composed of a feedforward term (T/R) and a feedback part (S/R). In this situation it is possible to obtain the closed-loop transfer function and derive some properties such as stability and robustness. First, the general GPC control scheme of figure 4.3 must be rearranged to take the form of Equation (4.15).

The control law of Equation (4.9) gives the future control sequence **u**. As a receding strategy is being used, only the first element of that sequence $\triangle u(t \mid t)$ is actually sent to the process, therefore the control action is given by

$$\triangle u(t) = \mathbf{K}(\mathbf{w} - \mathbf{f}) = \sum_{i=N_1}^{N_2} k_i[w(t+i) - f(t+i)] \tag{4.16}$$

where **K** is the first row of matrix $(\mathbf{G}^T\mathbf{G} + \lambda\mathbf{I})^{-1}\mathbf{G}^T$.

The general case in which the $C(z^{-1})$ polynomial is not equal to zero will be considered to obtain the free response. In many situations this polynomial is not identified, since identification is not easy due to its time-varying characteristics and the difficulty of the CARIMA model to describe general deterministic disturbances. In these cases it is substituted by the so-called T polynomial that can be regarded as a fixed observer or a prefilter, as will be discussed later.

Then the plant model is given by:

$$A(z^{-1})y(t) = B(z^{-1})\,u(t-1) + T(z^{-1})\frac{e(t)}{\triangle}$$

The Diophantine equation that must be solved now includes the T polynomial:

$$T(z^{-1}) = E_j(z^{-1})\,\triangle\,A(z^{-1}) + z^{-j}F_j(z^{-1}) \tag{4.17}$$

Using this equation and the plant model, the future output value is given by

$$y(t+j) = \frac{B(z^{-1})}{A(z^{-1})}u(t+j-1) + E_j(z^{-1})e(t+j) + \frac{F_j(z^{-1})}{A(z^{-1})\triangle}e(t)$$

replacing the $e(t)$ from the plant model and using (4.17):

$$y(t+j) = \frac{F_j}{T} y(t) + \frac{E_j B}{T} \triangle u(t+j-1) + E_j e(t+j)$$

The best prediction is obtained by replacing $e(t+j)$ by its expected value (zero):

$$\hat{y}(t+j \mid t) = \frac{F_j}{T} y(t) + \frac{E_j B}{T} \triangle u(t+j-1)$$

This expression is a function of known values and future control actions. The control actions can be separated into past ones (those taken before instant t) and future ones (which must be calculated by the controller) using the Diophantine equation[1]

$$E_j(z^{-1})B(z^{-1}) = H_j(z^{-1})T(z^{-1}) + z^{-j}I_j(z^{-1}) \tag{4.18}$$

that leads to the prediction equation:

$$\hat{y}(t+j \mid t) = H_j \triangle u(t+j) + \frac{I_j}{T} \triangle u(t-1) + \frac{F_j}{T} y(t)$$
$$= H_j \triangle u(t+j) + I_j \triangle u^f(t-1) + F_j y^f(t) \tag{4.19}$$

Using the filtered variables $y^f(t) = \frac{y(t)}{T}$ and $\triangle u^f(t-1) = \frac{\triangle u(t-1)}{T}$, this equation provides the same prediction along the horizon as given by (4.6) when $T(z^{-1}) = 1$, where the coefficients of H_j are the elements of matrix $\mathbf{G}$ and I_j are the rows of vector $\mathbf{G}'$.

Now the free response of the system (the one needed for the control law) is given by:

$$\mathbf{f} = \mathbf{I}(z^{-1}) \triangle u^f(t-1) + \mathbf{F}(z^{-1}) y^f(t) = \mathbf{I}(z^{-1}) \frac{\triangle u(t-1)}{T(z^{-1})} + \mathbf{F}(z^{-1}) \frac{y(t)}{T(z^{-1})}$$

Once the free response has been obtained when the T polynomial is considered, it can be included in the expression of the control law given by (4.16):

$$\triangle u(t) = \mathbf{K}(\mathbf{w} - \mathbf{f}) = \sum_{i=N_1}^{N_2} k_i[w(t+i) - f(t+i)]$$

$$= \sum_{i=N_1}^{N_2} k_i w(t+i) - \sum_{i=N_1}^{N_2} k_i \frac{I_i(z^{-1})}{T(z^{-1})} \triangle u(t-1) - \sum_{i=N_1}^{N_2} k_i \frac{F_i(z^{-1})}{T(z^{-1})} y(t)$$

Omitting the term z^{-1} and reordering the last equation

$$\left[T + z^{-1} \sum_{i=N_1}^{N_2} k_i I_i \right] \triangle u(t) = T \sum_{i=N_1}^{N_2} k_i w(t) - \sum_{i=N_1}^{N_2} k_i F_i y(t)$$

---

[1] Notice that this equation with $T(z^{-1}) = 1$ is implicitly used to derive $\mathbf{G}$ and $\mathbf{G}'$ in (4.6).

where it has been considered that the future reference trajectory keeps constant along the horizon or its evolution is unknown and therefore $w(t + i)$ is taken as equal to $w(t)$. In the other case, the first term of the right hand side should be expressed as $T \sum_{i=N_1}^{N_2} k_i z^i w(t)$ and therefore the following relations could change slightly.

The values of polynomials R and S can be obtained by comparing the last equation with (4.15), and are given by:

$$R(z^{-1}) = \frac{T(z^{-1}) + z^{-1} \sum_{i=N_1}^{N_2} k_i I_i}{\sum_{i=N_1}^{N_2} k_i}$$

$$S(z^{-1}) = \frac{\sum_{i=N_1}^{N_2} k_i F_i}{\sum_{i=N_1}^{N_2} k_i}$$

The closed-loop characteristic equation comes from inclusion of the control action given by (4.15) in the plant model expressed as:

$$A \bigtriangleup y(t) = B \bigtriangleup u(t-1) + Te(t)$$

Therefore, if the control action

$$\bigtriangleup u(t) = \frac{T}{R} w(t) - \frac{S}{R} y(t)$$

is replaced in the plant model, the following expression is obtained:

$$A \bigtriangleup y(t) = B z^{-1}(\frac{T}{R} w(t) - \frac{S}{R} y(t)) + Te(t)$$

Extracting $y(t)$ from this equation provides the closed-loop relation that gives the output as a function of the reference and the disturbance

$$y(t) = \frac{BTz^{-1}}{RA \bigtriangleup + BSz^{-1}} w(t) + \frac{TR}{RA \bigtriangleup + BSz^{-1}} e(t) \qquad (4.20)$$

and consequently the characteristic equation is given by:

$$RA \bigtriangleup + BSz^{-1} = 0$$

With a few manipulations and using (4.18), the characteristic polynomial can be decomposed as:

$$RA \bigtriangleup + BSz^{-1} = \frac{1}{\sum_{i=N_1}^{N_2} k_i} (T\tilde{A} + T \sum_{i=N_1}^{N_2} k_i z^{i-1}(B - \tilde{A} H_i)) = TP_c$$

Therefore Equation (4.20) turns to

$$y(t) = \frac{Bz^{-1}}{P_c} w(t) + \frac{R}{P_c} e(t)$$

where it is shown that the T polynomial is cancelled in the closed-loop transfer function between output and reference, as is the case in any observer, and that stability and performance are driven by the roots of polynomial P_c. However, it is difficult to establish clear dependencies of these roots on the tuning parameters N_1, N_2, N_u and λ. It is interesting to note that $P_c(1) = B(1)$, which guarantees offset-free response since the static gain of the transfer function between output and reference is always one.

When the GPC is written in the general pole-placement structure, some stability properties can be derived from the transfer function. A paper by Clarke and Mohtadi [57] presents some properties related to stability. It is proven that stability can be guaranteed if the tuning parameters (horizons and control-weighting factor) are correctly chosen. In the following sections, two formulations related to GPC with guaranteed stability will be treated in more detail.

4.6 The Role of the T Polynomial

Although the T polynomial does not appear in the transfer function between the output and the reference, this is not the case for the transfer function between the output and the disturbance. From Equation (4.19) it can be seen that both the output $y(t)$ and the control move $\triangle u(t)$ appear in the prediction, and therefore in the control law, filtered by $1/T$. Thus, from a practical point of view, it means that the T polynomial can be treated as a filter. By ensuring that the degree of T is big enough, the roll-off of the filter attenuates the component of prediction error caused by model mismatch, which is particularly important at high frequencies. Notice that low-frequency disturbances can be removed by the $\triangle$ term that appears in the prediction.

The high-frequency disturbances are mainly due to the presence of high-frequency unmodelled dynamics and unmeasurable load disturbances. If there are no unmodelled dynamics, the effect of T is the rejection of disturbances, with no influence on reference tracking. In this case T can be used to *detune* the response to unmeasurable high-frequency load disturbances, preventing excessive control actions.

On the other hand, T is used as a design parameter that can influence robust stability. In this case the predictions will not be optimal but robustness in the face of uncertainties can be achieved, in a similar interpretation to that used by Ljung [128]. Then this polynomial can be considered as a prefilter as well as an observer. The effective use of observers is known to play an essential role in the robust realization of predictive controllers (see [57] for the effect of prefiltering on robustness).

4.6.1 Selection of the T Polynomial

The selection of the filter polynomial T is not a trivial matter. Although some guidelines are given in [183] for mean-level and deadbeat GPC, a systematic design strategy for the T filter has not been completely established. Usually it is assumed that the stronger filtering (considered as stronger that filtering with smaller bandwidth or bigger slope if the bandwidth is the same) has better robustness properties against high-frequency uncertainties. But this is not always true, as is shown with some counterexamples in [210]. In this paper and in [209] Yoon and Clarke present guidelines for the selection of T. These guidelines are based upon the robustness margin improvement at high frequencies and conclude stating that, for open-loop stable processes, the best choice is

$$T(z^{-1}) = A(z^{-1})(1 - \beta z^{-1})^{N_1 - \delta(P)}$$

Where β is close to the dominant root of A, N_1 is the minimum prediction horizon and $\delta(P)$ is the degree of polynomial P (the filter used to generate a reference trajectory with specified dynamics, see Section 2.1.2).

Notice that this idea of filtering for improving robustness also lies in Internal Model Control (IMC)[141] where, once the controller that provides the desired performance is obtained, it is detuned with a filter to improve robustness.

4.6.2 Relationships with Other Formulations

The prefiltering with T can be compared to $\mathcal{H}_\infty$ optimization based on the Q parameterization [141], obtaining equivalent robustness results [208]. It implies that prefiltering with polynomial T is an alternative to the optimal Q whose computation is demanding, especially in the adaptive case.

Robustness is improved by the introduction of polynomial T but, on the other hand, this fact implies that the prediction is no longer optimal. In a certain way, this idea is similar to the one used in the Linear Quadratic Gaussian (LQG) regulator to recover the good robustness properties that the Linear Quadratic Regulator (LQR) loses with the inclusion of the observer. This recovery is achieved by means of the LQG/LTR method, that consists of the Loop Transfer Recovery, (LTR), in such a way that it approaches the open-loop transfer function of the LQR method. This can be done by acting on the Kalman filter parameters, working with fictitious covariances (see [81]). In this way robustness is gained although prediction deteriorates. In both cases, the loss of optimality in the prediction or estimation is not considered a problem, since the controller works and is robust.

4.7 The P Polynomial

In the presentation of Generalized Predictive Control Clarke [58] points out the possibility of the use of an additional polynomial $P(z^{-1})$ as a design element in a similar way as employed in the Minimum Variance Controller [55] as a weighting polynomial that would be used for model following.

The $P(z^{-1})$ polynomial can appear when defining an auxiliary output as happens, for instance, in EPSAC (see Chapter 2)

$$\psi(t) = P(z^{-1})y(t)$$

in such a way that it affects the output. This polynomial allows the control objectives to expand by using its roots as design parameters. In this way *deadbeat*, pole-placement or LQ control can be achieved.

This can easily be incorporated into the standard GPC formulation by considering the augmented plant

$$A \triangle P(z^{-1})y(t) = BA \triangle P(z^{-1})u(t-1)$$

with P(1)=1 to guarantee $\psi(t) = y(t)$ in steady state.

This is equivalent to defining filtered auxiliary signals $\nu(t) = P(z^{-1})u(t)$ and $\psi(t) = P(z^{-1})y(t)$. Thus the plant is given by:

$$A \triangle \psi(t) = BA \triangle \nu(t-1)$$

Now the error that appears in the cost function is defined by $w(t+j) - \psi(t+j)$, which is equivalent to considering a reference trajectory generated by $1/P(z^{-1})$. A *deadbeat* control of $\psi(t)$ can be achieved by acting on $\nu(t)$, whose closed-loop transfer function is given by:

$$\psi(t) = \frac{B(z^{-1})}{B(1)}w(t)$$

This means that

$$y(t) = \frac{B(z^{-1})}{B(1)P(z^{-1})}w(t) \qquad (4.21)$$

and the GPC algorithm is solved to provide the auxiliary control increment $\triangle\nu(t)$, from which the system input is calculated as:

$$u(t) = u(t-1) + \frac{\triangle\nu(t)}{P(z^{-1})}$$

As can be observed (4.21) is the typical response of a pole-placement method, with poles placed at zeros of the chosen $P(z^{-1})$. That is, the output is made to track the dynamics specified by $P(z^{-1})$.

4.8 Consideration of Measurable Disturbances

Many processes are affected by external disturbances caused by the variation of variables that can be measured. This situation is typical in processes whose outputs are affected by variations of the load regime. Consider, for instance, a cooled jacket continuous reactor where the temperature is controlled by manipulating the water flow entering the cooling jacket. Any variation of the reactive flows will influence the reactor temperature. These types of perturbations, also known as load disturbances, can easily be handled by the use of feedforward controllers. Known disturbances can be taken explicitly into account in MPC, as will be seen in the following.

Consider a process described by the following In this case the CARIMA model must be changed to include the disturbances:

$$A(z^{-1})y(t) = B(z^{-1})u(t-1) + D(z^{-1})v(t) + \frac{1}{\triangle}C(z^{-1})e(t) \qquad (4.22)$$

where the variable $v(t)$ is the measured disturbance at time t and $D(z^{-1})$ is a polynomial defined as:

$$D(z^{-1}) = d_0 + d_1 z^{-1} + d_2 z^{-2} + \ldots + d_{n_d} z^{-n_d}$$

Multiplying Equation (4.22) by $\triangle E_j(z^{-1})z^j$:

$$E_j(z^{-1})\tilde{A}(z^{-1})y(t+j) = E_j(z^{-1})B(z^{-1}) \triangle u(t+j-1)$$
$$+ E_j(z^{-1})D(z^{-1}) \triangle v(t+j) + E_j(z^{-1})e(t+j)$$

By using (4.3), and after some manipulation, we get:

$$y(t+j) = F_j(z^{-1})y(t) + E_j(z^{-1})B(z^{-1}) \triangle u(t+j-1)$$
$$+ E_j(z^{-1})D(z^{-1}) \triangle v(t+j) + E_j(z^{-1})e(t+j)$$

Notice that because the degree of $E_j(z^{-1})$ is $j-1$, the noise terms are all in the future. By taking the expectation operator and considering that $E[e(t)] = 0$, the expected value for $y(t+j)$ is given by:

$$\hat{y}(t+j|t) = E[y(t+j)] = F_j(z^{-1})y(t) + E_j(z^{-1})B(z^{-1}) \triangle u(t+j-1)$$
$$+ E_j(z^{-1})D(z^{-1}) \triangle v(t+j)$$

By making the polynomial $E_j(z^{-1})D(z^{-1}) = H_j(z^{-1}) + z^{-j}H_j'(z^{-1})$, with $\delta(H_j(z^{-1})) = j - 1$, the prediction equation can now be written as:

$$\hat{y}(t+j|t) = G_j(z^{-1}) \triangle u(t+j-1) + H_j(z^{-1}) \triangle v(t+j)$$
$$+G_j'(z^{-1}) \triangle u(t-1) + H_j'(z^{-1}) \triangle v(t) + F_j(z^{-1})y(t) \qquad (4.23)$$

Notice that the last three terms of the right-hand side of this equation depend on past values of the process output, measured disturbances and input variables and correspond to the free response of the process considered if the control signals and measured disturbances are kept constant; while the first term only depends on future values of the control signal and can be interpreted as the forced response, that is, the response obtained when the initial conditions are zero $y(t-j) = 0$, $\triangle u(t-j-1) = 0$, $\triangle v(t-j)$ for $j > 0$.

The second term of Equation (4.23) depends on the future deterministic disturbances. In some cases, when they are related to the process load, future disturbances are known. In other cases, they can be predicted using trends or other means. If this is the case, the term corresponding to future deterministic disturbances can be computed. If the future load disturbances are supposed to be constant and equal to the last measured value (i.e., $v(t+j) = v(t)$), then $\triangle v(t+j) = 0$ and the second term of this equation vanishes.

Equation (4.23) can be rewritten as

$$\hat{y}(t+j|t) = G_j(z^{-1})\,\triangle u(t+j-1) + H_j(z^{-1})\,\triangle v(t+j) + f_j$$

with $f_j = G_j'(z^{-1})\,\triangle u(t-1) + H_j'(z^{-1})\,\triangle v(t) + F_j(z^{-1})y(t)$.

Let us now consider a set of N j ahead predictions:

$$\hat{y}(t+1|t) = G_1(z^{-1})\,\triangle u(t) + H_j(z^{-1})\,\triangle v(t+1) + f_1$$
$$\hat{y}(t+2|t) = G_2(z^{-1})\,\triangle u(t+1) + H_j(z^{-1})\,\triangle v(t+2) + f_2$$
$$\vdots$$
$$\hat{y}(t+N|t) = G_N(z^{-1})\,\triangle u(t+N-1) + H_j(z^{-1})\,\triangle v(t+N) + f_N$$

Because of the recursive properties of the E_j polynomial, these expressions can be rewritten as

$$
\begin{bmatrix}
\hat{y}(t+1|t) \\
\hat{y}(t+2|t) \\
\vdots \\
\hat{y}(t+j|t) \\
\vdots \\
\hat{y}(t+N|t)
\end{bmatrix}
=
\begin{bmatrix}
g_0 & 0 & \cdots & 0 & \cdots & 0 \\
g_1 & g_0 & \cdots & 0 & \cdots & 0 \\
\vdots & \vdots & \ddots & \vdots & \vdots & \vdots \\
g_{j-1} & g_{j-2} & \cdots & g_0 & \vdots & 0 \\
\vdots & \vdots & \vdots & \vdots & \ddots & \vdots \\
g_{N-1} & g_{N-2} & \cdots & \cdots & \cdots & g_0
\end{bmatrix}
\begin{bmatrix}
\triangle u(t) \\
\triangle u(t+1) \\
\vdots \\
\triangle u(t+j-1) \\
\vdots \\
\triangle u(t+N-1)
\end{bmatrix}
$$

$$
+
\begin{bmatrix}
h_0 & 0 & \cdots & 0 & \cdots & 0 \\
h_1 & h_0 & \cdots & 0 & \cdots & 0 \\
\vdots & \vdots & \ddots & \vdots & \vdots & \vdots \\
h_{j-1} & \cdots & h_1 & h_0 & \vdots & 0 \\
\vdots & \vdots & \vdots & \vdots & \ddots & \vdots \\
h_{N-1} & \cdots & \cdots & \cdots & h_1 & h_0
\end{bmatrix}
\begin{bmatrix}
\triangle v(t+1) \\
\triangle v(t+2) \\
\vdots \\
\triangle v(t+j-1) \\
\vdots \\
\triangle v(t+N)
\end{bmatrix}
+
\begin{bmatrix}
f_1 \\
f_2 \\
\vdots \\
f_j \\
\vdots \\
f_N
\end{bmatrix}
$$

where $H_j(z^{-1}) = \sum_{i=1}^{j} h_i z^{-i}$ and h_i are the coefficients of the system step response to the disturbance.

By making $\mathbf{f}' = \mathbf{H}\mathbf{v} + \mathbf{f}$, the prediction equation is now

$$\mathbf{y} = \mathbf{G}\mathbf{u} + \mathbf{f}'$$

which has the same shape as the general prediction equation used in the case of zero measured disturbances. The future control signal can be found in the same way, simply using as free response the process response due to initial conditions (including external disturbances) and future "known" disturbances.

4.9 Use of a Different Predictor in GPC

In this section it is shown that a GPC is equivalent to a structure based on an optimal predictor plus a classical two-degree-of-freedom controller. If the optimal predictor is replaced by a Smith predictor [193], a new controller with similar nominal performance and more robust properties is obtained. This has great interest in the case of time-delay systems. The controller uses the same procedure to compute the control signal, although future outputs are calculated using the Smith predictor instead of the optimal predictor.

4.9.1 Equivalent Structure

In order to show the equivalence between the GPC and a structure composed of an optimal predictor and a classical controller, a CARIMA model with white integrated noise is used to compute the prediction. Let us consider a process with a dead time d, $T(z^{-1}) = 1$, $N_1 = d + 1$, $N_2 = d + N$, $N_u = N$, and the weighting sequences $\delta(j) = 1$, $\lambda(j) = \lambda$. Thus it is possible to write:

$$
\begin{aligned}
\hat{y}(t+d+j \mid t) &= (1 - a_1)\hat{y}(t+d+j-1 \mid t) \\
&+ (a_1 - a_2)\hat{y}(t+d+j-2 \mid t) + \ldots + a_{na}\hat{y}(t+d+j-n_a-1 \mid t) \\
&+ b_0 \triangle u(t+j-1) + \ldots + b_{nb} \triangle u(t+j-1-nb)
\end{aligned} \tag{4.24}
$$

If this equation is applied recursively for $j = 1, 2, \cdots, N$ we get

$$
\begin{bmatrix} \hat{y}(t+d+1 \mid t) \\ \hat{y}(t+d+2 \mid t) \\ \vdots \\ \hat{y}(t+d+N \mid t) \end{bmatrix} = \mathbf{G} \begin{bmatrix} \triangle u(t) \\ \triangle u(t+1) \\ \vdots \\ \triangle u(t+N-1) \end{bmatrix} + \mathbf{H} \begin{bmatrix} \triangle u(t-1) \\ \triangle u(t-2) \\ \vdots \\ \triangle u(t-nb) \end{bmatrix}
$$

$$
+ \mathbf{S} \begin{bmatrix} \hat{y}(t+d \mid t) \\ \hat{y}(t+d-1 \mid t) \\ \vdots \\ \hat{y}(t+d-na \mid t) \end{bmatrix}
$$

where $\mathbf{G}$, $\mathbf{H}$ and $\mathbf{S}$ are constant matrices of dimension $N \times N$, $N \times n_b$ and $N \times n_a + 1$, respectively. This equation can be written in a vector form as follows:

$$
\hat{\mathbf{y}} = \mathbf{Gu} + \mathbf{Hu'} + \mathbf{Sy'} = \mathbf{Gu} + \mathbf{f}
$$

where it is clear that $\mathbf{f} = \mathbf{H}\mathbf{u'} + \mathbf{S}\mathbf{y'}$ is composed of the terms in the past and correspond to the free response of the system.

If $\hat{\mathbf{y}}$ is introduced in the cost function, $J(N)$ is a function of $\mathbf{y'}$, $\mathbf{u}$, $\mathbf{u'}$ and the reference sequence. Minimizing $J(N)$ with respect to $\mathbf{u}$, that is, $\triangle u(t)$, $\triangle u(t+1)$ ... $\triangle u(t+N-1))$ leads to

$$
\mathbf{M} \begin{bmatrix} \triangle u(t) \\ \triangle u(t+1) \\ \vdots \\ \triangle u(t+N-1) \end{bmatrix} = \mathbf{P}_0 \begin{bmatrix} \hat{y}(t+d \mid t) \\ \hat{y}(t+d-1 \mid t) \\ \vdots \\ \hat{y}(t+d-na \mid t) \end{bmatrix} + \mathbf{P}_1 \begin{bmatrix} \triangle u(t-1) \\ \triangle u(t-2) \\ \vdots \\ \triangle u(t-nb) \end{bmatrix}
$$
$$
+ \mathbf{P}_2 \begin{bmatrix} w(t+d+1) \\ w(t+d+2) \\ \vdots \\ w(t+d+N) \end{bmatrix}
$$

where $\mathbf{M} = \mathbf{G}^T\mathbf{G} + \lambda\mathbf{I}$ and $\mathbf{R} = \mathbf{G}^T$ are of dimension $N \times N$, $\mathbf{P}_0 = -\mathbf{G}^T S$ of dimension $N \times n_a + 1$ and $\mathbf{P}_1 = -\mathbf{G}^T\mathbf{H}$ of dimension $N \times n_b$. As in a receding horizon algorithm only the value of $\triangle u(t)$ is computed, if $\mathbf{q}$ is the first row of matrix $\mathbf{M}^{-1}$, $\triangle u(t)$ is given by:

$$
\triangle u(t) = \mathbf{q}\mathbf{P}_0 \begin{bmatrix} \hat{y}(t+d \mid t) \\ \hat{y}(t+d-1 \mid t) \\ \vdots \\ \hat{y}(t+d-na \mid t) \end{bmatrix} + \mathbf{q}\mathbf{P}_1 \begin{bmatrix} \triangle u(t-1) \\ \triangle u(t-2) \\ \vdots \\ \triangle u(t-nb) \end{bmatrix}
$$
$$
+ \mathbf{q}\mathbf{P}_2 \begin{bmatrix} w(t+d+1) \\ w(t+d+2) \\ \vdots \\ w(t+d+N) \end{bmatrix}
$$

Therefore the control increment $\triangle u(t)$ can be written as:

$$
\triangle u(t) = \mathbf{q}\mathbf{P}_0\mathbf{y'} + \mathbf{q}\mathbf{P}_1\mathbf{u'} + \mathbf{q}\mathbf{P}_2\mathbf{w}
$$

The resulting control scheme (see Figure 4.4) is a linear feedback of predictions $\hat{y}(t+d \mid t)$, ..., $\hat{y}(t+d-na \mid t)$ generated by an optimal predictor. That is, prediction over a time horizon equal to the process dead time. The controller coefficients are computed for each choice of N and λ.

The block diagram presented in Figure 4.4 can easily be transformed into the one shown in Figure 4.5, which shows that a GPC is equivalent to a classical controller plus an optimal predictor. This classical controller is composed of a reference filter and a cascade block. If the plant can be modelled by a first-order system with a dead time then the classical controller results in a

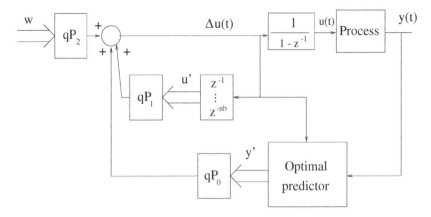

Fig. 4.4. Control scheme

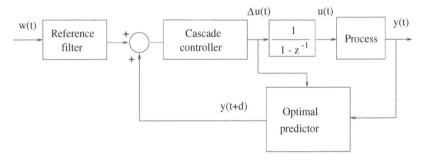

Fig. 4.5. Equivalent control structure of the GPC

simple PI [153]. In general, the primary controller is of the same order as the model of the plant.

This relation can be used to study how to improve the robustness of the GPC using a different predictor structure. Note that the computation of the classical controller is done independently of the predictor structure, so different predictors can be compared using the same controller parameters of the GPC.

The prediction can be obtained directly from polynomials A and B and the delay d considering Equation (4.24) for $j = 1$:

$$\hat{y}(t+1 \mid t) = (1-\tilde{A}(z^{-1}))zy(t)+\tilde{B}(z^{-1})u(t-d) \quad \text{with } \tilde{B}(z^{-1}) = (1-z^{-1})B(z^{-1})$$

Using the same procedure for $j = 2,\ldots$, $\hat{y}(t+j \mid t)$ is computed as:

$$\hat{y}(t+j \mid t) = ((1 - \tilde{A}(z^{-1}))z)^j y(t) + \sum_{i=1}^{j}(1 - \tilde{A}(z^{-1}))^{i-1}\tilde{B}(z^{-1})u(t-d+j-1)$$

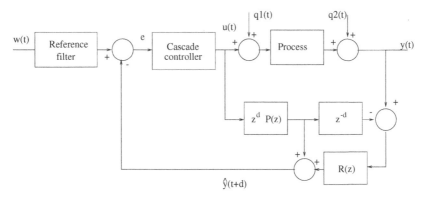

Fig. 4.6. Equivalent structure for the OP in the GPC

The predicted output at instant $t + d$ is therefore

$$\hat{y}(t + d \mid t) = ((1 - \tilde{A}(z^{-1}))z)^d y(t) + (1 - (1 - \tilde{A}(z^{-1}))^d)z^d P u(t)$$

where $P(z) = \frac{B(z^{-1})z^{-1}}{A(z^{-1})} z^{-d}$ is the plant model.

Defining $R(z) = (1 - \tilde{A}(z^{-1}))^d z^d$, the prediction can be written as:

$$\hat{y}(t + d \mid t) = R(z)y(t) + (z^d - R(z))P(z)u(t)$$

The closed-loop block diagram of the whole control system is shown in Figure 4.6. Now it is clear that the GPC has a structure similar to the well-known dead-time compensators like the Smith predictor (obtained when $R(z) = 1$) and that in the absence of a dead time the final control law is a classical two-degree-of-freedom controller.

The theoretical comparative results about the robustness and performance of GPC and the GPC which uses a Smith predictor can be found in [155], where it is shown that for stable processes, the Smith predictor (SP) based control structure has similar performance and more robustness than the one based on the optimal predictor (OP) when using the same primary controller. An application of this controller to mobile robot path tracking can be found in [157].

It is easy to show that the norm-bound uncertainty of the controller is:

$$\mid \delta P \mid \leq \left| \frac{1 + C(z)z^d P(z)}{C(z)R(z)} \right|$$

Notice that the norm-bound uncertainty is inversely proportional to $\mid R(z) \mid$ and that for the GPC the block $R(z)$ has high pass characteristics, so the controller has low values of the norm-bound uncertainty at high frequencies. On the other hand, the Smith predictor based controller has $R(z) = 1$ and consequently a higher robustness index.

As has been mentioned in this chapter, the robustness of the GPC can be improved by the use of a prefilter in the prediction equations. The effect of this filter (the T polynomial) can also be analyzed in Figure 4.6 because, when T is included in the predictor $R(z)$ is a function of $T(z)$. If T is chosen appropriately then $R(z)$ could have low pass characteristic and the robustness of the controller could be increased. On the other hand, the disturbance rejection response deteriorates when the robustness increases [8]. Note that for the Smith predictor based controller a filter $F(z)$ could also be included, but in this case the tuning of F is simpler than the tuning of T in the GPC case as for the new structure $R(z) = F(z)$. Note that to improve the robustness F must be chosen as a low pass filter [156].

For the computation of the controller the following steps must be taken:

- compute the prediction of the output (from t to $t+d$) using the open-loop model of the plant without considering disturbances
- correct each open-loop prediction adding the mismatch between the output and the prediction:

$$\hat{y}(t+d-i\mid t) = \hat{y}(t+d-i\mid t) + y(t-i) - \hat{y}(t-i)$$

- compute the control law as in the normal GPC using the coefficients of $\mathbf{q}$, $\mathbf{P}_0$, $\mathbf{P}_1$ and $\mathbf{P}_2$.

Note that to compute the control law it is also possible to use the forced and free response concepts used in standard GPC. In this case the forced response can be computed as done in GPC but the free response must be computed using the Smith predictor from t to $t + d$ and the optimal predictor from $t + d + 1$ to $t + N$.

4.9.2 A Comparative Example

In order to illustrate the robustness properties of the proposed GPC an example comparing the SPGPC and the GPC is presented. It corresponds to a temperature control of a heat exchanger, where the model used in the predictor is a first-order system with dead time, such as the one presented in [154]. This system represents a typical industrial process, and because of this it is normally used to evaluate the performance of industrial controllers. In the example ARIMA disturbances are considered.

The relation between the output temperature and the input flow in the heat exchanger was obtained using the reaction curve method and is given by:

$$P(s) = \frac{0.12e^{-3s}}{1 + 6s}$$

Using a sample time $T_s = 0.6s$ the obtained discrete plant is

$$P(z) = \frac{bz^{-1}}{1 - az^{-1}} z^{-d}$$

where the nominal values of the parameters are $d_n = 5, a_n = 0.905$ and $b_n = 0.0114$. In practice there are errors in the estimation of all parameters but for this example only dead time uncertainty is considered, with a maximum value of $\delta d = 2$.

The GPC is computed in order to obtain (for the nominal case) a step response faster than the open-loop one. Thus, the GPC parameters were chosen as $N = 15, \lambda = 0.8$. The SPGPC has the same parameters.

The closed-loop behaviour of the GPC and the SPGPC for the nominal case are shown in Figure 4.7(a). At $t = 0$ a step change in the reference is performed and at $t = 100$ samples a 10% step load disturbance is applied to the system. The noise is generated with an ARIMA model with uniform distribution in ± 0.005. As can be observed, both systems have similar setpoint tracking and disturbance rejection behaviour for the nominal case.

In the next simulation the delay of the plant is set to $d = 7$ and again a step change in the reference is considered at $t = 0$. Figure 4.7(b) shows that the SPGPC based control system is stable and the GPC one is unstable.

It must be stated that the comparison has been made with a GPC without T polynomial and the introduction of this in the formulation could improve its response when the dead time uncertainty appears, but at the same time the nominal disturbance rejection response will be deteriorated.

4.10 Constrained Receding Horizon Predictive Control

In spite of the great success of GPC in industry, there was an original lack of theoretic results about the properties of predictive control and an initial gap in important questions such as stability and robustness. In fact, the majority of stability results are limited to the infinite horizon case and there is a lack of a clear theory relating the closed-loop behaviour to design parameters, such as horizons and weighting sequences.

Bearing in mind the need to solve some of these drawbacks, a variation of the standard formulation of GPC appears developed by Clarke and Scattolini called Constrained Receding-Horizon Predictive Control (CRHPC) [61], [144],[151], which allows stability and robustness results to be obtained for small horizons. The idea basically consists of deriving a future control sequence so that the predicted output over some future time range is constrained to be at the reference value exactly, as shown in Figure 4.8. Some degrees of freedom of the future control signals are employed to force the

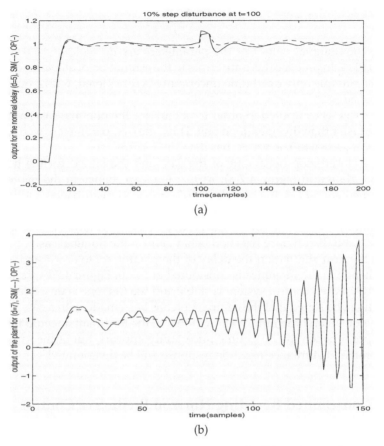

Fig. 4.7. Behaviour of the GPC(solid) and SPGPC (dashed) based control systems: (a) nominal case, (b) dead time uncertainty case

output, whilst the rest is available to minimize the cost function over a certain interval.

4.10.1 Computation of the Control Law

The computation of the control law is performed in a similar way to GPC, although the calculations become a little more complicated. Control signals must minimize the objective function

$$J(N_y, N_u) = \sum_{i=1}^{N_y} \mu(i)[\hat{y}(t+i \mid t) - y^o(t+i)]^2 + \sum_{i=0}^{N_u} \lambda(i)[\triangle u(t+j)]^2 \quad (4.25)$$

where $y^o(t+i)$ is the reference, subject to constraints

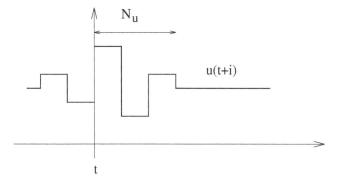

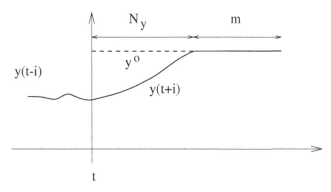

Fig. 4.8. Constrained Receding Horizon Predictive Control

$$y(t + N_y + i) = y^o(t + N_y + 1) \qquad i = 1 \ldots m$$
$$\triangle u(t + N_u + j) = 0 \qquad j > 0 \tag{4.26}$$

The design parameters are the values of the horizons, the weighting sequences $\mu(i)$ and $\lambda(i)$ and the value of m, which defines the instants of coincidence between output and reference. The system is modelled by:

$$\triangle A(z^{-1})y(t) = B(z^{-1}) \triangle u(t - d) \tag{4.27}$$

In order to solve the optimization problem, it is necessary to calculate the output prediction, which is done by defining polynomials $E_i(z^{-1})$, $F_i(z^{-1})$ and $G_i(z^{-1})$ where E_i is of degree $i - 1$ such that

$$1 = E_i(z^{-1}) \triangle A(z^{-1}) + z^{-i}F_i(z^{-1})$$

$$z^{-d}E_i(z^{-1})B(z^{-1}) = \sum_{h=1}^{i} s_h z^{-h} + z^{-i+1}G_i(z^{-1})$$

where s_h are the coefficients of the process step response.

Multiplying Equation (4.27) by $E_i(z^{-1})$ and taking the last equations into account leads to:

$$y(t+i) = \sum_{h=1}^{i} s_h \, \triangle \, u(t+i-h) + f(t+i)$$

$$f(t+i) = F_i(z^{-1})y(t) + G_i(z^{-1}) \, \triangle \, u(t-1)$$

Now defining the following sequences of future variables

$$Y(t) = [\, y(t+1) \; y(t+2) \; \ldots \; y(t+N_y) \,]^T$$
$$Y^o(t) = [\, y^o(t+1) \; y^o(t+2) \; \ldots \; y^o(t+N_y) \,]^T$$
$$\triangle U(t) = [\, \triangle u(t) \; \triangle \, u(t+1) \; \ldots \; \triangle \, u(t+N_u) \,]^T$$
$$F(t) = [\, f(t+1) \; f(t+2) \; \ldots \; f(t+N_y) \,]^T$$
$$\overline{Y}(t) = [\, y(t+N_y+1) \; y(t+N_y+2) \; \ldots \; y(t+N_y+m) \,]^T$$
$$\overline{Y}^o(t) = [\, y^o(t+N_y+1) \; y^o(t+N_y+2) \; \ldots y^o(t+N_y+m) \,]^T$$
$$\overline{F}^o(t) = [\, f(t+N_y+1) \; f(t+N_y+2) \; \ldots \; f(t+N_y+m) \,]^T$$

the following matrices, of dimension $N_y \times (N_u+1)$ and $m \times (N_u+1)$, respectively;

$$G = \begin{bmatrix} s_1 & 0 & 0 & \ldots & 0 \\ s_2 & s_1 & 0 & \ldots & 0 \\ \vdots & \vdots & \vdots & \vdots & \vdots \\ s_{Ny} & s_{Ny-1} & s_{Ny-2} & \cdots & s_{Ny-Nu} \end{bmatrix}$$

$$\overline{G} = \begin{bmatrix} s_{Ny+1} & s_{Ny} & \cdots & s_{Ny-Nu+1} \\ s_{Ny+2} & s_{Ny+1} & \cdots & s_{Ny-Nu+2} \\ \vdots & \vdots & \vdots & \vdots \\ s_{Ny+m} & s_{Ny+m-1} & \cdots & s_{Ny-Nu+m} \end{bmatrix}$$

and the weighting sequences

$$M(t) = \text{diag} \, \{\mu(1), \mu(2), \ldots \mu(N_y)\}$$
$$\Lambda(t) = \text{diag} \, \{\lambda(0), \lambda(1), \ldots \lambda(N_u)\}$$

the following relations hold:

$$Y(t) = G \, \triangle U(t) + F(t)$$
$$\overline{Y}(t) = \overline{G} \, \triangle U(t) + \overline{F}(t)$$

Then the cost function (Equation 4.25) and the constraints (4.26) can be written as:

$$J = [Y(t) - Y^o(t)]^T M(t) [Y(t) - Y^o(t)] + \triangle U^T(t) \Lambda(t) \triangle U(t)$$
$$\overline{G} \triangle U(t) + \overline{F}(t) = \overline{Y}^o(t)$$

The solution can be obtained by the use of Lagrange multipliers. If the common case of constant weighting sequence is considered, that is, $M(t) = \mu I$, $\Lambda(t) = \lambda I$, the solution can be written as:

$$\triangle U(t) = (\mu G^T G + \lambda I)^{-1}[\mu G^T(Y^o(t) - F(t)) + \overline{G}^T(\overline{G}(\mu G^T G + \lambda I)^{-1}\overline{G}^T)^{-1}$$
$$\times (\overline{Y}^o(t) - \overline{F}(t) - \mu\overline{G}(\mu G^T G + \lambda I)^{-1}G^T(Y^o(t) - F(t)))] \qquad (4.28)$$

As it is a receding horizon strategy, only the first element of vector $\triangle U(t)$ is used, repeating the calculation at the next sampling time. This method provides an analytical solution that, as can be observed in (4.28) proves to be more complex than the standard GPC solution. Computational burden can be considerable since various matrix operations, including inversion, must be made, although some calculations can be optimized knowing that G is triangular and the matrices to be inverted are symmetrical. This factor can be decisive in the adaptive case, where all vectors and matrices can change at every sampling time.

Obtaining the control signal requires inversion of matrix $\overline{G}(\mu G^T G + \lambda I)^{-1}\overline{G}^T)$. ¿From the definition of matrix $\overline{G}$, it can be derived that condition $m \leq N_u + 1$ must hold, which can be interpreted as the number m of output constraints cannot be bigger than the number of control signal variations $N_u + 1$. Another condition for invertibility is that $m \leq n + 1$, since the coefficient s_i of the step response is a linear combination of the previous $n + 1$ values (where n is the system order). Notice that this last condition constrains the order of the matrix to invert $(m \times m)$ to relative small values, as in the majority of situations the value of m will not be bigger than two or three.

4.10.2 Properties

As has been stated, one of the advantages of this method is the availability of stability results for finite horizons, compensating in a certain way the computational burden it carries. The following results present the outstanding properties of CRHPC, whose demonstration can be found in [151], based upon a state-space formulation of the control law (4.28). The fact that the output follows the reference over some range guarantees the monotonicity of the associated Riccati equation and in consequence stability, based upon the results by Kwon and Pearson [115].

Property 1. If $N_y = N_u > n + d + 1$ and $m = n + 1$ then the closed-loop system is asymptotically stable.

Property 2. If $N_u = n + d$ and $m = n + 1$ the control law results in a stable dead-beat control.

Property 3. If the system is asymptotically stable, $\mu = 0, m = 1$ and there exists ν such that either

$$s_\nu \leq s_{\nu+1} \leq \ldots \leq s_\infty, \quad s_\nu > s_\infty/2, \quad s_\infty > 0$$

or

$$s_\nu \geq s_{\nu+1} \geq \ldots \geq s_\infty, \quad s_\nu < s_\infty/2, \quad s_\infty < 0$$

Then, for any $N_y = N_u \geq \nu - 1$ the closed-loop system is asymptotically stable.

Property 4. Under the latter conditions, the closed-loop system is asymptotically stable for $N_u = 0$ and $N_y \geq \nu - 1$.

Property 5. If the open-loop system is asymptotically stable, $\mu = 0, m = 1, N_u = 0$ and constants $K > 0$ and $0 < \eta < 1$ exist such that

$$\mid s_i - s_\infty \mid \leq K\eta^i, \quad i \geq 0$$

Then, for any N_y such that

$$\mid s_{Ny+1} \mid > K\frac{1+\eta}{1-\eta}\eta^{Ny+1}$$

the closed-loop system is asymptotically stable.

So it is seen that the method is able to stabilize any kind of process, such as unstable or nonminimum phase ones. As in GPC, the use of filtered inputs and outputs can result in a pole-placement control. Using the $P(z^{-1})$ polynomial introduced in [59] shows the close relationship between predictive control and pole placement [61]. This polynomial appears when defining an auxiliary output $\psi(t) = P(z^{-1})y(t)$, which substitutes $y(t)$ in (4.25) and (4.26) in such a way that the desired closed-loop poles are the roots of $z^{n+1}P(z^{-1})$.

4.11 Stable GPC

To overcome the lack of stability results for GPC, a new formulation has been developed by Rossiter and Kouvaritakis which ensures that the associated cost function is monotonically decreasing, guaranteeing closed-loop stability. For this reason, this algorithm is called Stable Generalized Predictive Control SGPC [112], [186] and it is based on stabilizing the loop before the application of the control strategy. Now the control actions that must be calculated are the future values of the reference that are sent to the closed loop, instead of the system inputs, which are functions of these. The stabilizing inner-loop controller is designed to obtain a finite closed-loop impulse response; this fact simplifies the implementation of the algorithm at the same time as ensuring the monotonicity of the cost function.

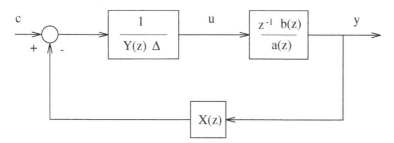

Fig. 4.9. Stable loop

4.11.1 Formulation of the Control Law

The model considered for obtaining the control law is:

$$G(z) = \frac{z^{-1}b(z)}{a(z)} = \frac{b_1 z^{-1} + b_2 z^{-2} + \ldots + b_n z^{-n}}{1 + a_1 z^{-1} + a_2 z^{-2} + \ldots + a_n z^{-n}} \tag{4.29}$$

where, for the sake of simplicity and without any loss of generality, the order of numerator and denominator is considered to be the same; if the delay is bigger than one, it is enough to set the first terms of $b(z)$ to zero. As has been stated, before optimizing the cost function, the loop is stabilized by means of polynomials $X(z)$ and $Y(z)$ as shown in Figure 4.9. Signal c is the closed-loop reference and is the value that will appear in the cost function.

Polynomial $X(z)$ and $Y(z)$ satisfy the following relations:

$$a(z)Y(z) \triangle (z) + z^{-1}b(z)X(z) = 1$$

$$K(z) = \frac{X(z)}{Y(z) \triangle (z)}$$

with $\triangle(z) = 1 - z^{-1}$ and $K(z)$ the overall feedback controller. With these relations, it can be deduced that:

$$y(z) = z^{-1}b(z)c(z)$$

$$\triangle u(z) = A(z)c(z)$$

where $A(z) = a(z)\triangle$.

The cost function that, like standard GPC, measures the discrepancies between future outputs and reference as well as the necessary control effort can be obtained making use of these relations. The cost must be expressed as a function of the future values of c in such a way that the reference signals for the stabilized system can be obtained from its minimization. In order to achieve this objective, the following vectors are considered:

$$y^+ = [\, y(t+1)\, y(t+2)\, \ldots\, y(t+n_y)\,]^T$$

$$c^+ = [\, c(t+1)\, c(t+2)\, \ldots\, c(t+n_c)\,]^T$$

$$\triangle u^+ = [\, \triangle u(t)\, \triangle u(t+1)\, \ldots\, \triangle u(t+n_y-1)\,]^T$$

$$c^- = [\, c(t)\, c(t-1)\, \ldots\, c(t-n)\,]^T$$

where n_y and n_c are the output and reference horizons.

Simulating Equation (4.29) forward in time it can be written

$$y^+ = \Gamma_b c^+ + H_b c^- + M_b c^\infty$$
$$\triangle u^+ = \Gamma_a c^+ + H_A c^- + M_A c^\infty$$

where the last terms of each equation represent the free response y_f and $\triangle u_f$ of y and $\triangle u$, respectively. Matrices Γ_A, Γ_b, H_A, H_b, M_A and M_b can easily be derived as shown in [112] while c^∞ is the vector of $n_y - n_c$ rows that contains the desired future values of the command input c. The elements of this vector are chosen to ensure offset-free steady state.

Usually, if $r(t + i)$, $i = 1, \ldots, n_y$ is the setpoint to be followed by the system output, c^∞ is chosen to be $[1, 1, \ldots, 1]^T \times r(t+n_y)/b(1)$. If step setpoint changes are assumed, c^∞ can be chosen as the vector of future references premultiplied by a matrix E formed out of the last $n_y - n_c$ rows of I_{ny}.

Then the cost function can be written as:

$$J = \|r^+ - y^+\|_2 + \lambda\|\triangle u^+\|_2 = [c^+ - c_o]^T S^2 [c^+ - c_o] + \gamma$$

where

$$S^2 = \Gamma_b^T \Gamma_b + \lambda \Gamma_A^T \Gamma_A$$
$$c_o = S^{-2}[\Gamma_b^T(r^+ - y_f) - \lambda\Gamma_A^T \triangle u_f]$$
$$\gamma = \|r^+ - y_f\|_2 + \lambda\|\triangle u_f\|_2 - \|c_o\|_2$$

As γ is a known constant, the control law of SGPC comes from the minimization of cost $J = \|S(c^+ - c_o)\|$, and is defined by

$$\triangle u(t) = A(z)c(t) \qquad c(t) = \frac{p_r(z)}{p_c(z)} r(t + ny)$$

where $p_r(z)$ and $p_c(z)$ are polynomials defined as:

$$p_r = e^T T[\Gamma_b^T(I - M_b E) - \Gamma_b^T \Gamma_A E]$$
$$p_c = e^T T[\Gamma_b^T H_b + \lambda\Gamma_b H_A]$$

and $T = (\Gamma_b^T \Gamma_b + \lambda\Gamma_A\Gamma_A)^{-1}$; e is the first standard basis vector and the polynomials are in descending order of z.

The stability of the algorithm is a consequence of the fact that y and $\triangle u$ are related to c by means of finite impulse response operators ($z^{-1}b(z)$ and $A(z)$), which can be used to establish that the cost is monotonically decreasing [186] and can therefore be interpreted as a Lyapunov function guaranteeing stability.

4.12 Exercises

4.1. Given the process

$$y(t) - 1.5y(t-1) + 0.56y(t-2) = 0.9u(t-1) - 0.6u(t-2)$$

and $N_1 = 1$ and $N_2 = 3$:

1. Solve the Diophantine equation.
2. Compute $G_i(z^{-1})$ for $i = 1, 2$ and 3.
3. Write the elements of the prediction vector as functions of $\triangle u(t-1)$, $y(t-1)$, $y(t-2)$ and the future control signals.
4. Simulate the response when the setpoint changes from 0 to 1.

4.2. Given the system $G(s) = \frac{0.5}{1+10s}e^{-2s}$:

1. Obtain the discrete model for a sampling time of 1 second.
2. Use a GPC to steer the output from 0 to 1.5.
3. Consider that there is a model mismatch and the model used by the controller has a time constant of 12 seconds (that is, the first model is used to simulate the process and the second is used to design the controller). Simulate the results with the nominal controller.
4. Add a T polynomial and simulate the results.
5. Use a Smith predictor and simulate the results.

4.3. Design a GPC for the system described by

$$\frac{(0.6445 - 0.7176z^{-1} - 0.0906z^{-2})z^{-2}}{1 - 0.9183z^{-1} + 0.084z^{-2} - 0.002z^{-3}}$$

Try different values for the tuning parameters (horizons and weights) and simulate the results for a setpoint change of 1 unit and a step disturbance at the output of 0.2.

4.4. A DC motor can be modelled by

$$\frac{0.00489z^{-1} + 0.00473z^{-2}}{(1-z^{-1})(1-0.935z^{-1})}$$

where the input is the voltage applied to the motor and the output is the shaft angle.

1. Design a GPC and simulate a setpoint move from 0 to 3.
2. Try to obtain the same results with a DMC. Justify the results.

4.5. Design a GPC for the system described by

$$G(z) = \frac{-1 + 1.2592z^{-1}}{1 - 0.7408z^{-1}}$$

with $N_1 = 1$, $N_2 = 30$, $N_u = 10$ and $\lambda = 0.1$. Change the value of λ to 0 and see the results (for the justification see Section 6.9).

4.6. Given an oscillatory system $G(s) = \frac{50}{s^2+25}$, find the parameters of a GPC such that the overshoot is less than 20%. Is it possible to control this process with a MAC?

5

Simple Implementation of GPC for Industrial Processes

One of the reasons for the success of the traditional PID controllers in industry is that PID are very easy to implement and tune using heuristic tuning rules such as the Ziegler-Nichols rules frequently used in practice. A Generalized Predictive Controller, as shown in the previous chapter, results in a linear control law which is very easy to implement once the controller parameters are known. The derivation of the GPC parameters requires, however, some mathematical complexities such as recursively solving the Diophantine equation, forming the matrices $\mathbf{G}, \mathbf{G}'$ and $\mathbf{f}$ and then solving a set of linear equations. Although this is not a problem for people in the research control community where mathematical packages are normally available, it may be discouraging for practitioners used to much simpler ways of implementing and tuning controllers.

The previously mentioned computation has to be carried out only once when dealing with processes with fixed parameters, but if the process parameters change, the GPC's parameters have to be derived again, perhaps in real time, at every sampling time if a self-tuning control is used. This may be difficult because, on one hand, some distributed control equipment has only limited mathematical computation capabilities for the controllers, and, on the other hand, the computation time required for the derivation of the GPC parameters may be excessive for the sampling time required by the process and the number of loops implemented.

The goal of this chapter is to show how a GPC can very easily be implemented and tuned for a wide range of processes in industry. It will be shown that a GPC can be implemented with a limited set of instructions, available in most control systems, and that the computation time required, even for tuning, is very short. The method to implement the GPC is based on the fact that a wide range of processes in industry can be described by a few parameters and that a set of simple Ziegler-Nichols type of functions relating GPC parameters to process parameters can be obtained. By using these functions the implementation and tuning of a GPC is almost as simple as the implementation and tuning of a PID.

The influence of modelling errors is also analyzed in this chapter, with a section dedicated to performing a robustness analysis of the method when modelling errors are taken into account.

5.1 Plant Model

Most processes in industry, when considering small changes around an operating point can be described by a linear model of, normally, very high order. This is because most industrial processes are composed of many dynamic elements, usually first order, so the full model is of an order equal to the number of elements. In fact, each mass or energy storage element in the process provides a first-order element in the model. Consider, for instance, a long pipe used for heat exchanging purposes, as the case of a cooler or a steam generator. The pipe can be modelled by breaking it into a set of small pieces, each of which can be considered a first-order system. The resulting model will have an order equal to the number of pieces used to model the pipe, that is, a very high-order model. These very high-order models would be difficult to use for control purposes but, fortunately, it is possible to approximate the behaviour of such high-order processes by a system with one time constant and a dead time.

As shown in [63], we may consider a process having N first-order elements in series, each having a time constant τ/N. That is, the resulting transfer function will be

$$G(s) = \frac{1}{(1 + \frac{\tau}{N}s)^N}$$

Changing the value of N from 1 to ∞ the response shifts from exact first order to pure dead time (equal to τ). When a time constant is much larger than the others (as in many processes) the smaller time constants work together to produce a lag that acts as pure dead time. In this situation the dynamical effects are mainly due to this larger time constant, as can be seen in Figure 5.1. It is therefore possible to approximate the model of a very high-order, complex, dynamic process with a simplified model consisting of a first-order process combined with a dead time element. This type of system can then be described by the following transfer function:

$$G(s) = \frac{K}{1 + \tau s}e^{-s\tau_d} \tag{5.1}$$

where K is the process static gain, τ is the time constant or process lag, and τ_d is the dead time or delay.

5.1.1 Plant Identification: The Reaction Curve Method

Once the model structure is defined, the next step is to choose the correct value for the parameters. In order to identify these parameters, a suitable

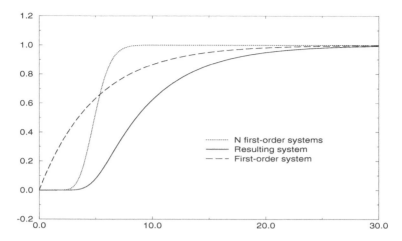

Fig. 5.1. System response

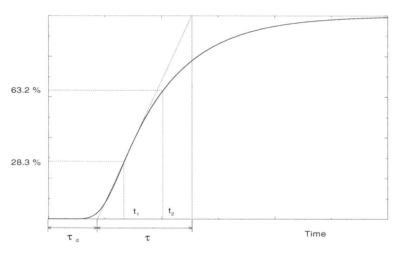

Fig. 5.2. Reaction curve

stimulus must be applied to the process input. In the Reaction Curve Method a step perturbation, that is, an input with a wide frequency content, is applied to the process and the output is recorded to fit the model to the data.

The step response or reaction curve of the process looks like Figure 5.2. Here, a step of magnitude $\triangle u$ is produced in the manipulated variable and the time response of the process variable $y(t)$ is shown. The process parameters of Equation (5.1) can be obtained by measuring two times: t_1, the time when the output reaches 28.3 % of its steady-state value $\triangle y$, and t_2, when the output reaches 63.2 %. Using these values, the process parameters are given by:

$$K = \frac{\triangle y}{\triangle u}$$
$$\tau = 1.5(t_2 - t_1)$$
$$\tau_d = 1.5(t_1 - \tfrac{1}{3}t_2)$$

A similar and perhaps more intuitive way of obtaining the process parameters consists of finding the inflection point of the response and drawing the line that represents the slope at that point [95]. The static gain is obtained by the former expression and the times τ and τ_d come directly from the response, as can be seen in the figure. The values obtained are similar in the two approaches.

The Reaction Curve method is probably one of the most popular methods used in industry for tuning regulators, as in the Ziegler-Nichols method for tuning PIDs, and it is used in the pretune stage of some commercial adaptive and auto-tuning regulators.

5.2 The Dead Time Multiple of the Sampling Time Case

5.2.1 Discrete Plant Model

When the dead time τ_d is an integer multiple of the sampling time T ($\tau_d = dT$), the corresponding discrete transfer function of Equation (5.1) has the form

$$G(z^{-1}) = \frac{bz^{-1}}{1 - az^{-1}} z^{-d} \tag{5.2}$$

where discrete parameters a, b and d can easily be derived from the continuous parameters by discretization of the continuous transfer function, resulting in the following expressions:

$$a = e^{-\frac{T}{\tau}} \qquad\qquad b = K(1 - a) \qquad\qquad d = \frac{\tau_d}{T}$$

If a CARIMA (Controlled Auto-Regressive and Integrated Moving Average) model is used to model the random disturbances in the system and the noise polynomial is chosen to be 1 , the following equation is obtained

$$(1 - az^{-1})y(t) = bz^{-d}u(t - 1) + \frac{\varepsilon(t)}{\triangle}$$

where $u(t)$ and $y(t)$ are the control and output sequences of the plant, $\varepsilon(t)$ is a zero mean white noise and $\triangle = 1 - z^{-1}$. This equation can be transformed into:

$$y(t + 1) = (1 + a)y(t) - ay(t - 1) + b \triangle u(t - d) + \varepsilon(t + 1) \tag{5.3}$$

5.2.2 Problem Formulation

As was shown in the previous chapter, the Generalized Predictive Control (GPC) algorithm consists of applying a control sequence that minimizes a multistage cost function of the form

$$J(N_1, N_2) = \sum_{j=N_1}^{N_2} \delta(j)[\hat{y}(t+j \mid t) - w(t+j)]^2$$

$$+ \sum_{j=1}^{N_2-d} \lambda(j)[\triangle u(t+j-1)]^2 \tag{5.4}$$

Notice that the minimum output horizon N_1 should be set to a value greater than the dead time d as the output for smaller time horizons cannot be affected by the first action $u(t)$. In the following N_1 and N_2 will be considered to be $N_1 = d+1$ and $N_2 = d+N$, where N is the control horizon.

If $\hat{y}(t+d+j-1 \mid t)$ and $\hat{y}(t+d+j-2 \mid t)$ are known, it is clear, from Equation (5.3) that the best expected value for $\hat{y}(t+d+j \mid t)$ is given by:

$$\hat{y}(t+d+j \mid t) = (1+a)\hat{y}(t+d+j-1 \mid t) - a\hat{y}(t+d+j-2 \mid t)$$
$$+ b \triangle u(t+j-1) \tag{5.5}$$

If Equation (5.5) is applied recursively for $j = 1, 2, \ldots, i$, we get

$$\hat{y}(t+d+i \mid t) = G_i(z^{-1})\hat{y}(t+d \mid t) + D_i(z^{-1}) \triangle u(t+i-1) \tag{5.6}$$

where $G_i(z^{-1})$ is of degree 1 and $D_i(z^{-1})$ is of degree $i-1$. Notice that when $\delta(i) = 1$ and $\lambda(i) = \lambda$, the polynomials $D_i(z^{-1})$ are equal to the polynomials $G_i(z^{-1})$ given in [58] for the case of $d = 0$ and the terms $f(t+i)$ given in that reference are equal to $G_i(z^{-1})y(t)$ of equation (5.6).

If $\hat{y}(t+d+i \mid t)$ is introduced in Equation (5.4), $J(N)$ is a function of $\hat{y}(t+d \mid t)$, $\hat{y}(t+d-1 \mid t)$, $\triangle u(t+N_2-d-1)$, $\triangle u(t+N_2-d-2)$ ... $\triangle u(t)$ and the reference sequence.

Minimizing $J(N)$ with respect to $\triangle u(t), \triangle u(t+1) \ldots \triangle u(t+N-1)$ leads to

$$\mathbf{M\,u} = \mathbf{P\,y} + \mathbf{R\,w} \tag{5.7}$$

where

$$\mathbf{u} = [\triangle u(t) \ \triangle u(t+1) \ \cdots \ \triangle u(t+N-1)]^T$$
$$\mathbf{y} = [\hat{y}(t+d \mid t) \ \hat{y}(t+d-1 \mid t)]^T$$
$$\mathbf{w} = [w(t+d+1) \ w(t+d+2) \ \cdots \ w(t+d+N)]^T$$

$\mathbf{M}$ and $\mathbf{R}$ are matrices of dimension $N \times N$ and $\mathbf{P}$ of dimension $N \times 2$. Let us call $\mathbf{q}$ the first row of matrix $\mathbf{M}^{-1}$. Then $\triangle u(t)$ is given by

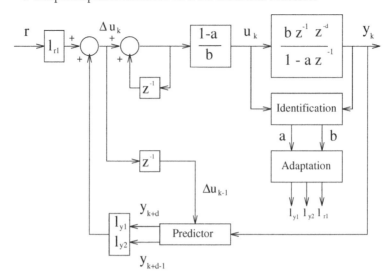

Fig. 5.3. Control Scheme

$$\triangle u(t) = \mathbf{q\,P\,y} + \mathbf{q\,R\,w} \qquad (5.8)$$

When the future setpoints are unknown, $w(t + d + i)$ is supposed to be equal to the current reference $r(t)$. The reference sequence can be written as:

$$\mathbf{w} = [1 \cdots 1] r(t)$$

The control increment $\triangle u(t)$ can be written as

$$\triangle u(t) = l_{y1}\hat{y}(t + d \mid t) + l_{y2}\hat{y}(t + d - 1 \mid t) + l_{r1}r(t) \qquad (5.9)$$

where $\mathbf{q\,P} = [l_{y1}l_{y2}]$ and $l_{r1} = \sum\limits_{i=1}^{N} q_i \sum\limits_{j=1}^{N} r_{ij}$. The coefficients l_{y1}, l_{y2} and l_{r1} are functions of a, b, $\delta(i)$ and $\lambda(i)$. If the GPC is designed considering the plant to have a unit static gain, the coefficients in (5.9) will only depend on $\delta(i)$ and $\lambda(i)$ (which are supposed to be fixed) and on the pole of the plant which will change for the adaptive control case. Notice that by doing this, a normalized weighting factor λ is used and it should be corrected for systems with different static gains.

The resulting control scheme is shown in Figure 5.3. The estimated plant parameters are used to compute the controller coefficients (l_{y1}, l_{y2}, l_{r1}). The values $\hat{y}(t + d \mid t)$, $\hat{y}(t + d - 1 \mid t)$ are obtained by the use of the predictor given by equation (5.5). The control signal is divided by the process static gain in order to get a system with a unitary static gain.

Notice that the controller coefficients do not depend on the dead time d and for fixed values of $\delta(i)$ and $\lambda(i)$ they will be a function of the estimated pole ($\hat{a}$). The standard way of computing the controller coefficients would be by computing the matrices $\mathbf{M}$, $\mathbf{P}$ and $\mathbf{R}$ and solving Equation (5.7) followed by the generation of the control law of Equation (5.9). This involves the triangularization of an $N \times N$ matrix, which could be prohibitive for some real-time applications.

As suggested in [43], the controller coefficients can be obtained by interpolating in a set of previously computed values as shown in Figure 5.4. Notice that this can be accomplished in this case because the controller coefficients only depend on one parameter. The number of points of the set used depends on the variability of the process parameters and on the accuracy needed. The set does not need to be uniform and more points can be computed in regions where the controller parameters vary substantially in order to obtain a better approximation or to reduce the computer memory needed.

The predictor needed in the algorithm to calculate $\hat{y}(t+d \mid t)$, $\hat{y}(t+d-1 \mid t)$ is obtained by applying Equation (5.5) sequentially for $j = 1 - d \cdots 0$. Notice that it basically consists of a model of the plant which is projected towards the future with the values of past inputs and outputs, and it only requires straightforward computation.

5.2.3 Computation of the Controller Parameters

The algorithm just described can be used to compute controller parameters of GPC for plants which can be described by Equation (5.2) (most industrial plants can be described this way) over a set covering the region of interest.

Notice that the sampling time of a digital controller is chosen in practice according to the plant time response. Sampling time between $1/15$ and $1/4$ of T_{95} (the time needed by the system to reach 95 % of the final output value) is recommended in [95]. The pole of the discrete form of the plant transfer function is therefore going to vary between 0.5 and 0.95 for most industrial processes when sampled at appropriate rates.

The curves shown in Figure 5.4 correspond to the controller parameters (l_{y1}, l_{y2}, l_{r1}) obtained for $\delta(i) = \delta^i$ and $\lambda(i) = \lambda^i$ with $\delta = 1$, $\lambda = 0.8$ and $N = 15$. The pole of the system has been changed with a 0.0056 step from 0.5 to 0.99. Notice that due to the fact that the closed-loop static gain must be equal to 1, the sum of the three parameters equals zero. This result implies that only two of the three parameters need to be known.

By looking at Figure 5.4 it can be seen that the functions relating the controller parameters to the process pole can be approximated by functions of the form:

$$l_{yi} = k_{1i} + k_{2i} \frac{a}{k_{3i} - a} \qquad i = 1, 2 \tag{5.10}$$

The coefficients k_{ji} can be calculated by a least squares adjustment using the set of known values of l_{yi} for different values of a. Equation (5.10) can be written as:

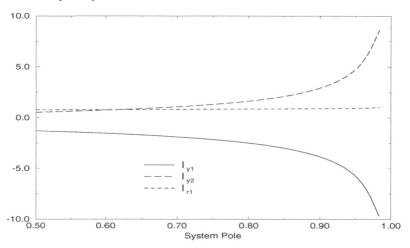

Fig. 5.4. Controller parameters

$$al_{yi} = l_{yi}k_{3i} - k_{1i}k_{3i} + a(k_{i1}a - k_{2i})$$

Repeating this equation for the N_p points used to obtain the approximation, we get

$$\begin{bmatrix} a^1 l_{yi}^1 \\ a^2 l_{yi}^2 \\ \vdots \\ a^{N_p} l_{yi}^{N_p} \end{bmatrix} = \begin{bmatrix} l_{yi}^1 & 1 & a^1 \\ l_{yi}^2 & 1 & a^2 \\ \vdots & \vdots & \vdots \\ l_{yi}^{N_p} & 1 & a^{N_p} \end{bmatrix} \begin{bmatrix} x_1 \\ x_2 \\ x_3 \end{bmatrix} + \begin{bmatrix} e^1 \\ e^2 \\ \vdots \\ e^{N_p} \end{bmatrix} \tag{5.11}$$

where $l_i^j, a^j, e^j, j = 1...N_p$ are the N_p values of the system pole, the precalculated parameters, and the approximation errors, $x_1 = k_{3i}$, $x_2 = -k_{1i}k_{3i}$ and $x_3 = k_{1i} - k_{2i}$.

Equation (5.11) can be written in matrix form

$$\mathbf{Y} = \mathbf{M}\,\mathbf{X} + \mathbf{E}$$

In order to calculate the optimum values of X we minimize $E^T E$ obtaining

$$\mathbf{X} = (\mathbf{M}^T\,\mathbf{M})^{-1}\,\mathbf{M}^T\,\mathbf{Y}$$

The desired coefficients can now be evaluated as:

$$\begin{aligned} k_{3i} &= x_1 \\ k_{1i} &= -x_2/k_{3i} \\ k_{2i} &= k_{1i} - x_3 \end{aligned}$$

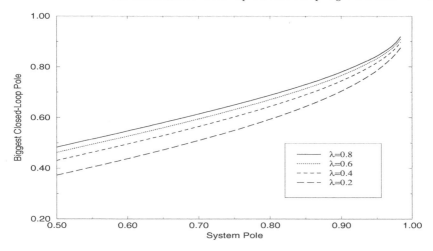

Fig. 5.5. Influence of the control-weighting factor

In the case of $\lambda = 0.8$ and for a control horizon of 15, the controller coefficients are given by:

$$l_{y1} = -0.845 - 0.564\frac{a}{1.05 - a}$$
$$l_{y2} = 0.128 + 0.459\frac{a}{1.045 - a}$$
$$l_{r1} = -l_{y1} - l_{y2}$$

These expressions give a very good approximation to the true controller parameters and fit the set of computed data with a maximum error of less than 1 % of the nominal value for the range of interest of the open-loop pole.

5.2.4 Role of the Control-weighting Factor

The control-weighting factor λ affects the control signal in Equation (5.9). The bigger this value is, the smaller the control effort is allowed to be. If it is given a small value, the system response will be fast since the controller tends to minimize the error between the output and the reference, forgetting about control effort. The controller parameters l_{y1}, l_{y2} and l_{r1} and therefore the closed-loop poles depend on the values of λ.

Figure 5.5 shows the value of the modulus of the biggest closed-loop pole when changing λ from 0 to 0.8 by increments of 0.2. As can be seen, the modulus of the biggest closed-loop pole decreases with λ, indicating faster systems. For a value of λ equal to 0, the closed-loop poles are zero, indicating deadbeat behaviour.

A set of functions was obtained by making λ change from 0.3 to 1.1 by increments of 0.1. It was found that the values of the parameters $k_{ij}(\lambda)$ of

Equation (5.10) obtained could be approximated by functions with the form: $\text{sgn}(k_{ij})e^{c_0+c_1\lambda+c_2\lambda^2}$.

By applying logarithm the coefficients c_1, c_2 and c_3 can be adjusted using a polynomial fitting procedure. The following expressions were obtained for the grid of interest:

$$
\begin{aligned}
k_{11} &= -e^{0.3598-0.9127\lambda+0.3165\lambda^2} \\
k_{21} &= -e^{0.0875-1.2309\lambda+0.5086\lambda^2} \\
k_{31} &= 1.05 \\
k_{12} &= e^{-1.7383-0.40403\lambda} \\
k_{22} &= e^{-0.32157-0.8192\lambda+0.3109\lambda^2} \\
k_{32} &= 1.045
\end{aligned}
\tag{5.12}
$$

The values of the control parameters l_{y1} and l_{y2} obtained when introducing the k_{ij} given in Equation (5.10) are a very good approximation to the real ones. The maximum relative error for $0.55 < a < 0.95$ and $0.3 \leq \lambda \leq 1.1$ is less than 3 %.

5.2.5 Implementation Algorithm

Once the λ factor has been decided, the values k_{ij} can very easily be computed by Expressions (5.12) and the approximate adaptation laws given by Equation (5.10) can easily be employed. The proposed algorithm in the adaptive case is:

---

0. Compute k_{ij} with Expressions (5.12).
1. Perform an identification step.
2. Make $l_i = k_{1i} + k_{2i}\frac{\hat{a}}{k_{3i}-\hat{a}}$ for $i = 1, 2$ and $l_{r1} = -l_{y1} - l_{y2}$.
3. Compute $\hat{y}(t + d \mid t)$ and $\hat{y}(t + d - 1 \mid t)$ using equation (5.5) recursively.
4. Compute control signal $u(t)$ with
   $\triangle u(t) = l_{y1}\hat{y}(t + d \mid t) + l_{y2}\hat{y}(t + d - 1 \mid t) + l_{r1}r(t)$
5. Divide the control signal by the static gain.
6. Go to step 1.

Notice that in a fixed-parameter case the algorithm is simplified since the controller parameters need to be computed only once (unless the control-weighting factor λ is changed) and only steps 3 and 4 have to be carried out at every sampling time.

5.2.6 An Implementation Example

In order to show the straightforwardness of this method, an application to a typical process such as a simple furnace is presented. First, identification

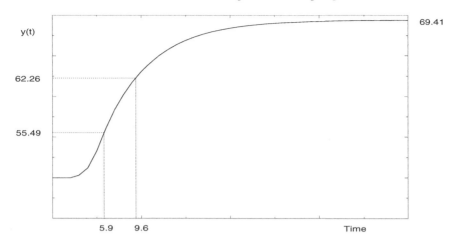

Fig. 5.6. Outlet temperature response

by means of the Reaction Curve method is performed and then the precalculated GPC is applied. The process basically consists of a water flow being heated by fuel which can be manipulated by a control valve. The output variable is the coil outlet temperature whereas the manipulated variable is the fuel flow.

The stationary values of the variables are: inlet temperature 20 °C, outlet temperature 50 °C and fuel valve at 18.21 %. Under these conditions, the fuel rate is changed to a value of 30 % and the outlet temperature response is shown in Figure 5.6, reaching a final value of $y = 69.41$ °C. The plant parameters can be obtained directly from this response as was explained in Section 3.1.1. First, the times t_1 (when the response reaches 28.3 % of its final value) and t_2 (63.2 %) are obtained, resulting in $t_1 = 5.9$ and $t_2 = 9.6$ seconds. Then the plant parameters are

$$K = \frac{\triangle y}{\triangle u} = \frac{69.411 - 50}{30 - 18.21} = 1.646$$
$$\tau = 1.5(t_2 - t_1) = 1.5(9.6 - 5.9) = 5.55$$
$$\tau_d = 1.5(t_1 - \tfrac{1}{3}t_2) = 1.5(5.9 - \tfrac{9.6}{3}) = 4.05$$

with these values, and a sampling time of one second, the equivalent discrete transfer function results in

$$G(z^{-1}) = \frac{0.2713z^{-1}}{1 - 0.8351z^{-1}}z^{-4}$$

As the process is considered to have fixed parameters, the controller coefficients l_{y1}, l_{y2} and l_{r1} can be calculated offline. In this case, choosing $\lambda = 0.8$, the coefficients are obtained from Equation (5.12).

```
/* Predictor */ for (i=2; i<=5; i++)
y[i]=1.8351*y[i-1]-0.8351*y[i-2]+0.2713*(u[5-i]-u[6-i]);

/* Control Law */
u[0]=u[1]+(-3.0367*y[5]+1.9541*y[4]+1.08254*r)/1.646;

/* Updating */ for (i=5; i>0; i--)    u[i] = u[i-1];
y[0] = y[1];
```

Fig. 5.7. C code of implementation algorithm

$$k_{11} = -0.845$$
$$k_{21} = -0.564$$
$$k_{31} = 1.05$$
$$k_{12} = 0.128$$
$$k_{22} = 0.459$$
$$k_{32} = 1.045$$

In consequence the controller coefficients are:

$$l_{y1} = -3.0367 \qquad\qquad l_{y2} = 1.9541 \qquad\qquad l_{r1} = 1.0826$$

Therefore, at every sampling time it will only be necessary to compute the predicted outputs and the control law. The predictions needed are $\hat{y}(t+4 \mid t)$ and $\hat{y}(t+3 \mid t)$, that will be calculated from the next equation with $i = 1 \cdots 4$

$$\hat{y}(t+i \mid t) = (1+a)\hat{y}(t+i-1 \mid t) - a\hat{y}(t+i-2 \mid t) + b \, \triangle \, u(t+i-5)$$

and the control law, where G is the static gain

$$u(t) = u(t-1) + (l_{y1}\,\hat{y}(t+4 \mid t) + l_{y2}\,\hat{y}(t+3 \mid t) + l_{r1}\,r)/G$$

The implementation of the controller in a digital computer will result in a simple program, a part of whose code written in C is shown in Figure 5.7. Two arrays, u and y, are used. The first is used to store the past values of the control signal and the second to store the values of the outputs. In this process with a dead time of 4, the arrays are:

$$\mathbf{y} = [\, y(t-1), y(t), \hat{y}(t+1), \hat{y}(t+2), \hat{y}(t+3), \hat{y}(t+4)\,]$$
$$\mathbf{u} = [\, u(t), u(t-1), u(t-2), u(t-3), u(t-4), u(t-5)\,]$$

Notice that $\mathbf{y}$ contains the predicted outputs and the outputs in t and $t-1$, because these last two values are needed in the first two predictions $\hat{y}(t+1), \hat{y}(t+2)$.

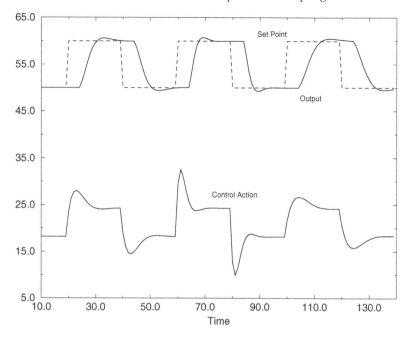

Fig. 5.8. System response

The closed-loop response to a setpoint change of $+10\ ^0$C is plotted in Figure 5.8, where the evolution of the control signal can also be seen . Additionally, the control-weighting factor λ was changed at $t = 60$ from the original value of 0.8 to a smaller 0.3; notice that the control effort increases and the output tends to be faster. On the other hand, if λ takes a bigger value, such as 1.3, the behaviour tends to be slower; this change was performed at $t = 100$.

5.3 The Dead Time Nonmultiple of the Sampling Time Case

5.3.1 Discrete Model of the Plant

When the dead time τ_d of the process is not an integer multiple of the sampling time T ($dT \leq \tau_d \leq (d+1)T$), Equation (5.2) cannot be employed. In this case the fractional delay time can be approximated [66] by the first two terms of the Padé expansion and the plant discrete transfer function can be written as:

$$G(z^{-1}) = \frac{b_0 z^{-1} + b_1 z^{-2}}{1 - az^{-1}} z^{-d} \tag{5.13}$$

As can be seen this transfer function is slightly different from the previous model (Equation(5.2)), presenting an additional zero; a new parameter appears in the numerator. Using the same procedure as in the previous case, a similar implementation of GPC can be obtained for this family of processes.

To obtain the discrete parameters a, b_0 and b_1, the following relations can be used [66]: first, the dead time is decomposed as $\tau_d = dT + \epsilon T$ with $0 < \epsilon < 1$. Then the parameters are:

$$a = e^{-\frac{T}{\tau}} \qquad b_0 = K(1-a)(1-\alpha) \qquad b_1 = K(1-a)\alpha \qquad \alpha = \frac{a(a^{-\epsilon} - 1)}{1 - a}$$

Since the derivation of the control law is very similar in this case to the case in the previous section, some steps will be omitted here for simplicity.

The function J to be minimized is also that of Equation (5.4). Using the CARIMA model with the noise polynomial equal to 1, the system can be written as

$$(1 - az^{-1})y(t) = (b_0 + b_1 z^{-1})z^{-d}u(t - 1) + \frac{\varepsilon(t)}{\triangle}$$

which can be transformed into:

$$y(t+1) = (1+a)y(t) - ay(t-1) + b_0 \triangle u(t-d) + b_1 \triangle u(t-d-1) + \varepsilon(t+1) \quad (5.14)$$

If $\hat{y}(t + d + i - 1 \mid t)$ and $\hat{y}(t + d + i - 2 \mid t)$ are known, it is clear, from Equation (5.14), that the best expected value for $\hat{y}(t + d + i \mid t)$ is given by:

$$\hat{y}(t + d + i \mid t) = (1 + a)\hat{y}(t + d + i - 1 \mid t) - a\hat{y}(t + d + i - 2 \mid t) \\ + b_0 \triangle u(t + i - 1) + b_1 \triangle u(t + i - 2)$$

If $\hat{y}(t + d + i \mid t)$ is introduced in the function to be minimized, $J(N)$ is a function of $\hat{y}(t + d \mid t)$, $\hat{y}(t + d - 1 \mid t)$, $\triangle u(t + N_2 - d - 1)$, $\triangle u(t + N_2 - d - 2)$ ... $\triangle u(t)$, $\triangle u(t - 1)$ and the reference sequence.

Minimizing $J(N)$ with respect to $\triangle u(t)$, $\triangle u(t+1)$ ... $\triangle u(t+N-1)$ leads to

$$\mathbf{M}\,\mathbf{u} = \mathbf{P}\,\mathbf{y} + \mathbf{R}\,\mathbf{w} + \mathbf{Q}\,\triangle u(t-1) \qquad (5.15)$$

where

$$\mathbf{u} = [\,\triangle u(t)\;\triangle u(t+1)\;\cdots\;\triangle u(t+N-1)\,]^T$$
$$\mathbf{y} = [\,\hat{y}(t+d\mid t)\,\hat{y}(t+d-1\mid t)\,]^T$$
$$\mathbf{w} = [\,w(t+d+1)\,w(t+d+2)\;\cdots\;w(t+d+N)\,]^T$$

$\mathbf{M}$ and $\mathbf{R}$ are matrices of dimension $N \times N$, $\mathbf{P}$ of dimension $N \times 2$ and $\mathbf{Q}$ of $N \times 1$.

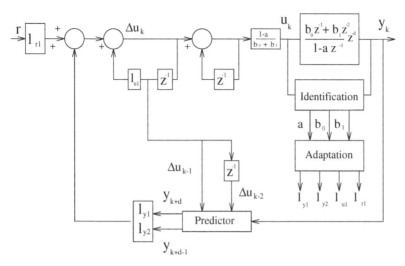

Fig. 5.9. Control scheme

Notice that the term $\mathbf{Q} \triangle u(t-1)$ did not appear in the simpler plant because of the different plant parameters; hence the control law will not be the same. Let us call $\mathbf{q}$ the first row of matrix $\mathbf{M}^{-1}$. Then $\triangle u(t)$ is given by

$$\triangle u(t) = \mathbf{q\,P\,y} + \mathbf{q\,R\,w} + \mathbf{q\,Q} \triangle u(t-1) \tag{5.16}$$

If the reference sequence is considered to be $\mathbf{w} = [1\cdots 1]r(t)$, the control increment $\triangle u(t)$ can be written as:

$$\triangle u(t) = l_{y1}\,\hat{y}(t+d\mid t) + l_{y2}\,\hat{y}(t+d-1\mid t) + l_{r1}\,r(t) + l_{u1} \triangle u(t-1) \tag{5.17}$$

where $\mathbf{q\,P} = [l_{y1}\ l_{y2}]$, $l_{r1} = \sum_{i=1}^{N} q_i \sum_{j=1}^{N} r_{ij}$ and $l_{u1} = \mathbf{q\,Q}$. The resulting control scheme is shown in Figure 5.9, where the values $\hat{y}(t+d\mid t)$, $\hat{y}(t+d-1\mid t)$ are obtained by use of the predictor previously described. Notice that the predictor basically consists of a model of the plant projected towards the future with the values of past inputs and outputs.

5.3.2 Controller Parameters

The plant estimated parameters can be used to compute the controller coefficients (l_{y1}, l_{y2}, l_{r1} and l_{u1}). These coefficients are functions of the plant parameters $a, b_0, b_1, \delta(i)$ and $\lambda(i)$. If the GPC is designed considering the plant to have a unit static gain, there exists a relationship between the plant parameters

$$1 - a = b_0 + b_1$$

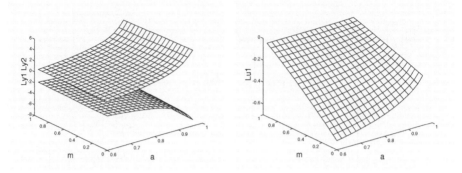

Fig. 5.10. Controller parameters l_{y1}, l_{y2} and l_{u1} as functions of a and m

so that only two of the three parameters will be needed to calculate the coefficients in (5.17). One parameter will be the system pole a and the other will be:

$$m = \frac{b_0}{b_0 + b_1}$$

Parameter m indicates how close the true dead time is to parameter d in the model used in Equation (5.13). That is, if $m = 1$ the plant dead time is d and if $m = 0$ it is $d + 1$; so a range of m between 1 and 0 covers the fractional dead times between d and $d + 1$.

Once the values of $\delta(i)$ and $\lambda(i)$ have been chosen and the plant parameters are known, the controller coefficients can easily be derived. The control signal is divided by the process static gain in order to get a system with a unitary static gain and reduce the number of parameters.

In order to avoid the heavy computational requirements needed to compute matrices **M**, **P**, **R** and **Q** and solve Equation (5.15), the coefficients can be obtained by interpolating in a set of previously computed values as shown in Figure 5.10. Notice that this can be accomplished in this case because the controller coefficients only depend on two parameters. As they have been obtained considering a unitary static gain, they must be corrected dividing the coefficients l_{y1}, l_{y2} and l_{r1} by this value.

The algorithm just described can be used to compute controller parameters of GPC for plants which can be described by Equation (5.13) over a set covering the region of interest. This region is defined by values of the pole in the interval $[0.5, 0.95]$ and the other plant parameter m that will vary between 0 and 1.

The curves shown in Figure 5.10 correspond to the controller parameters l_{y1}, l_{y2} and l_{u1} for $\delta(i) = \delta^i$ and $\lambda(i) = \lambda^i$ with $\delta = 1$, $\lambda = 0.8$ and $N = 15$. Notice that because the closed-loop static gain must equal the value 1, the sum of parameters l_{y1}, l_{y2} and l_{r1} equals zero. This result implies that only three of the four parameters need to be known.

The expressions relating the controller parameters to the process parameters can be approximated by functions of the form:

$$k_{1i}(m) + k_{2i}(m)\frac{a}{k_{3i}(m) - a} \tag{5.18}$$

The coefficients $k_{ji}(m)$ depend on the value of m and can be calculated by a least squares fitting using the set of known values of l_{yi} for different values of a and m. Low-order polynomials that give a good approximation for $k_{ji}(m)$ have been obtained by Bordons [30].

In the case of $m = 0.5$, $\lambda = 0.8$ and for a control horizon of 15, the controller coefficients are given by:

$$l_{y1} = -0.9427 - 0.5486\frac{a}{1.055 - a}$$
$$l_{y2} = 0.1846 + 0.5082\frac{a}{1.0513 - a}$$
$$l_{u1} = -0.3385 + 0.0602\frac{a}{1.2318 - a}$$
$$l_{r1} = -l_{y1} - l_{y2}$$

These expressions give a very good approximation of the true controller parameters and fit the set of computed data with a maximum error of less than 2 % of the nominal values for the range of interest of plant parameters.

The influence of the control-weighting factor λ on the controller parameters can also be taken into account. For small values of λ, the parameters are bigger so that they produce a bigger control effort, thus this factor has to be considered in the approximate functions. With a procedure similar to that in previous sections, the values of $k_{ij}(m)$ in Expressions (5.18) can be approximated as functions of λ, obtaining a maximum error of around 3 % (with the worst cases at the limits of the region).

The algorithm in the adaptive case will consider the plant parameters and the control law and can be seen here:

1. Perform an identification step.
2. Compute $k_{ij}(m, \lambda)$.
3. Calculate l_{y1}, l_{y2} and l_{u1}. Make $l_{r1} = -l_{y1} - l_{y2}$.
4. Compute $\hat{y}(t + d \mid t)$ and $\hat{y}(t + d - 1 \mid t)$ using equation (5.5) recursively.
5. Compute $u(t)$ with:
   $$\triangle u(t) = (l_{y1}\,\hat{y}(t + d \mid t) + l_{y2}\,\hat{y}(t + d - 1 \mid t) + l_{r1}\,r(t))/G + l_{u1}\,\triangle u(t - 1)$$
6. Go to step 1.

5.3.3 Example

This example is taken from [196] and corresponds to the distillate composition loop of a binary distillation column. The manipulated variable is the
reflux flow rate and the controlled variable is the distillate composition. Although the process is clearly nonlinear, it can be modelled by a first-order
model plus a dead time of the form $G(s) = Ke^{-s\tau_d}/(1 + \tau s)$ at different operating points. Notice that this is a reasonable approach since, in fact, a distillation column is composed of a number of plates, each being a first-order
element.

As the process is nonlinear, the response varies for different operating
conditions, so different values for the parameters K, τ and τ_d were obtained,
changing the reflux flow rate from 3.5 to 4.5 mol/min (see Table 5.1). By these
tests, it was seen that variations in the process parameters as $0.107 \le K \le$
0.112, $15.6 \le \tau_d \le 16.37$, and $40.49 \le \tau \le 62.8$ should be considered (where
τ and τ_d are in minutes).

Considering a sample time of five minutes the dead time is not an integer
and the discrete transfer function must be that of equation (5.13). The discrete
parameters can be seen in the same table. The system pole can vary between
0.8838 and 0.9234, while parameter m is going to move between 0.7381 and
0.8841.

Table 5.1. Process parameters for different operating conditions

Flow	K	τ	τ_d	a	$b_0(\times10^{-3})$	$b_1(\times10^{-3})$	d
4.5	0.107	62.8	15.6	0.9234	7.2402	0.9485	3
4	0.112	46.56	15.65	0.8981	9.9901	1.4142	3
3.5	0.112	40.49	16.37	0.8838	9.6041	3.4067	3

To cope with the variations of the system dynamics depending on the
operating point, a gain-scheduling predictive controller can be developed.
To do this, a set of controller parameters is obtained for each operating point,
and depending on the reflux flow, the parameters in each condition will be
calculated by interpolating in these sets.

For a fixed value of λ, the coefficients $k_{ij}(m)$ are used to calculate the
values $l_i(a, m)$. For $\lambda = 0.8$, the following expressions can be used:

$$k_{11} = 0.141 * m^2 - 0.125 * m - 0.920$$
$$k_{21} = -0.061 * m^2 + 0.202 * m - 0.625$$
$$k_{31} = -0.015 * m + 1.061$$
$$k_{12} = -0.071 * m^2 + 0.054 * m + 0.180$$
$$k_{22} = -0.138 * m + 0.575$$
$$k_{32} = -0.015 * m + 1.058$$
$$k_{14} = -0.115 * m^2 + 0.847 * m - 0.729$$
$$k_{24} = -0.113 * m + 0.112$$
$$k_{34} = -0.071 * m^2 + 0.091 * m + 1.181$$

The GPC parameters for the different operating conditions are given in Table 5.2.

Table 5.2. GPC parameters for different operating conditions

Flow	l_{y1}	l_{y2}	l_{r1}	l_{u1}
4.5	-4.577	3.629	0.948	-0.035
4	-3.881	2.954	0.927	-0.039
3.5	-3.637	2.745	0.892	-0.091

As the dead time is of three sampling periods, the predictor is

$$\hat{y}(t+i \mid t) = (1+a)\hat{y}(t+i-1 \mid t) - a\hat{y}(t+i-2 \mid t) + b_0 \triangle u(t+i-3) + b_1 \triangle u(t+i-4)$$

and the control law is

$$u(t) = u(t-1) + (l_{y1}\hat{y}(t+3 \mid t) + l_{y2}\hat{y}(t+2 \mid t) + l_{r1}r)/G + l_{u1} \triangle u(t-1)$$

where G is the static gain and the controller and predictor parameters are obtained by interpolating the flow, once filtered by a low-pass filter, in Table 5.2.

To show the system behaviour, changes of the setpoint covering the complete region are produced, and it can be observed in Figure 5.11 that the output follows the reference, in spite of the changes in the system dynamics due to the changing operating point.

5.4 Integrating Processes

In industrial practice it is easy to find some processes including an integral effect. The output of one of these processes grows infinitely when excited by a step input. This is the case of a tank where the level increases, provided there is an input flow and a constant output. Also the angle of an electrical

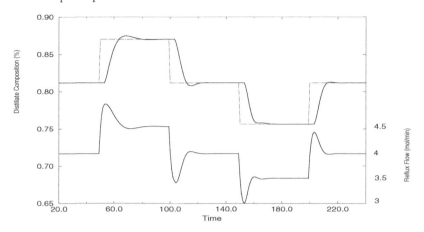

Fig. 5.11. Changes in operating conditions

motor shaft grows while being powered until the torque equals the load. The behaviour of these processes differs drastically from that of those considered up to now in this chapter.

These processes cannot be modelled by a first-order-plus-delay transfer function, but they need the addition of an $1/s$ term to model the integrating effect. Hence, the transfer function for this kind of process will be:

$$G(s) = \frac{K}{s(1 + \tau s)} e^{-\tau_d s} \tag{5.19}$$

In the general case of dead time being nonmultiple of the sampling time the equivalent discrete transfer function when a zero-order hold is employed is given by:

$$G(z) = \frac{b_0 z^{-1} + b_1 z^{-2} + b_2 z^{-3}}{(1 - z^{-1})(1 - az^{-1})} z^{-d} \tag{5.20}$$

In the simpler case of the dead time being an integer multiple of the sampling time, the term b_2 disappears.

The GPC control law for processes described by (5.19) will be calculated in this section. Notice that some formulations of MPC are unable to deal with these processes since they use the truncated impulse or step response, which is not valid for unstable processes. As GPC makes use of the transfer function, there is no problem about unstable processes.

5.4.1 Derivation of the Control Law

The procedure for obtaining the control law is analogous to the one used in previous sections, although logically the predictor will be different and the final expression will change slightly.

Using a CARIMA model with the noise polynomial equal to 1, the system can be written as

$$(1 - z^{-1})(1 - az^{-1})y(t) = (b_0 + b_1 z^{-1} + b_2 z^{-2})z^{-d}u(t - 1) + \frac{\varepsilon(t)}{\triangle}$$

which can be transformed into:

$$y(t + 1) = (2 + a)y(t) - (1 + 2a)y(t - 1) + ay(t - 2)$$
$$+ b_0 \triangle u(t - d) + b_1 \triangle u(t - d - 1) + b_2 \triangle u(t - d - 2) + \varepsilon(t + 1)$$

If the values of $\hat{y}(t+d+i-1 \mid t)$, $\hat{y}(t+d+i-2 \mid t)$ and $\hat{y}(t+d+i-3 \mid t)$ are known, then the best predicted output at instant $t + d + i$ will be:

$$\hat{y}(t + d + i \mid t) = (2 + a)\hat{y}(t + d + i - 1 \mid t) - (1 + 2a)\hat{y}(t + d + i - 2 \mid t) +$$
$$a\hat{y}(t + d + i - 3 \mid t) + b_0 \triangle u(t + i - 1) + b_1 \triangle u(t + i - 2) + b_2 \triangle u(t + i - 3)$$

With these expressions of the predicted outputs, the cost function to be minimized will be a function of $\hat{y}(t+d \mid t)$, $\hat{y}(t+d-1 \mid t)$ and $\hat{y}(t+d-2 \mid t)$, as well as the future control signals $\triangle u(t+N-1)$, $\triangle u(t+N-2)$ ... $\triangle u(t)$, and past inputs $\triangle u(t-1)$ and $\triangle u(t-2)$ and, of course, of the reference trajectory.

Minimizing $J(N_1, N_2, N_3)$ leads to the following matrix equation for calculating $\mathbf{u}$

$$\mathbf{M}\,\mathbf{u} = \mathbf{P}\,\mathbf{y} + \mathbf{R}\,\mathbf{w} + \mathbf{Q}_1 \triangle u(t - 1) + \mathbf{Q}_2 \triangle u(t - 2)$$

where $\mathbf{M}$ and $\mathbf{R}$ are matrices of dimension $N \times N$, $\mathbf{P}$ of dimension $N \times 2$ and $\mathbf{Q}_1$ and $\mathbf{Q}_2$ of $N \times 1$. As in the previous section, $\mathbf{u}$ are the future input increments and $\mathbf{y}$ the predicted outputs.

The first element of vector $\mathbf{u}$ can be obtained by

$$\triangle u(t) = \mathbf{q}\,\mathbf{P}\,\mathbf{y} + \mathbf{q}\,\mathbf{R}\,\mathbf{w} + \mathbf{q}\,\mathbf{Q}_1 \triangle u(t - 1) + \mathbf{q}\,\mathbf{Q}_2 \triangle u(t - 2)$$

where $\mathbf{q}$ is the first row of matrix $\mathbf{M}^{-1}$.

If the reference is considered to be constant over the prediction horizon and equal to the current setpoint

$$\mathbf{w} = [1 \ldots 1]r(t + d)$$

the control law results as

$$\triangle u(t) = l_{y1}\hat{y}(t + d \mid t) + l_{y2}\hat{y}(t + d - 1 \mid t) + l_{y3}\hat{y}(t + d - 2 \mid t)$$
$$+ l_{r1}r(t + d) + l_{u1} \triangle u(t - 1) + l_{u2} \triangle u(t - 2) \qquad (5.21)$$

where $\mathbf{q}\,\mathbf{P} = [l_{y1}\ l_{y2}\ l_{y3}]$, $l_{r1} = \sum_{i=1}^{N}(q_i \sum_{j=1}^{N} r_{ij})$, $l_{u1} = \mathbf{q}\,\mathbf{Q}_1$ and $l_{u2} = \mathbf{q}\,\mathbf{Q}_2$.

Therefore the control law results in a linear expression depending on six coefficients which depend on the process parameters (except on the dead time) and on the control-weighting factor λ. Furthermore, one of these coefficients is a linear combination of the others, since the following relation must hold to get a closed loop with unitary static gain:

$$l_{y1} + l_{y2} + l_{y3} + l_{r1} = 0$$

5.4.2 Controller Parameters

The control law (5.21) is very easy to implement provided the controller parameters l_{y1}, l_{y2}, l_{y3}, l_{r1}, l_{u1} and l_{u2} are known. The existence of available relationships of these parameters with process parameters is of crucial importance for a straightforward implementation of the controller. In a similar way to the previous sections, simple expressions for these relationships will be obtained.

As the process can be modelled by (5.20) four parameters (a, b_0, b_1 and b_2) are needed to describe the plant. Expressions relating the controller coefficients to these parameters can be obtained as earlier, although the resulting functions are not as simple, due to the number of plant parameters involved. As the dead time can often be considered as a multiple of the sampling time, simple functions will be obtained for this case from now on. Then b_2 will be considered equal to 0.

In a similar way to the process without integrator case, the process can be considered to have $(b_0 + b_1)/(1 - a) = 1$ in order to work with normalized plants. Then the computed parameters must be divided by this value that will not equal 1 in general.

The controller coefficients will be obtained as a function of the pole a and a parameter:

$$n = \frac{b_0}{b_0 + b_1}$$

This parameter has a short range of variability for any process. As b_0 and b_1 are related to the continuous parameters by (see [11])

$$b_0 = K(T + \tau(-1 + e^{-\frac{T}{\tau}})) \qquad b_1 = K(\tau - e^{-\frac{T}{\tau}}(T + \tau))$$

then

$$n = \frac{a - 1 - \log a}{(a - 1)\log a}$$

that for the usual values of the system pole is going to vary between $n = 0.5$ and $n = 0.56$. Therefore the controller parameters can be expressed as functions of the system pole, and n for a fixed value of λ.

The shape of the parameters is displayed in Figure 5.12 for a fixed value of $\lambda = 1$. It can be seen that the coefficients depend mainly on the pole a, being almost independent of n except in the case of l_{u1}. Functions of the form

$$f(a, n, \lambda) = k_1(n, \lambda) + k_2(n, \lambda)\frac{a}{k_3(n, \lambda) - a}$$

where k_i can be approximated by

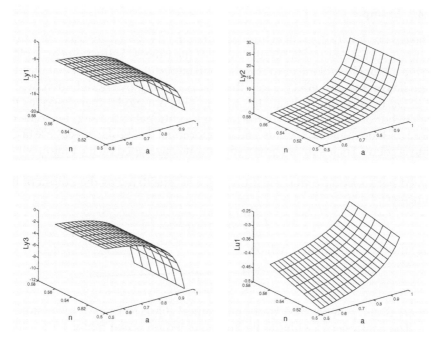

Fig. 5.12. Controller coefficients l_{y1}, l_{y2}, l_{y3} and l_{u1}

$$k_{y1,1} = -e^{0.955-0.559\lambda+0.135\lambda^2}$$
$$k_{y1,2} = -e^{0.5703-0.513\lambda+0.138\lambda^2}$$
$$k_{y1,3} = 1.0343$$
$$k_{y2,1} = e^{0.597-0.420\lambda+0.0953\lambda^2}$$
$$k_{y2,2} = e^{1.016-0.4251\lambda+0.109\lambda^2}$$
$$k_{y2,3} = 1.0289$$
$$k_{y3,1} = -e^{-1.761-0.422\lambda+0.071\lambda^2}$$ \hfill (5.22)
$$k_{y3,2} = -e^{0.103-0.353\lambda+0.089\lambda^2}$$
$$k_{y3,3} = 1.0258$$
$$k_{u1,1} = 1.631n - 1.468 + 0.215\lambda - 0.056\lambda^2$$
$$k_{u1,2} = -0.124n + 0.158 - 0.026\lambda + 0.006\lambda^2$$
$$k_{u1,3} = 1.173 - 0.019\lambda$$

provide good approximations for l_{y1}, l_{y2}, l_{y3}, l_{r1} and l_{u1} in the usual range of the plant parameter variations. Notice that an approximate function for l_{r1} is not supplied since it is linearly dependent on the other coefficients. The functions fit the set of computed data with a maximum error of less than 1.5 % of the nominal values. Notice that closer approximations can be obtained if developed for a concrete case where the range of variability of the process parameters is smaller.

5.4.3 Example

The control law (5.21) will be implemented in an extensively used system as a direct-current motor. When the input of the process is the voltage applied to the motor (U) and the output is the shaft angle (θ) it is obvious that the process has an integral effect, given that the position grows indefinitely whilst it is fed by a certain voltage. In order to obtain a model that describes the behaviour of the motor the inertia load (proportional to the angular acceleration) and the dynamic friction load (proportional to angular speed) are taken into account. Their sum is equal to the torque developed by the motor, which depends on the voltage applied to it. It is a first-order system with regards to speed but a second-order one if the angle is considered as the output of the process:

$$J\frac{d^2\theta}{dt^2} + f\frac{d\theta}{dt} = M_m$$

and the transfer function will be:

$$\frac{\theta(s)}{U(s)} = \frac{K}{s(1 + \tau s)}$$

where K and τ depend on electromechanical characteristics of the motor.

The controller is going to be implemented on a real motor with a feed voltage of 24 V and nominal current of 1.3 A, subjected to a constant load. The Reaction Curve Method is used to obtain experimentally the parameters of the motor, applying a step in the feed voltage and measuring the evolution of the angular speed (which is a first-order system). The parameters obtained are

$$K = 2.5 \qquad \tau = 0.9 \text{ second}$$

and zero dead time. Taking a sampling time of $T = 0.06$ second one gets the discrete transfer function:

$$G(z) = \frac{0.004891z^{-1} + 0.004783z^{-2}}{(1 - z^{-1})(1 - 0.935507z^{-1})}$$

If a high value of the control-weighting factor is taken to avoid overshooting ($\lambda = 2$) the control parameters (5.21) can be calculated using expressions (5.22):

$$l_{y1} = -11.537$$
$$l_{y2} = 19.242$$
$$l_{y3} = -8.207$$
$$l_{u1} = -0.118$$
$$l_{r1} = 0.502$$

The evolution of the shaft angle when some steps are introduced in the reference can be seen in Figure 5.13. It can be observed that there is no overshooting due to the high value of λ chosen. The system has a dead zone such

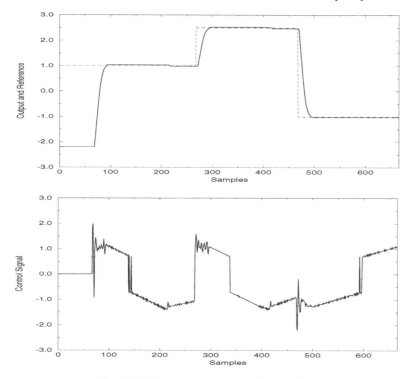

Fig. 5.13. Motor response for setpoint changes

that it is not sensitive to control signals less than 0.7 V; in order to avoid this a nonlinearity is added.

It is important to remember that the sampling time is very small (0.06 second) which could make the implementation of the standard GPC algorithm impossible. However, due to the simple formulation used here, the implementation is reduced to the calculation of Expression (5.21) and hardly takes any time in a computer.

The process is disturbed by the addition of an electromagnetic break that changes the load and the friction constant. The model parameters used for designing the GPC do not coincide with the process parameters, but in spite of this, as can be seen in Figure 5.14, GPC is able to control the motor reasonably well even though a slight overshoot appears.

5.5 Consideration of Ramp Setpoints

It is usual for a process reference signal to keep a certain constant value for a time and to move to other constant values by step changes during normal plant operation. This is what has been considered up to now, that is,

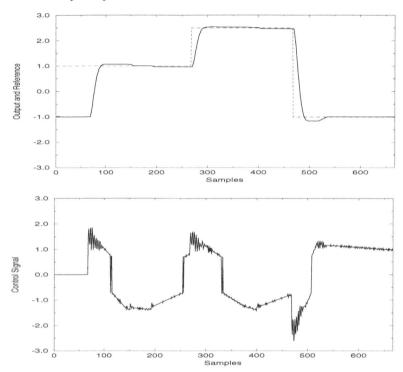

Fig. 5.14. Motor response with electromagnetic break

$w(t + d + 1) = w(t + d + 2) \ldots = r(t)$, where $r(t)$ is the setpoint at instant t which is going to maintain a fixed value.

But the reference evolution will not behave like this in all circumstances. On many occasions it can evolve as a ramp, which changes smoothly to another constant setpoint. In general it would be desirable for the process output to follow a mixed trajectory composed of steps and ramps.

This situation frequently appears in different industrial processes. In the food and pharmaceutical industries some thermal processes require the temperature to follow a profile given by ramps and steps. It is also of interest that in the control of motors and in robotics applications, the position or velocity follows evolutions of this type.

GPC will be reformulated when the reference is a ramp, defined by a parameter α indicating the increment at each sampling time. The reference trajectory is therefore:

$$w(t + d + 1) = r(t + d) + \alpha$$
$$w(t + d + 2) = r(t + d) + 2\alpha$$
$$\cdots \qquad \cdots$$
$$w(t + d + N) = r(t + d) + N\alpha$$

Employing the procedure used throughout this chapter, and for first-order systems with dead time, we get

$$\mathbf{M}\,\mathbf{u} = \mathbf{P}\,\mathbf{y} + \mathbf{R}\,\mathbf{w} + \mathbf{Q}\,\triangle u(t-1)$$

If $\mathbf{q}$ is the first row of matrix $\mathbf{M}^{-1}$ then $\triangle u(t)$ can be expressed as

$$\triangle u(t) = \mathbf{q}\,\mathbf{P}\,\mathbf{y} + \mathbf{q}\,\mathbf{R}\,\mathbf{w} + \mathbf{q}\,\mathbf{Q}\,\triangle u(t-1)$$

By making $\mathbf{h} = \mathbf{q}\,\mathbf{R}$ the term of the preceding expression including the reference ($\mathbf{h}\,\mathbf{w}$) takes the form:

$$\mathbf{h}\,\mathbf{w} = \sum_{i=1}^{N} h_i\ r(t+d+i) = \sum_{i=1}^{N} h_i\ (r(t+d)+\alpha\ i) = \sum_{i=1}^{N} h_i\ r(t+d)+\alpha \sum_{i=1}^{N} h_i\ i$$

Therefore

$$\mathbf{h}\,\mathbf{w} = l_{r1}\ r(t+d) + \alpha\ l_{r2}$$

The control law can now be written as

$$\triangle u(t) = l_{y1}\hat{y}(t+d\mid t) + l_{y2}\hat{y}(t+d-1\mid t) + l_{r1}r(t+d) + \alpha\, l_{r2} + l_{u1}\triangle u(t-1)$$
$$(5.23)$$

where $\mathbf{q}\,\mathbf{P} = [l_{y1}\ \ l_{y2}]$, $l_{u1} = \mathbf{q}\,\mathbf{Q}$, $l_{r1} = \sum_{i=1}^{N}(q_i \sum_{j=1}^{N} r_{ij})$ and $l_{r2} = \alpha \sum_{i=1}^{N} h_i\ i$.

The control law is therefore linear. The new coefficient l_{r2} is due to the ramp. It can be noticed that when the ramp becomes a constant reference, the control law coincides with the one developed for the constant reference case. The only modification that needs to be made because of the ramps is the term $l_{r2}\alpha$. The predictor is the same and the resolution algorithm does not differ from the one used for the constant reference case. The new parameter l_{r2} is a function of the process parameters (a, m) and of the control weighting factor (λ). As in the previous cases an approximating function can easily be obtained. Notice that the other parameters are exactly the same as in the constant reference case, meaning that the previously obtained expressions can be used.

In what has been seen up to now (nonintegrating processes, integrating processes, constant reference, ramp reference), a new coefficient appeared in the control law with each new situation. All these situations can be described by the following control law:

Table 5.3. Coefficients that may appear in the control law. The $\times$ indicates that the coefficient exists

Process	Reference	l_{y1}	l_{y2}	l_{y3}	l_{u1}	l_{u2}	l_{r1}	l_{r2}
$\frac{k}{1+\tau s}e^{-\tau_d s}$	Constant	$\times$	$\times$	0	0	0	$\times$	0
τ_d integer	Ramp	$\times$	$\times$	0	0	0	$\times$	$\times$
$\frac{k}{1+\tau s}e^{-\tau_d s}$	Constant	$\times$	$\times$	0	$\times$	0	$\times$	0
τ_d non integer	Ramp	$\times$	$\times$	0	$\times$	0	$\times$	$\times$
$\frac{k}{s(1+\tau s)}e^{-\tau_d s}$	Constant	$\times$	$\times$	$\times$	$\times$	0	$\times$	0
τ_d integer	Ramp	$\times$	$\times$	$\times$	$\times$	0	$\times$	$\times$
$\frac{k}{s(1+\tau s)}e^{-\tau_d s}$	Constant	$\times$	$\times$	$\times$	$\times$	$\times$	$\times$	0
τ_d non integer	Ramp	$\times$	$\times$	$\times$	$\times$	$\times$	$\times$	$\times$

$$\triangle u(t) = l_{y1}\hat{y}(t+d\mid t) + l_{y2}\hat{y}(t+d-1\mid t) + l_{y3}\hat{y}(t+d-2\mid t)$$
$$+ l_{r1}r(t+d) + \alpha\, l_{r2} + l_{u1}\triangle u(t-1) + l_{u2}\triangle u(t-2)$$

Table 5.3 shows which coefficients of this control law may be zero depending on the particular situation.

5.5.1 Example

As an application example, a GPC with ramp following capability is going to be designed for the motor described earlier. The reference trajectory is composed of a series of steps and ramps defined by the value of α ($\alpha = 0$ for the case of constant reference).

The same controller parameters as in the previous example are used, with the addition of the new parameter $l_{r2} = 2.674$. Considering that $(b_0 + b_1)/(1 - a) = 0.15$, the control law is given by:

$$\triangle u(t) = -76.92\, y(t) + 128.29\, y(t-1) - 54.72\, y(t-2)$$
$$+ 3.35r(t) + 17.82\, \alpha - 0.12\triangle u(t-1)$$

As the dead time is zero, the predicted outputs are known at instant t.

The results obtained are shown in Figure 5.15, where it can be seen that the motor is able to follow the ramp reference quite well.

5.6 Comparison with Standard GPC

The approximations made in the method can affect the quality of the controlled performance. Some simulation results are presented that compare the results obtained with the proposed method with those when the standard GPC algorithm as originally proposed by Clarke et al. [58] is used.

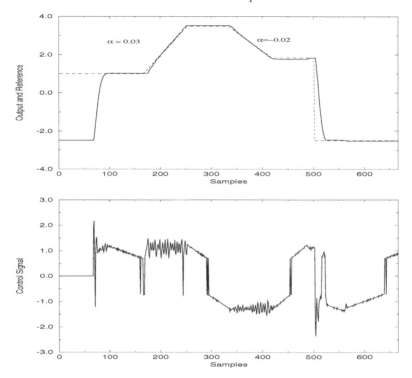

Fig. 5.15. Combined steps and ramps setpoint

Two indices are used to measure the performance: ISE (sum of the square errors during the transient) and ITAE (sum of the absolute error multiplied by discrete time). Also the number of floating-point operations and the computing time needed to calculate the control law are analyzed.

First, the performance of the proposed algorithm is compared with that of the standard GPC with no modelling errors. In this situation the error is only caused by the approximative functions of the controller parameters. For the system $G(s) = \frac{1.5}{1+10s}e^{-4s}$ with a sampling time of one second, the values for the proposed algorithm when the process is perturbed by a white noise uniformly distributed in the interval ± 0.015 are ISE= 7.132, ITAE= 101.106 and for the standard controller ISE= 7.122, ITAE=100.536. The plot comparing the two responses is not shown because there is practically no difference.

The plant model is supposed to be first-order plus deadtime. If the process behaviour can be reasonably described by this model, there will not be a substantial loss of performance. Consider, for instance, the process modelled by:

$$G_p(z) = \frac{0.1125z^{-1} - 0.225z^{-2}}{1 - z^{-1} + 0.09z^{-2}}z^{-3}$$

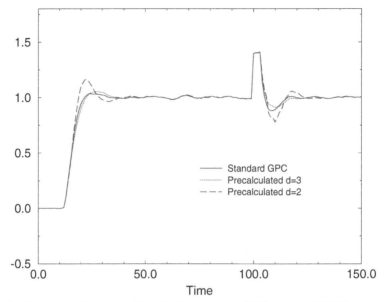

Fig. 5.16. System performance. Standard GPC: ISE = 6.809, ITAE = 395.608; proposed algorithm $d = 3$ ISE = 6.987, ITAE = 402.544; proposed algorithm $d = 2$ ISE = 7.064, ITAE = 503.687

For control purposes, it is approximated by the following first-order model, obtained from data generated by the process G_p:

$$G_m(z) = \frac{0.1125z^{-1}}{1 - 0.8875z^{-1}} z^{-3}$$

That is, the precomputed GPC is working in the presence of unmodelled dynamics. From the previous studies of robust stability, it can be deduced that the closed-loop system is going to be stable. The performance in this situation is shown in Figure 5.16 where the system response for both controllers is shown; notice that for $t = 100$ a disturbance is added to the output. The figure also shows the behaviour of the precomputed GPC when an additional deadtime mismatch is included, that is, the controller uses a model with $d = 2$ instead of the true value $d = 3$.

Logically, there is a slight loss of performance due to the uncertainties, that must be considered in conjunction with the benefits in the calculation. Besides, consider that in a real case the uncertainties (such as deadtime mismatch) can also affect the standard GPC since high-frequency effects are usually very difficult to model.

The computational requirements of the method are compared with the standard in Table 5.4 for this example, working with a control horizon of $N = 15$ and $\lambda = 0.8$. The table shows the computation needed for the calculation of the control law in both floating-point operations and CPU time on a personal computer.

Table 5.4. Computational requirements for the standard and precalculated GPC

Algorithm	Calculation	Operations (flops)
Standard	Build matrices	1057
	Compute $G^T G + \lambda I$	10950
	Inversion	7949
	Rest	1992
	TOTAL	**21948**
Proposed	**TOTAL**	79

As can be seen, these examples show that although a little performance is lost, there is a great improvement in real-time implementability, reaching a computing effort around 275 times smaller. This advantage can represent a crucial factor for the implementation of this strategy in small controllers with low computational facilities, considering that the impact on the performance is negligible. The simulations also show the robustness of the controller in the presence of structured uncertainties.

5.7 Stability Robustness Analysis

The elaboration of mathematical models of processes in real life requires simplifications to be adopted. In practice no mathematical model capable of exactly describing a physical process exists. It is always necessary to bear in mind that modelling errors may adversely affect the behaviour of the control system. The aim is that the controller should be insensitive to these uncertainties in the model, that is, that it should be robust. The aim here is to deal with the robustness of the controller presented. In any case, developments of predictive controller design using robust criteria can be found, for instance, in [44] and [208].

The modelling errors, or uncertainties, can be represented in different forms, reflecting in certain ways the knowledge of the physical mechanisms which cause the discrepancy between the model and the process as well as the capacity to formalize these mechanisms so that they can be handled. Uncertainties can, in many cases, be expressed in a structured way, as expressions in function of determined parameters which can be considered in the transfer function [69]. However, there are usually residual errors particularly dominant at high frequencies which cannot be modelled in this way, which constitute unstructured uncertainties [65]. In this section a study of the precalculated GPC stability in the presence of both types of uncertainties is made; that is, the stability robustness of the method will be studied.

This section aims to study the influence of uncertainties on the behaviour of the process working with a controller which has been developed for the nominal model. That is, both the predictor and the controller parameters are calculated for a model which does not exactly coincide with the real process to be controlled. The following question is asked: what discrepancies are permissible between the process and the model for the controlled system to be stable?

The controller parameters l_{y1}, l_{y2}, l_{r1} and l_{u1} that appear in the control law

$$\triangle u(t) = l_{y1}\hat{y}(t + d \mid t) + l_{y2}\hat{y}(t + d - 1 \mid t) + l_{r1}r(t) + l_{u1} \triangle u(t - 1)$$

have been precalculated for the model (not for the process as this is logically unknown). Likewise the predictor works with the parameters of the model, although it keeps up to date with the values taken from the output produced by the real process.

5.7.1 Structured Uncertainties

A first-order model with pure delay, in spite of its simplicity, describes the dynamics of most plants in the process industry. However, it is fundamental to consider the case where the model is unable to completely describe all the dynamics of the real process. Two types of structured uncertainties are considered: parametric uncertainties and unmodelled dynamic uncertainties. In the first case, the order of the control model is supposed to be identical to the order of the plant but the parameters are considered to be within an uncertainty region around the nominal parameters (these parameters will be the pole, the gain, and the coefficient $m = b_0/(b_0 + b_1)$ that measures the fractional delay between d and $d+1$). The other type of uncertainty will take into account the existence of process dynamics not included in the control model as an additional unmodelled pole and delay estimation error. This will be reflected in differences between the plant and model orders.

The uncertainty limits have been obtained numerically for the range of variation of the process parameters $(0.5 < a < 0.98, 0 \leq m \leq 1)$ with a delay $0 \leq d \leq 10$ obtaining the following results (for more details, see [44]):

- **uncertainty at the pole:** for a wide working zone $(a < 0.75)$ and for normal values of the delay an uncertainty of more than $\pm 20\%$ is allowed. For higher poles the upper limit decreases due almost exclusively to the fact that the open loop would now be unstable. The stable area only becomes narrower for very slow systems with large delays. Notice that this uncertainty refers to the time constant (τ) uncertainty of the continuous process $(a = \exp(-T/\tau))$ and thus the time constant can vary around 500% of the nominal one in many cases.

- **gain uncertainty:** When the gain of the model is G_e and that of the process is $\gamma \times G_e$, γ will be allowed to move between 0.5 and 1.5, that is, uncertainties in the value of the gain of about 50% are permitted. For small delays $(1, 2)$ the upper limit is always above the value $\gamma = 2$ and only comes close to the value 1.5 for delays of about 10. It can thus be concluded that the controller is very robust when faced with this type of error.
- **uncertainty in m:** The effect of this parameter can be ignored since a variation of 300% is allowed without reaching instability.
- **unmodelled pole:** The real process has another less dominant pole $(k \times a)$ apart from the one appearing in the model (a), and the results show that the system is stable even for values of k close to 1; stability is only lost for systems with very large delays.
- **delay estimation error:** From the results obtained in a numerical study, it is deduced that for small delays stability is guaranteed for errors of up to two units through the range of the pole, but when bigger poles are dealt with this only happens for small values of a, and even for delay 10 only a delay mismatch of one unit is permitted. It can be concluded, therefore, that a good delay estimation is fundamental to GPC, because for errors of more than one unit the system can become unstable if the process delay is high.

5.7.2 Unstructured Uncertainties

In order to consider unstructured uncertainties, it will be assumed from now on that the dynamic behaviour of a determined process is described not by an invariant time linear model but by a family of linear models. Thus the real possible processes (G) will be in a vicinity of the nominal process $(\widetilde{G})$, which will be modelled by a first-order-plus-delay system.

A family $\mathcal{F}$ of processes in the frequency domain will therefore be defined which in the Nyquist plane will be represented by a region about the nominal plant for each ω frequency. If this family is defined as

$$\mathcal{F} = \{G : | G(i\omega) - \widetilde{G}(i\omega) | \leq \bar{l}_a(\omega)\}$$

the region consists of a disc with its centre at $G(i\omega)$ and radius $\bar{l}_a(\omega)$. Therefore any member of the family fulfils the condition

$$G(i\omega) = \widetilde{G}(i\omega) + l_a(i\omega) \qquad | l_a(i\omega) | \leq \bar{l}_a(\omega)$$

This region will change with ω because l_a does and, therefore, in order to describe family $\mathcal{F}$ we will have a zone formed by the discs at different frequencies. If one wishes to work with multiplicative uncertainties the family of processes can be described by

$$\mathcal{F} = \{G : \left| \frac{G(i\omega) - \widetilde{G}(i\omega)}{\widetilde{G}(i\omega)} \right| \leq \bar{l}_m(\omega)\} \tag{5.24}$$

simply considering

$$l_m(i\omega) = l_a(i\omega)/\widetilde{G}(i\omega) \qquad \bar{l}_m(i\omega) = \bar{l}_a(\omega)/|\widetilde{G}(i\omega)|$$

Therefore any member of family $\mathcal{F}$ satisfies

$$G(i\omega) = \widetilde{G}(i\omega)(1 + l_m(i\omega)) \qquad |l_m(i\omega)| \leq \bar{l}_m(i\omega)$$

This representation of uncertainties in the Nyquist plane as a disc around the nominal process can encircle any set of structured uncertainties, although some times it can result in a rather conservative attitude [141].

The measurement of the robustness of the method can be tackled using the robust stability theorem [141], that for discrete systems states:

Suppose that all processes G of family $\mathcal{F}$ have the same number of unstable poles, which do not become unobservable for the sampling, and that a controller $C(z)$ stabilizes the nominal process $\widetilde{G}(s)$. Then the system has robust stability with controller C if and only if the complementary sensitivity function for the nominal process satisfies the following relation:

$$|\widetilde{T}(e^{i\omega T})|\,\bar{l}_m(\omega) < 1 \qquad 0 \leq \omega \leq \pi/T \tag{5.25}$$

Using this condition the robustness limits will be obtained for systems that can be described by (5.13), and for all the values of the parameters that describe the system (a, m and d). For each value of the frequency ω the limits can be calculated as:

$$\bar{l}_m = \left| \frac{1 + \widetilde{G}C}{\widetilde{G}C} \right| \qquad \bar{l}_a = \bar{l}_m\,|\widetilde{G}|$$

In Figure 5.17 the form taken by the limits in function of ωT can be seen for some values of a, and fixed values of m and d. Both limits are practically constant and equal to unity at low frequencies and change (the additive limit $\bar{l}_a$ decreases and the multiplicative $\bar{l}_m$ increases) at a certain point. Notice that these curves show the great degree of robustness that the GPC possesses since $\bar{l}_m$ is relatively big at high frequencies, where multiplicative uncertainties are normally smaller than unity, and increases with frequency as uncertainties do. The small value of $\bar{l}_a$ at high frequencies is due to the fact that the process itself has a small gain at those frequencies; remember that both limits are dependent and related by $\bar{l}_a = \bar{l}_m\,|\widetilde{G}|$.

Figure 5.18 shows the frequency response of the nominal process alone and with the controller, as well as the discs of radius $\bar{l}_a$ and $\bar{l}_m\,|\widetilde{G}C|$ for a certain frequency. All the G processes belonging to the $\mathcal{F}$ family maintaining

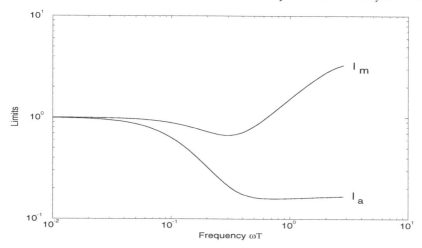

Fig. 5.17. General shape of $\bar{l}_a$ and $\bar{l}_m$

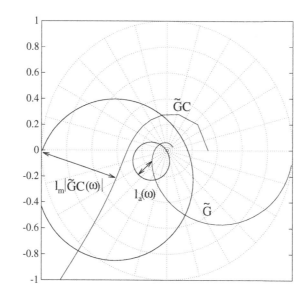

Fig. 5.18. Polar diagram of the process $\widetilde{G}$ and $\widetilde{G}C$ showing the limits for a given frequency

the stability of the closed loop can be found inside the disc of radius $\bar{l}_a$. The shape of the frequency response leads to limits $\bar{l}_a$ and $\bar{l}_m$. Thus, $\widetilde{G}C(\omega)$ has a big modulus (due to the integral term) at low frequencies, leading to a value

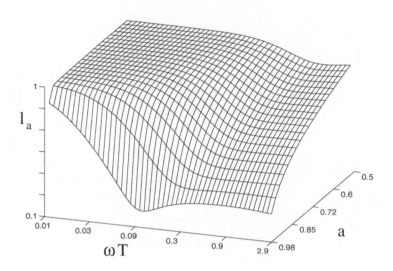

Fig. 5.19. Limit $\bar{l}_a$ for dead time 1

of $\bar{l}_m$ close to unity. When ω increases, $\widetilde{G}C(\omega)$ separates from -1 (without decreasing in modulus) and therefore the limit can safely grow.

It can be seen that the most influential parameters are pole a and delay d. The evolution of limit $\bar{l}_a$ with frequency (ωT) for parameter a changing between 0.5 and 0.98 is presented in Figure 5.19 for a concrete value of delay, $d = 1$, and for an average value of m, $m = 0.5$. As was to be expected, the limit decreases for greater poles because with open-loop poles near the limit of the unit circle the uncertainty allowed is smaller, as it would be easier to enter the open-loop unstable zone.

5.7.3 General Comments

The results obtained for both types of uncertainties are qualitatively the same. It can be concluded that the factor that mainly affects robustness is delay uncertainty, because of its effect at high frequencies. The robustness zone decreases when the open-loop pole increases whilst the parameter m hardly has any influence. As the analysis has been performed based on a particular choice of parameters in the GPC formulation the conclusions depend on these values. The influence of the choice of these parameters on the closed-loop stability is studied in [153].

In any case, the GPC algorithm presented has shown itself to be very robust against the types of uncertainties considered. For small delays the closed loop is stable for static gain mismatch of more than 100% and time constant mismatch of more than 200%.

The stability robustness of GPC can be improved with the use of an observer polynomial, the so-called $T(z^{-1})$ polynomial. In [57] a reformulation of the standard GPC algorithm including this polynomial can be found. In order to do this, the CARIMA model is expressed in the form:

$$A(z^{-1})y(t) = B(z^{-1})u(t-1) + \frac{T(z^{-1})}{\triangle}\xi(t)$$

Up to now the $T(z^{-1})$ has been considered equal to 1, describing the most common disturbances or as the colouring polynomial $C(z^{-1})$. But it can also be considered as a design parameter. In consequence the predictions will not be optimal but on the other hand robustness in the face of uncertainties can be achieved, in a similar interpretation as that used by Ljung [128]. Then this polynomial can be considered as a prefilter as well as an observer. The effective use of observers is known to play an essential role in the robust realization of predictive controllers (see [57] for the effect of prefiltering on robustness and [209] for guidelines for the selection of T).

This polynomial can easily be added to the proposed formulation, computing the prediction with the values of inputs and outputs filtered by $T(z^{-1})$. Then, the predictor works with $y^f(t) = y(t)/T(z^{-1})$ and $u^f(t) = u(t)/T(z^{-1})$. The actual prediction for the control law is computed as $\hat{y}(t+d) = T(z^{-1})\hat{y}^f(t+d)$.

5.8 Composition Control in an Evaporator

This chapter ends with an application of the method to a typical process. An evaporator, a very common process in industry, has been chosen as a testing bed for the GPC. This process involves a fair number of interrelated variables and although it may appear to be rather simple compared to other processes of greater dimension, it allows the performance of any control technique to be checked. The results presented in this section have been obtained by simulation on a nonlinear model of the process.

5.8.1 Description of the Process

The process in question is a forced circulation evaporator in which the raw material is mixed with an extraction of the product and pumped through a vertical heat exchanger through which water steam is circulating, which condenses in the tubes. The mix evaporates and passes through a separating vessel where the liquid and the vapour are separated. The former is made to recycle and a part is extracted as the final product whilst the vapour is condensed with cooling water. Figure 5.20 shows the diagram of this process, used in many production sectors, such as the sugar industry.

The process behaviour can be modelled by a series of equations obtained from the mass and energy balance equations, as well as by making some

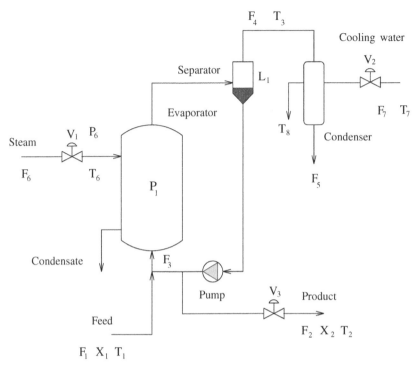

Fig. 5.20. Diagram of the evaporator

realistic assumptions. The equations describing the process behaviour can be found in [149]. The main variables, together with their values at the point of operation, are grouped in table 5.5.

Table 5.5. Process variables and values at operating point

Variable	Description	Value	Units
F_1	Feed flow rate	10.0	kg/min
F_2	Product flow rate	2.0	kg/min
F_4	Vapour flow rate	8.0	kg/min
F_5	Condensate flow rate	8.0	kg/min
X_1	Feed composition	5.0	%
X_2	Product composition	25.0	%
L_1	Separator level	1.0	m
P_1	Operating pressure	50.5	kPa
P_6	Steam pressure	194.7	kPa
F_6	Steam flow rate	9.3	kg/min
F_7	Cooling water flow rate	208.0	kg/min

The system dynamics are mainly dictated by the differential equations modelling the mass balances:

- mass balance in the liquid:

$$\rho A \frac{dL_1}{dt} = F_1 - F_4 - F_2 \tag{5.26}$$

where ρ is the density of the liquid and A the section of the separator, whose product can be considered constant.
- mass balance in the solute:

$$M \frac{dX_2}{dt} = F_1 X_1 - F_2 X_2 \tag{5.27}$$

where M is the total quantity of liquid in the evaporator.
- mass balance in the process vapour, the total amount of water vapour can be expressed in a function of the pressure existing in the system according to

$$C \frac{dP_1}{dt} = F_4 - F_5 \tag{5.28}$$

where C is a constant that converts the steam mass into an equivalent pressure.

The dynamics of the interchanger and the condenser can be considered very fast compared to previous ones.

Degrees of Freedom

Twelve equations can be found for twenty variables so there are eight degrees of freedom. Eight more equations must therefore be considered to close the problem; these will be the ones which provide the values of the manipulated variables and the disturbances:

- three manipulated variables: the steam pressure P_6, which depends on the opening of valve V_1, the cooling water flow rate F_7, controlled by valve V_2 and the product flow rate F_2 with V_3.
- five disturbances: feed flow rate F_1, circulating flow rate F_3, composition and temperature of feed X_1 and T_1, and cooling water temperature T_7.

A single solution can be obtained with these considerations which allows the value of the remaining variables to be calculated.

5.8.2 Obtaining the Linear Model

As can be deduced from Equations (5.26)-(5.28) the process is a nonlinear system with a strong interrelationship amongst the variables. Even so, a linear model with various independent loops will be used to design the controller.

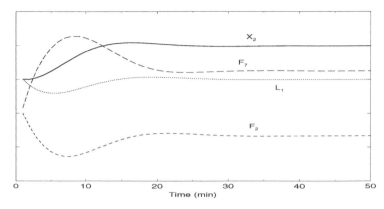

Fig. 5.21. Evaporator response to step input

It is clear that these hypotheses of work are incorrect, as will be made obvious when putting the control to work. The control of the evaporator includes maintaining certain stable working conditions as well as obtaining a product of a determined quality. To achieve the first objective it is necessary to control the mass and energy of the system, which can be achieved by keeping the level in the separator L_1 and the process pressure P_1 constant . In order to do this, two PI-type local controllers will be used, so that the level L_1 is controlled by acting on the product flow rate F_2 and the process pressure P_1 is controlled by the cooling water F_7. The justification of the choice of the couplings and the tuning of these loops can be found in [149].

The other objective is to obtain a determined product composition. This is achieved by acting on the remaining manipulated variable, the steam pressure which supplies the energy to the evaporator, P_6. The interaction amongst the variables is very strong, as can be seen by using Bristol method [36] and it would even be possible to have coupled X_2 with F_7 and P_1 with P_6. To obtain the linear model of the composition loop, a step is applied at the input (P_6) and the effect on the output (X_2) is studied. As was expected and as can be seen from Figure 5.21 which shows the evolution of the more significant variables for a 10% step, the interaction amongst the variables is considerable.

The evolution of the composition does not therefore follow the pattern of a first-order system, due mainly to the fact that the experiment was not done in open loop because the level and pressure regulators are functioning which, as has been indicated, need to be activated for stable functioning of the evaporator and indirectly affect the composition. The approximation of loop $X_2 - P_6$ to that of a first-order model with delay of the form

$$G(s) = \frac{K}{1 + \tau s} e^{-\tau_d s}$$

that, as is known, in spite of its simplicity is much used in practice, will be attempted. The reaction curve method will be used to obtain the model parameters; this provides the values of K, τ and τ_d starting from the graph of the system response at a step input. Due to the nonlinearity the system behaviour will be different for inputs of different value and different sign. By conducting various experiments for steps of different signs and magnitudes the following parameters can be considered to be appropriate:

$$K = 0.234\,\%/\text{KPa} \qquad \tau = 4.5\,\text{min} \qquad \tau_d = 3.5\,\text{min}$$

By taking a sampling time of one minute the delay is not integer so that the discrete transfer function about the working point will be (see conversion expressions in Chapter 3):

$$G(z^{-1}) = \frac{0.02461z^{-1} + 0.02202z^{-2}}{1 - 0.8007374z^{-1}}z^{-3}$$

This transfer function will be used for the design of the controller in spite of its limitations because of the existence of the previously mentioned phenomena.

5.8.3 Controller Design

Once a linear model of the process is obtained, the design of the controller is direct if the precalculated GPC is used. It is only necessary to calculate the parameters which appear in the control law:

$$\triangle u(t) = (l_{y1}\hat{y}(t+d \mid t) + l_{y2}\hat{y}(t+d-1 \mid t) + l_{r1}r(t))/K + l_{u1}\triangle u(t-1) \quad (5.29)$$

If one wants to design a fixed (nonadaptive) regulator, it is only necessary to calculate these parameters once. In the case of the evaporator with $a = 0.8007374$, $m = b_0/(b_0 + b_1) = 0.528$ and for a value of λ of 1.2 one has:

$$l_{y1} = -2.2748$$
$$l_{y2} = 1.5868$$
$$l_{r1} = 0.6879$$
$$l_{u1} = -0.1862$$

The control signal at each instant is therefore

$$u(t) = 0.814u(t-1) + 0.186u(t-2) - 9.721\hat{y}(t+d \mid t)$$
$$+6.781\hat{y}(t+d-1 \mid t) + 2.939r(t)$$

In order to complete the computations of the control law (5.29) the predicted values of the output at instants $t + d$ and $t + d - 1$ are necessary. This computation is easy to do given the simplicity of the model. It is enough to project the equation of the model towards the future

$$\hat{y}(t+i) = (1+a)\hat{y}(t+i-1) - a\hat{y}(t+i-2)$$
$$+b_0(u(t-d+i-1) - u(t-d+i-2))$$
$$+b_1(u(t-d+i-2) - u(t-d+i-3))   i = 1 \ldots d$$

where the elements $\hat{y}(t) = y(t)$ and $\hat{y}(t-1) = y(t-1)$ are known values at instant t. Note that if one wanted to make the controller adaptive it would be enough to just calculate the new value of l_i when the parameters of the system change. The simplicity of the control law obtained is obvious, being comparable to that of a digital PID, and it is therefore easy to implant in any control system.

5.8.4 Results

In the following, some results of applying the previous control law to the evaporator are presented (simulated to a nonlinear model). Even though the simplifications which were employed in the design phase (monovariable system, first-order linear model) were not very realistic, a reasonably good behaviour of the closed-loop system is obtained.

In Figure 5.22 the behaviour of the process in the presence of changes in the reference of the composition is shown. It can be observed that the output clearly follows the reference although with certain initial overshoot. It should be taken into account that the loops considered to be independent are greatly interrelated amongst themselves and in particular that the composition is very disturbed by the variations in the cooling water flow rate F_7, which is constantly changing to keep the process pressure constant.

In spite of the overshoot, the behaviour can be considered good. It can be compared to that obtained with a classical controller such as PI. Some good values for adjusting this controller are those calculated in [149]:

$$K = 1.64 \text{ kPa}/\%   T_I = 3.125 \text{ min}$$

In Figure 5.23 both regulators are compared for a change in the reference from 28 to 25 %. The GPC is seen to be faster and overshoots less than the PI and does not introduce great complexity in the design, as was seen in the previous section. The responses of both controllers in the presence of changes in the feed flow rate are reflected in Figure 5.24, where at instant $t = 50$ the feed flow rate changes from 10 to 11 kg/min and at $t = 200$ the composition at the input brusquely changes from 5 to 6 %. It can be seen that these changes considerably affect the composition of the product and although both controllers return the output to the reference value, the GPC does it sooner and with less overshoot, reducing the peaks by about 30%.

Tests can also be made with regard to the study of robustness. It is already known that the model used does not correspond to the real one (which

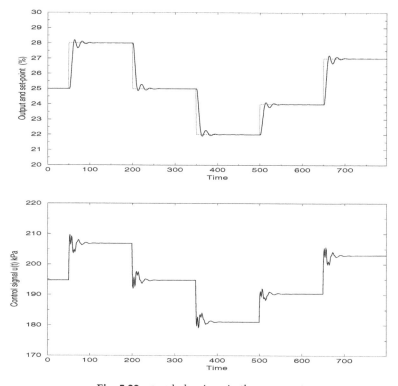

Fig. 5.22. GPC behaviour in the evaporator

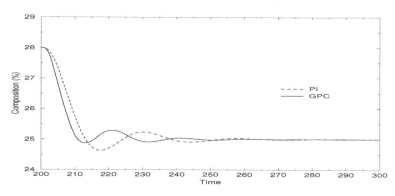

Fig. 5.23. Comparison of GPC and PI for a setpoint change

is neither first order nor linear) and therefore the controller already possesses certain robustness. However, to corroborate the robustness results
previously presented, instead of using the linear model that best fits the
nonlinear process as a control model , a model with estimation errors is
going to be used. For example, when working in a model with an error
on the pole estimation of the form that $\hat{a} = \alpha \times a$ with $\alpha = 0.9$, that is

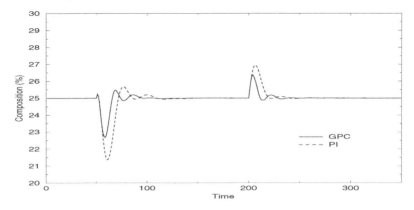

Fig. 5.24. Comparison of GPC and PI for feed changes

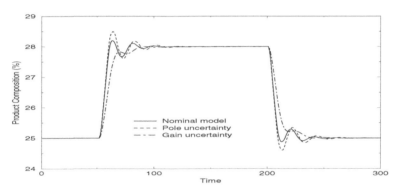

Fig. 5.25. Influence of errors on the estimation of gain and delay

an error of 10%, it is necessary to calculate the new l_i parameters using $\hat{a} = 0.9 \times a = 0.9 \times 0.8007374 = 0.7206$ (supposing that $a = 0.8007374$ is the *real* value). Thus, by recalculating the control law (including the predictor) for this new value and leaving the gain unaltered, the response shown in Figure 5.25 is obtained. As can be seen, the composition is hardly altered by this modelling error and similar response to the initial model is obtained. The same can be done by changing the system gain. For the model values considered to be *good*, uncertainties of up to 100% in the gain can be seen to be permissible without any problem. Thus by doubling the gain of the model used and calculating the new control law, slower but not less satisfactory behaviour is obtained, as is shown in the same figure.

Knowledge of delay is a fundamental factor of model based predictive methods, to such an extent that large errors in estimation can give rise to instability. Whilst for a difference of one unit the response hardly varies, the same is not true if the discrepancy is two or more units. Figure 5.26 shows the effect of using a model with delay 1 the *real* being equal to 3. The response is

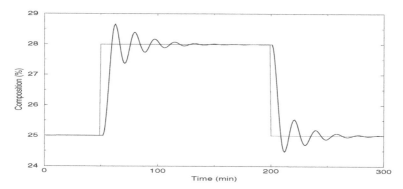

Fig. 5.26. Influence of error on the estimation of delay

defective but does not reach instability. It should be noted that once a mistake in the value of the delay is detected it is very easy to correct as it is enough to calculate higher or lower values of $\hat{y}(t+d)$ using the model equation whose other parameters are unchanged. Furthermore, it is not necessary to change the values of the l_i coefficients as they are independent of the delay.

5.9 Exercises

5.1. Given a system described by a static gain $K = 0.25$, a time constant $\tau = 10.5$ and a dead time of $\tau_d = 10$, compute l_{y1}, l_{y2} and l_{r1} with the given formula and simulate the process output to a step setpoint change:

1. Compare the results with those obtained by the *standard* algorithm.
2. Simulate the response to a setpoint composed of a ramp with slope 0.5 unit/second until time $t = 50$ and constant from this time to $t = 100$.
3. Add a white noise of mean 0.005 and compute the results.

5.2. Use the general procedure described in Chapter 3 to compute the values of l_{y1}, l_{y2} and l_{r1} for the system $G(z) = \frac{0.4z^{-1}}{1-0.8z^{-1}}$ with $N = 3$ and $\lambda = 0.8$.

5.3. Use the method described in this chapter to simulate the response of the process $G(s) = \frac{0.41e^{-50s}}{s(1+50s)}$ to a setpoint change from 0 to 2 when the sampling time is 10 seconds. Try different values for λ.

5.4. Control the following process $G(s) = \frac{1.12e^{-45s}}{1+87s}$ when the sampling time is 10 seconds and when it is sampled every 5 seconds. Compare the control law and results in the two cases.

5.5. Consider that the distillation column in Example 5.3.3 can work in a new operating regime at 5 mol/min. In this situation, $K = 0.102$, $\tau = 65.3$ and $\tau_d = 16.7$ (time in minutes).

1. Obtain the discrete model when $T = 5$ minutes.
2. Obtain the controller parameters for $\lambda = 0.8$.
3. Simulate the response to a setpoint that takes the value 0.1 at time $t = 20$, 0 at time $t = 120$ and -0.1 at time $t = 220$ (values taken around the nominal values of the variables).

6

Multivariable Model Predictive Control

Most industrial plants have many variables that have to be controlled (outputs) and many manipulated variables or variables used to control the plant (inputs). In certain cases a change in one of the manipulated variables mainly affects the corresponding controlled variable, and each the input-output pair can be considered as a single-input single-output (SISO) plant and controlled by independent loops. In many cases, when one of the manipulated variables is changed, it not only affects the corresponding controlled variable but also upsets the other controlled variables. These interactions between process variables may result in poor performance of the control process or even instability. When the interactions are not negligible, the plant must be considered to be a process with multiple inputs and outputs (MIMO) instead of a set of SISO processes. The control of MIMO processes has been extensively treated in literature; perhaps the most popular way of controlling MIMO processes is by designing decoupling compensators to suppress or diminish the interactions and then designing multiple SISO controllers. This requires first determining how to pair the input and output variables, that is, which manipulated variable will be used to control which output variables, and that the plant have the same number of manipulated and controlled variables. Total decoupling is very difficult to achieve for processes with complex dynamics or exhibiting dead times.

One of the advantages of Model Predictive Control is that multivariable processes can be handled in a straightforward manner [191], [204]. This chapter is dedicated to showing how MPC can be implemented on MIMO processes. The controllers that are obtained when transfer function matrices, state space models or convolution models are used are analyzed here.

6.1 Derivation of Multivariable GPC

A CARIMA model for an n-output, m-input multivariable process can be expressed as

$$\mathbf{A}(z^{-1})y(t) = \mathbf{B}(z^{-1})u(t-1) + \frac{1}{\triangle}\mathbf{C}(z^{-1})e(t) \tag{6.1}$$

where $\mathbf{A}(z^{-1})$ and $\mathbf{C}(z^{-1})$ are $n \times n$ monic polynomial matrices and $\mathbf{B}(z^{-1})$ is an $n \times m$ polynomial matrix defined as:

$$\mathbf{A}(z^{-1}) = I_{n \times n} + A_1 z^{-1} + A_2 z^{-2} + \cdots + A_{n_a} z^{-n_a}$$
$$\mathbf{B}(z^{-1}) = B_0 + B_1 z^{-1} + B_2 z^{-2} + \cdots + B_{n_b} z^{-n_b}$$
$$\mathbf{C}(z^{-1}) = I_{n \times n} + C_1 z^{-1} + C_2 z^{-2} + \cdots + C_{n_c} z^{-n_c}$$

The operator $\triangle$ is defined as $\triangle = 1 - z^{-1}$. The variables $y(t)$, $u(t)$ and $e(t)$ are the $n \times 1$ output vector, the $m \times 1$ input vector and the $n \times 1$ noise vector at time t. The noise vector is supposed to be a white noise with zero mean.

Let us consider the following finite horizon quadratic criterion

$$J(N_1, N_2, N_3) = \sum_{j=N_1}^{N_2} \|\hat{y}(t+j \mid t) - w(t+j)\|_R^2 + \sum_{j=1}^{N_3} \|\triangle u(t+j-1)\|_Q^2 \tag{6.2}$$

where $\hat{y}(t+j \mid t)$ is an optimum j step ahead prediction of the system output on data up to time t; that is, the expected value of the output vector at time t if the past input and output vectors and the future control sequence are known. N_1 and N_2 are the minimum and maximum prediction horizons and $w(t+j)$ is a future setpoint or reference sequence for the output vector. R and Q are positive definite weighting matrices.

6.1.1 White Noise Case

We shall first consider the most usual case when matrix $\mathbf{C}(z^{-1}) = I_{n \times n}$. The reason for this is that the colouring polynomials are very difficult to estimate with sufficient accuracy in practice, especially in the multivariable case. In fact, many predictive controllers use colouring polynomials as design parameters. The optimal prediction for the output vector can be generated as in the monovariable case as follows:

Consider the following Diophantine equation:

$$I_{n \times n} = \mathbf{E}_j(z^{-1})\tilde{\mathbf{A}}(z^{-1}) + z^{-j}\mathbf{F}_j(z^{-1}) \tag{6.3}$$

where $\tilde{\mathbf{A}}(z^{-1}) = \mathbf{A}(z^{-1})\triangle$, $\mathbf{E}_j(z^{-1})$ and $\mathbf{F}_j(z^{-1})$ are unique polynomial matrices of order $j-1$ and n_a respectively. If (6.1) is multiplied by $\triangle\mathbf{E}_j(z^{-1})z^j$:

$$\mathbf{E}_j(z^{-1})\tilde{\mathbf{A}}(z^{-1})y(t+j) = \mathbf{E}_j(z^{-1})\mathbf{B}(z^{-1})\triangle u(t+j-1) + \mathbf{E}_j(z^{-1})e(t+j)$$

By using (6.3) and after some manipulation we get:

$$y(t+j) = \mathbf{F}_j(z^{-1})y(t) + \mathbf{E}_j(z^{-1})\mathbf{B}(z^{-1})\triangle u(t+j-1) + \mathbf{E}_j(z^{-1})e(t+j) \tag{6.4}$$

Notice that because the degree of $\mathbf{E}_j(z^{-1})$ is $j-1$, the noise terms of equation (6.4) are all in the future. By taking the expectation operator and considering that $E[e(t)] = 0$, the expected value for $y(t+j)$ is given by:

$$\hat{y}(t+j|t) = E[y(t+j)] = \mathbf{F}_j(z^{-1})y(t) + \mathbf{E}_j(z^{-1})\mathbf{B}(z^{-1}) \triangle u(t+j-1) \quad (6.5)$$

Notice that the prediction can easily be extended to the nonzero mean noise case by adding vector $\mathbf{E}_j(z^{-1})E[e(t)]$ to prediction $\hat{y}(t+j|t)$.

Recursion of the Diophantine Equation

Let us consider that a solution $(\mathbf{E}_j(z^{-1}), \mathbf{F}_j(z^{-1}))$ for the Diophantine equation has been obtained. That is,

$$I_{n\times n} = \mathbf{E}_j(z^{-1})\tilde{\mathbf{A}}(z^{-1}) + z^{-j}\mathbf{F}_j(z^{-1}) \quad (6.6)$$

with

$$\tilde{\mathbf{A}}(z^{-1}) = \mathbf{A}(z^{-1})\triangle = I_{n\times n} + \tilde{A}_1 z^{-1} + \tilde{A}_2 z^{-2} + \cdots + \tilde{A}_{n_a} z^{-n_a} + \tilde{A}_{n_a+1} z^{-(n_a+1)}$$

$$= I_{n\times n} + (A_1 - I_{n\times n})z^{-1} + (A_2 - A_1)z^{-2} + \cdots + (A_{n_a} - A_{n_a-1})z^{-n_a} - A_{n_a} z^{-(n_a+1)}$$

$$\mathbf{E}_j(z^{-1}) = E_{j,0} + E_{j,1}z^{-1} + E_{j,2}z^{-2} + \cdots + E_{j,j-1}z^{j-1}$$

$$\mathbf{F}_j(z^{-1}) = F_{j,0} + F_{j,1}z^{-1} + F_{j,2}z^{-2} + \cdots + F_{j,n_a}z^{-n_a}$$

Now consider the Diophantine equation corresponding to the prediction for $\hat{y}(t+j+1|t)$

$$I_{n\times n} = \mathbf{E}_{j+1}(z^{-1})\tilde{\mathbf{A}}(z^{-1}) + z^{-(j+1)}\mathbf{F}_{j+1}(z^{-1}) \quad (6.7)$$

Let us subtract Equation (6.6) from equation (6.7)

$$\mathbf{0}_{n\times n} = (\mathbf{E}_{j+1}(z^{-1}) - \mathbf{E}_j(z^{-1}))\tilde{\mathbf{A}}(z^{-1}) + z^{-j}(z^{-1}\mathbf{F}_{j+1}(z^{-1}) - \mathbf{F}_j(z^{-1})) \quad (6.8)$$

Matrix $(\mathbf{E}_{j+1}(z^{-1}) - \mathbf{E}_j(z^{-1}))$ is of degree j. Let us make

$$(\mathbf{E}_{j+1}(z^{-1}) - \mathbf{E}_j(z^{-1})) = \tilde{\mathbf{R}}(z^{-1}) + R_j z^{-j}$$

where $\mathbf{R}(z^{-1})$ is an $n \times n$ polynomial matrix of degree smaller or equal to $j-1$ and R_j is an $n \times n$ real matrix. By substituting in Equation (6.8):

$$\mathbf{0}_{n\times n} = \tilde{\mathbf{R}}(z^{-1})\tilde{\mathbf{A}}(z^{-1}) + z^{-j}(R_j\tilde{\mathbf{A}}(z^{-1}) + z^{-1}\mathbf{F}_{j+1}(z^{-1}) - \mathbf{F}_j(z^{-1})) \quad (6.9)$$

As $\tilde{\mathbf{A}}(z^{-1})$ is monic, it is easy to see that $\tilde{\mathbf{R}}(z^{-1}) = \mathbf{0}_{n\times n}$. That is, matrix $\mathbf{E}_{j+1}(z^{-1})$ can be computed recursively by:

$$\mathbf{E}_{j+1}(z^{-1}) = \mathbf{E}_j(z^{-1}) + R_j z^{-j}$$

The following expressions can easily be obtained from (6.9):

$$R_j = F_{j,0}$$
$$F_{j+1,i} = F_{j,i+1} - R_j \tilde{A}_{i+1} \text{ for } i = 0 \cdots \delta(\mathbf{F}_{j+1})$$

It can easily be seen that the initial conditions for the recursion equation are given by:

$$\mathbf{E}_1 = I$$
$$\mathbf{F}_1 = z(I - \tilde{\mathbf{A}})$$

By making the polynomial matrix $\mathbf{E}_j(z^{-1})\mathbf{B}(z^{-1}) = \mathbf{G}_j(z^{-1}) + z^{-j}\mathbf{G}_{jp}(z^{-1})$, with $\delta(\mathbf{G}_j(z^{-1})) < j$, the prediction equation can now be written as:

$$\hat{y}(t+j|t) = \mathbf{G}_j(z^{-1})\triangle u(t+j-1) + \mathbf{G}_{jp}(z^{-1})\triangle u(t-1) + \mathbf{F}_j(z^{-1})y(t) \quad (6.10)$$

Notice that the last two terms of the right-hand side of equation (6.10) depend on past values of the process output and input variables and correspond to the free response of the process considered if the control signals are kept constant, while the first term depends only on future values of the control signal and can be interpreted as the forced response. That is, the response obtained when the initial conditions are zero $y(t-j) = 0$, $\triangle u(t-j) = 0$ for $j = 0, 1 \cdots$. Equation (6.10) can be rewritten as:

$$\hat{y}(t+j|t) = \mathbf{G}_j(z^{-1})\triangle u(t+j-1) + \mathbf{f}_j$$

with $\mathbf{f}_j = \mathbf{G}_{jp}(z^{-1})\triangle u(t-1) + \mathbf{F}_j(z^{-1})y(t)$. Let us now consider a set of N j ahead predictions:

$$\hat{y}(t+1|t) = \mathbf{G}_1(z^{-1})\triangle u(t) + \mathbf{f}_1$$
$$\hat{y}(t+2|t) = \mathbf{G}_2(z^{-1})\triangle u(t+1) + \mathbf{f}_2$$
$$\vdots \qquad\qquad\qquad\qquad\qquad (6.11)$$
$$\hat{y}(t+N|t) = \mathbf{G}_N(z^{-1})\triangle u(t+N-1) + \mathbf{f}_N$$

Because of the recursive properties of the $\mathbf{E}_j$ polynomial matrix described earlier, Expressions (6.11) can be rewritten as

$$
\begin{bmatrix}
\hat{y}(t+1|t) \\
\hat{y}(t+2|t) \\
\vdots \\
\hat{y}(t+j|t) \\
\vdots \\
\hat{y}(t+N|t)
\end{bmatrix}
=
\begin{bmatrix}
G_0 & 0 & \cdots & 0 & \cdots & 0 \\
G_1 & G_0 & \cdots & 0 & \cdots & 0 \\
\vdots & \vdots & \ddots & \vdots & \vdots & \vdots \\
G_{j-1} & G_{j-2} & \cdots & G_0 & \vdots & 0 \\
\vdots & \vdots & \vdots & \vdots & \ddots & \vdots \\
G_{N-1} & G_{N-2} & \cdots & \cdots & \cdots & G_0
\end{bmatrix}
\begin{bmatrix}
\triangle u(t) \\
\triangle u(t+1) \\
\vdots \\
\triangle u(t+j-1) \\
\vdots \\
\triangle u(t+N-1)
\end{bmatrix}
+
\begin{bmatrix}
\mathbf{f}_1 \\
\mathbf{f}_2 \\
\vdots \\
\mathbf{f}_j \\
\vdots \\
\mathbf{f}_N
\end{bmatrix}
$$

where $\mathbf{G}_j(z^{-1}) = \sum_{i=0}^{j-1} G_i z^{-i}$. The predictions can be expressed in condensed form as:

$$y = \mathbf{G}u + \mathbf{f}$$

Notice that if all initial conditions are zero, the free response $\mathbf{f}$ is also zero. If a unit step is applied to the first input at time t; that is,

$$\triangle u(t) = [1, 0, \cdots, 0]^T, \triangle u(t+1) = 0, \cdots, \triangle u(t+N-1) = 0$$

the expected output sequence $[\hat{y}(t+1)^T, \hat{y}(t+2)^T, \cdots, \hat{y}(t+N)^T]^T$ is equal to the first column of matrix $\mathbf{G}$ or the first columns of matrices $G_0, G_1, \cdots, G_{N-1}$. That is, the first column of matrix $\mathbf{G}$ can be calculated as the step response of the plant when a unit step is applied to the first control signal. Column i can be obtained in a similar manner by applying a unit step to the i input. In general, matrix G_k can be obtained as follows

$$(G_k)_{i,j} = y_{i,j}(t+k+1)$$

where $(G_k)_{i,j}$ is the (i, j) element of matrix G_k and $y_{i,j}(t+k+1)$ is the i-output of the system when a unit step has been applied to control input j at time t.

The free response term can be calculated recursively by:

$$\mathbf{f}_{j+1} = z(I - \tilde{\mathbf{A}}(z^{-1}))\mathbf{f}_j + \mathbf{B}(z^{-1}) \triangle u(t+j)$$

with $\mathbf{f}_0 = y(t)$ and $\triangle u(t+j) = 0$ for $j \geq 0$.

Notice that if matrix $\mathbf{A}(z^{-1})$ is diagonal, matrices $\mathbf{E}_j(z^{-1})$ and $\mathbf{F}_j(z^{-1})$ are also diagonal matrices and the problem is reduced to the recursion of n scalar Diophantine equations, which are much simpler to program and require less computation. The computation of $\mathbf{G}_j(z^{-1})$ and $\mathbf{f}_j$ is also considerably simplified.

If the control signal is kept constant after the first N_3 control moves, the set of predictions affecting the cost function (6.2) $\mathbf{y}_{N_{12}} = [\hat{y}(t+N_1|t)^T \cdots \hat{y}(t+N_2|t)^T]$ can be expressed as

$$\mathbf{y}_{N_{12}} = \mathbf{G}_{N_{123}} \mathbf{u}_{N_3} + \mathbf{f}_{N_{12}}$$

where $\mathbf{u}_{N_3} = [\triangle u(t)^T \cdots \triangle u(t+N_3-1)^T]^T$, $\mathbf{f}_{N_{12}} = [\mathbf{f}_{N_1}^T \cdots \mathbf{f}_{N_2}^T]^T$ and $\mathbf{G}_{N_{123}}$ is the following submatrix of $\mathbf{G}$

$$\mathbf{G}_{N_{123}} = \begin{bmatrix} G_{N_1-1} & G_{N_1-2} & \cdots & G_{N_1-N_3} \\ G_{N_1} & G_{N_1-1} & \cdots & G_{N_1+1-N_3} \\ \vdots & \ddots & \ddots & \vdots \\ G_{N_2-1} & G_{N_2-2} & \cdots & G_{N_2-N_3} \end{bmatrix}$$

with $G_i = 0$ for $i < 0$. Equation (6.2) can be rewritten as

$$J = (\mathbf{G}_{N_{123}} \mathbf{u}_{N_3} + \mathbf{f}_{N_{12}} - \mathbf{w})^T \overline{R}(\mathbf{G}_{N_{123}} \mathbf{u}_{N_3} + \mathbf{f}_{N_{12}} - \mathbf{w}) + \mathbf{u}_{N_3}^T \overline{Q} \mathbf{u}_{N_3}$$

where $\overline{R} = diag(R, \cdots, R)$ and $\overline{Q} = diag(Q, \cdots, Q)$.

If there are no constraints, the optimum can be expressed as:

$$\mathbf{u} = (\mathbf{G}_{N_{123}}^T \overline{R} \mathbf{G}_{N_{123}} + \overline{Q})^{-1} \mathbf{G}_{N_{123}}^T \overline{R}(\mathbf{w} - \mathbf{f}_{N_{12}})$$

Because of the receding control strategy, only $\triangle u(t)$ is needed at instant t. Thus only the first m rows of $(\mathbf{G}_{N_{123}}^T \overline{R} \mathbf{G}_{N_{123}} + \overline{Q})^{-1} \mathbf{G}_{N_{123}}^T \overline{R}$, say K, have to be computed. This can be done beforehand for the nonadaptive case. The control law can then be expressed as $\triangle u(t) = K(\mathbf{w} - \mathbf{f})$. That is a linear gain matrix that multiplies the predicted errors between the predicted references and the predicted free response of the plant.

In the case of adaptive control, matrix $\mathbf{G}_{N_{123}}$ has to be computed every time the estimated parameters change and the way of computing the control action increment would be by solving the linear set of equations: $(\mathbf{G}_{N_{123}}^T \overline{R} \mathbf{G}_{N_{123}} + \overline{Q})\mathbf{u} = \mathbf{G}_{N_{123}}^T \overline{R}(\mathbf{w} - \mathbf{f}_{N_{12}})$. Again only the first m components of $\mathbf{u}$ have to be found and as matrix $(\mathbf{G}_{N_{123}}^T \overline{R} \mathbf{G}_{N_{123}} + \overline{Q})$ is positive definite, Cholesky's algorithm [197] can be used to find the solution.

6.1.2 Coloured Noise Case

When the noise is coloured, $\mathbf{C}(z^{-1}) \neq I$, and provided that the colouring matrix $\mathbf{C}(z^{-1})$ is stable, the optimal predictions needed can be generated as follows [80].

First solve the Diophantine equation

$$\mathbf{C}(z^{-1}) = \mathbf{E}_j(z^{-1})\tilde{\mathbf{A}}(z^{-1}) + z^{-j}\mathbf{F}_j(z^{-1}) \qquad (6.12)$$

where $\mathbf{E}_j(z^{-1})$ and $\mathbf{F}_j(z^{-1})$ are unique polynomial matrices of order $j - 1$ and n_a, respectively. Note that the Diophantine Equation (6.12) can be solved recursively. Consider the Diophantine equations for j and $j + 1$:

$$\mathbf{C}(z^{-1}) = \mathbf{E}_j(z^{-1})\tilde{\mathbf{A}}(z^{-1}) + z^{-j}\mathbf{F}_j(z^{-1})$$
$$\mathbf{C}(z^{-1}) = \mathbf{E}_{j+1}(z^{-1})\tilde{\mathbf{A}}(z^{-1}) + z^{-(j+1)}\mathbf{F}_{j+1}(z^{-1})$$

By differentiating them we get Equation (6.8), hence $\mathbf{E}_{j+1}(z^{-1})$ and $\mathbf{F}_{j+1}(z^{-1})$ can be computed recursively by using the same expressions obtained in the case of $\mathbf{C}(z^{-1}) = I$ with initial conditions $\mathbf{E}_1(z^{-1}) = I$ and $\mathbf{F}_1(z^{-1}) = z(\mathbf{C}(z^{-1}) - \tilde{\mathbf{A}}(z^{-1}))$.

Define the polynomial matrices $\overline{\mathbf{E}}_j(z^{-1})$ and $\overline{\mathbf{C}}_j(z^{-1})$ such that

$$\overline{\mathbf{E}}_j(z^{-1})\mathbf{C}(z^{-1}) = \overline{\mathbf{C}}_j(z^{-1})\mathbf{E}_j(z^{-1}) \qquad (6.13)$$

with $\overline{E}_0 = I$ and $\det(\overline{\mathbf{C}}_j(z^{-1})) = \det(\mathbf{C}_j(z^{-1}))$. Note that

$$\overline{\mathbf{E}}_j(z^{-1})^{-1}\overline{\mathbf{C}}_j(z^{-1}) = \mathbf{C}(z^{-1})\mathbf{E}_j(z^{-1})^{-1}$$

and thus matrices $\overline{\mathbf{E}}_j(z^{-1})$ and $\overline{\mathbf{C}}_j(z^{-1})$ can be interpreted (and computed) as a left fraction matrix description of the process having a right matrix fraction description given by matrices $\mathbf{E}(z^{-1})$ and $\mathbf{C}(z^{-1})$. Define

$$\overline{\mathbf{F}}_j(z^{-1}) = z^j(\overline{\mathbf{C}}_j(z^{-1}) - \overline{\mathbf{E}}_j(z^{-1})\tilde{\mathbf{A}}(z^{-1})) \tag{6.14}$$

Premultiplying Equation (6.1) by $\overline{\mathbf{E}}_j(z^{-1})\triangle$,

$$\overline{\mathbf{E}}_j(z^{-1})\tilde{\mathbf{A}}(z^{-1})y(t+j) = \overline{\mathbf{E}}_j(z^{-1})\mathbf{B}(z^{-1})\triangle u(t+j-1) + \overline{\mathbf{E}}_j(z^{-1})e(t+j)$$

Using Equations (6.13) and (6.14) we get

$$\overline{\mathbf{C}}_j(z^{-1})(y(t+j) - \overline{\mathbf{E}}_j(z^{-1})e(t+j)) = \overline{\mathbf{E}}_j(z^{-1})\mathbf{B}(z^{-1})\triangle u(t+j-1) + \overline{\mathbf{F}}_j(z^{-1})y(t)$$

By taking the expected value $E[\overline{\mathbf{C}}_j(z^{-1})y(t+j) - \overline{\mathbf{E}}_j(z^{-1})e(t+j)] = \hat{y}(t+j|t)$. The optimal predictions $\hat{y}(t+j|t)$ can be generated by the equation:

$$\overline{\mathbf{C}}_j(z^{-1})\hat{y}(t+j|t) = \overline{\mathbf{E}}_j(z^{-1})\mathbf{B}(z^{-1})\triangle u(t+j-1) + \overline{\mathbf{F}}_j(z^{-1})y(t)$$

Now solving the Diophantine equation

$$I = \mathbf{J}_j(z^{-1})\overline{\mathbf{C}}_j(z^{-1}) + z^{-j}\mathbf{K}_j(z^{-1}) \tag{6.15}$$

with $\delta(\mathbf{J}(z^{-1})) < j$. Multiplying by $\mathbf{J}_j(z^{-1})^{-1}$ and using Equation (6.15)

$$(I - z^{-j}\mathbf{K}_j(z^{-1}))y(t+j|t) = \mathbf{J}_j(z^{-1})\overline{\mathbf{E}}_j(z^{-1})\mathbf{B}(z^{-1})\triangle u(t+j-1)$$

$$+\mathbf{J}_j(z^{-1})\overline{\mathbf{F}}_j(z^{-1})y(t)$$

or

$$y(t+j|t) = \mathbf{J}_j(z^{-1})\overline{\mathbf{E}}_j(z^{-1})\mathbf{B}(z^{-1})\triangle u(t+j-1) + (\mathbf{K}_j(z^{-1}) + \mathbf{J}_j(z^{-1})\overline{\mathbf{F}}_j(z^{-1}))y(t)$$

If $\mathbf{J}_j(z^{-1})\overline{\mathbf{E}}_j(z^{-1})\mathbf{B}(z^{-1}) = \mathbf{G}_j(z^{-1}) + z^{-j}\mathbf{G}p_j(z^{-1})$, with $\delta(\mathbf{G}_j(z^{-1})) < j$, the optimal j-step ahead prediction can be expressed as:

$$y(t+j|t) = \mathbf{G}_j(z^{-1})\triangle u(t+j-1) + \mathbf{G}p_j(z^{-1})\triangle u(t-1) + (\mathbf{K}_j(z^{-1})$$

$$+\mathbf{J}_j(z^{-1})\overline{\mathbf{F}}_j(z^{-1}))y(t)$$

The first term of the prediction corresponds to the forced response due to future control increments, while the last two terms correspond to the free response $\mathbf{f}c_j$ and are generated by past input increments and past output.

The set of pertinent j ahead predictions can be written as: $\mathbf{y}_{N_{12}} = \mathbf{G}_{N_{123}}\mathbf{u}_{N_3} + \mathbf{f}c_{N_{12}}$ generated as before.

The objective function can be expressed as:

$$J = (\mathbf{G}_{N_{123}}\mathbf{u}_{N_3} + \mathbf{f}c_{N_{12}} - \mathbf{w})^T\overline{R}(\mathbf{G}_{N_{123}}\mathbf{u}_{N_3} + \mathbf{f}c_{N_{12}} - \mathbf{w}) + \mathbf{u}_{N_3}^T\overline{Q}\mathbf{u}_{N_3}$$

And the optimal solution can be found by solving a set of linear equations as in the white noise case but notice that the computation required is more complex.

It is in reality very difficult to obtain the colouring polynomial matrix $\mathbf{C}(z^{-1})$, and in most cases this matrix is chosen arbitrarily by the user in order to gain robustness. If matrices $\mathbf{C}(z^{-1})$ and $\mathbf{A}(z^{-1})$ are chosen to be diagonal, the problem is transformed into generating a set of optimal predictions for a series of multi-input single-output processes, which is an easier problem to solve and the computation required can be substantially simplified.

Consider a CARIMA multivariable process with $\mathbf{A}(z^{-1}) = \mathrm{diag}(A_{ii}(z^{-1}))$ and $\mathbf{C}(z^{-1}) = \mathrm{diag}(C_{ii}(z^{-1}))$. The model equation corresponding to the ith-output variable can be expressed as:

$$A_{ii}(z^{-1})y_i(t) = \sum_{j=1}^{m} B_{ij}(z^{-1})u_j(t-1) + C_{ii}(z^{-1})\frac{e_i(t)}{\triangle} \qquad (6.16)$$

Solve the scalar Diophantine equation

$$C_{ii}(z^{-1}) = E_{i_k}(z^{-1})\tilde{A}_{ii}(z^{-1}) + z^{-k}F_{i_k}(z^{-1}) \qquad (6.17)$$

with $\delta(E_{i_k}(z^{-1})) = k - 1$ and $\delta(F_{i_k}(z^{-1})) = \delta(\tilde{A}_{ii}(z^{-1})) - 1$. Multiplying Equation (6.16) by $\triangle E_{i_k}(z^{-1})$ and using (6.17),

$$C_{ii}(z^{-1})(y_i(t+j) - E_{i_k}(z^{-1})e_i(t+j)) =$$

$$= E_{i_k}(z^{-1}) \sum_{j=1}^{m} B_{ij}(z^{-1}) \triangle u_j(t+j-1) + F_{i_k}(z^{-1})y_i(t)$$

As the noise terms are all in the future, the expected value of the left-hand side of this equation is: $E[C_{ii}(z^{-1})(y_i(t+k) - E_{i_k}(z^{-1})e_i(t+k))] = C_{ii}(z^{-1})\hat{y}_i(t+k|t)$.

The expected value of the output can be generated by the equation:

$$C_{ii}(z^{-1})\hat{y}_i(t+k|t) = E_{i_k}(z^{-1}) \sum_{j=1}^{m} B_{ij}(z^{-1}) \triangle u_j(t+k-1) + F_{i_k}(z^{-1})y_i(t)$$

$$(6.18)$$

Notice that this prediction equation could be used to generate the predictions in a recursive way. An explicit expression for the optimal k step ahead prediction can be obtained by solving the Diophantine equation

$$1 = C_{ii}(z^{-1})M_{i_k}(z^{-1}) + z^{-k}N_{i_k}(z^{-1}) \qquad (6.19)$$

with $\delta(M_{i_k}(z^{-1})) = k - 1$ and $\delta(N_{i_k}(z^{-1})) = \delta(C_{ii}(z^{-1})) - 1$.
Multiplying Equation (6.18) by $M_{i_k}(z^{-1})$ and using (6.19),

$$\hat{y}_i(t + k|t) = M_{i_k} E_{i_k}(z^{-1}) \sum_{j=1}^{m} B_{ij}(z^{-1}) \triangle u_j(t + k - 1)$$

$$+ M_{i_k}(z^{-1}) F_{i_k}(z^{-1}) y_i(t) + N_{i_k}(z^{-1}) y_i(t)$$

which can be expressed as

$$\hat{y}_i(t + k|t) = \sum_{j=1}^{m} G_{ij}(z^{-1}) \triangle u_j(t + k - 1) + \sum_{j=1}^{m} Gp_{ij}(z^{-1}) \triangle u_j(t + k - 1)$$

$$+ (M_{i_k}(z^{-1}) F_{i_k}(z^{-1}) + N_{i_k}(z^{-1})) y_i(t)$$

with $\delta(G_{ij}(z^{-1})) < k$. These predictions can be substituted in the cost function which can be minimized as previously. Note that the amount of computation required has been considerably reduced in respect to the case of a nondiagonal colouring matrix.

6.1.3 Measurable Disturbances

The measurable disturbances can be handled for the MIMO case in the same way as for SISO processes. It will be seen that only the *free* response has to be changed to take into account the measurable disturbances. Consider a multivariable process described by the following CARIMA model

$$\mathbf{A}(z^{-1}) y(t) = \mathbf{B}(z^{-1}) u(t) + \mathbf{D}(z^{-1}) v(t) + \frac{1}{\triangle} \mathbf{C}(z^{-1}) e(t) \qquad (6.20)$$

where the variable $v(t)$ is an $n \times 1$ vector of measured disturbances at time t and $\mathbf{D}(z^{-1})$ is an $n \times n$ polynomial matrix defined as:

$$\mathbf{D}(z^{-1}) = D_0 + D_1 z^{-1} + D_2 z^{-2} + \cdots + D_{n_d} z^{-n_d}$$

Multiplying Equation (6.20) by $\triangle \mathbf{E}_j(z^{-1}) z^j$:

$$\mathbf{E}_j(z^{-1}) \tilde{\mathbf{A}}(z^{-1}) y(t + j) = \mathbf{E}_j(z^{-1}) \mathbf{B}(z^{-1}) \triangle u(t + j - 1)$$

$$+ \mathbf{E}_j(z^{-1}) \mathbf{D}(z^{-1}) \triangle v(t + j) + \mathbf{E}_j(z^{-1}) e(t + j)$$

Using (6.3) and after some manipulation we get:

$$y(t + j) = \mathbf{F}_j(z^{-1}) y(t) + \mathbf{E}_j(z^{-1}) \mathbf{B}(z^{-1}) \triangle u(t + j - 1)$$

$$+ \mathbf{E}_j(z^{-1}) \mathbf{D}(z^{-1}) \triangle v(t + j) + \mathbf{E}_j(z^{-1}) e(t + j) \qquad (6.21)$$

Notice that because the degree of $\mathbf{E}_j(z^{-1})$ is $j - 1$, the noise terms of equation (6.4) are all in the future. By taking the expectation operator and considering that $E[e(t)] = 0$, the expected value for $y(t + j)$ is given by:

$$\hat{y}(t+j|t) = E[y(t+j)] = \mathbf{F}_j(z^{-1})y(t) + \mathbf{E}_j(z^{-1})\mathbf{B}(z^{-1})\,\triangle\,u(t+j-1)$$
$$+\mathbf{E}_j(z^{-1})\mathbf{D}(z^{-1})\,\triangle\,v(t+j)$$

By making the polynomial matrix

$$\mathbf{E}_j(z^{-1})\mathbf{D}(z^{-1}) = \mathbf{H}_j(z^{-1}) + z^{-j}\mathbf{H}_{jp}(z^{-1}),$$

with $\delta(\mathbf{H}_j(z^{-1})) = j - 1$, the prediction equation can now be written as:

$$\hat{y}(t+j|t) = \mathbf{G}_j(z^{-1})\,\triangle\,u(t+j-1) + \mathbf{H}_j(z^{-1})\,\triangle\,v(t+j) + \mathbf{G}_{jp}(z^{-1})\,\triangle\,u(t-1)$$
$$+\mathbf{H}_{jp}(z^{-1})\,\triangle\,v(t) + \mathbf{F}_j(z^{-1})y(t) \qquad (6.22)$$

Notice that the last three terms of the right-hand side of Equation (6.22) depend on past values of the process output measured disturbances and input variables and correspond to the free response of the process considered if the control signals and measured disturbances are kept constant, while the first term depends only on future values of the control signal and can be interpreted as the forced response. That is, the response obtained when the initial conditions are zero $y(t - j) = 0$, $\triangle u(t - j - 1) = 0$, $\triangle v(t - j)$ for $j > 0$.

The second term of Equation (6.22) depends on the future deterministic disturbances. In some cases, when they are related to the process load, future disturbances are known. In other cases, they can be predicted using trends or other means. If this is the case, the term corresponding to future deterministic disturbances can be computed. If the future load disturbances are supposed to be constant and equal to the last measured value (i.e., $v(t+j) = v(t)$), then $\triangle v(t + j) = 0$ and the second term of this equation vanishes.

Equation (6.22) can be rewritten as

$$\hat{y}(t+j|t) = \mathbf{G}_j(z^{-1})\,\triangle\,u(t+j-1) + \mathbf{H}_j(z^{-1})\,\triangle\,v(t+j) + \mathbf{f}_j$$

with $\mathbf{f}_j = \mathbf{G}_{jp}(z^{-1})\,\triangle\,u(t-1) + \mathbf{H}_{jp}(z^{-1})\,\triangle\,v(t) + \mathbf{F}_j(z^{-1})y(t)$.

Let us now consider a set of N j ahead predictions:

$$\hat{y}(t+1|t) = \mathbf{G}_1(z^{-1})\,\triangle\,u(t) + \mathbf{H}_j(z^{-1})\,\triangle\,v(t+1) + \mathbf{f}_1$$
$$\hat{y}(t+2|t) = \mathbf{G}_2(z^{-1})\,\triangle\,u(t+1) + \mathbf{H}_j(z^{-1})\,\triangle\,v(t+2) + \mathbf{f}_2$$
$$\vdots \qquad\qquad\qquad\qquad\qquad\qquad (6.23)$$
$$\hat{y}(t+N|t) = \mathbf{G}_N(z^{-1})\,\triangle\,u(t+N-1) + \mathbf{H}_j(z^{-1})\,\triangle\,v(t+N) + \mathbf{f}_N$$

Because of the recursive properties of the $\mathbf{E}_j$ polynomial matrix described earlier, Expressions (6.23) can be rewritten as

$$
\begin{bmatrix}
\hat{y}(t+1|t) \\
\hat{y}(t+2|t) \\
\vdots \\
\hat{y}(t+j|t) \\
\vdots \\
\hat{y}(t+N|t)
\end{bmatrix}
=
\begin{bmatrix}
G_0 & 0 & \cdots & 0 & \cdots & 0 \\
G_1 & G_0 & \cdots & 0 & \cdots & 0 \\
\vdots & \vdots & \ddots & \vdots & \vdots & \vdots \\
G_{j-1} & G_{j-2} & \cdots & G_0 & \vdots & 0 \\
\vdots & \vdots & \vdots & \vdots & \ddots & \vdots \\
G_{N-1} & G_{N-2} & \cdots & \cdots & \cdots & G_0
\end{bmatrix}
\begin{bmatrix}
\triangle u(t) \\
\triangle u(t+1) \\
\vdots \\
\triangle u(t+j-1) \\
\vdots \\
\triangle u(t+N-1)
\end{bmatrix}
$$

$$+ \begin{bmatrix} H_0 & 0 & \cdots & 0 & \cdots & 0 \\ H_1 & H_0 & \cdots & 0 & \cdots & 0 \\ \vdots & \vdots & \ddots & \vdots & \vdots & \vdots \\ H_{j-1} & \cdots & H_1 & H_0 & \vdots & 0 \\ \vdots & \vdots & \vdots & \ddots & \ddots & \vdots \\ H_{N-1} & \cdots & \cdots & \cdots & H_1 & H_0 \end{bmatrix} \begin{bmatrix} \triangle v(t+1) \\ \triangle v(t+2) \\ \vdots \\ \triangle v(t+j-1) \\ \vdots \\ \triangle v(t+N) \end{bmatrix} + \begin{bmatrix} \mathbf{f}_1 \\ \mathbf{f}_2 \\ \vdots \\ \mathbf{f}_j \\ \vdots \\ \mathbf{f}_N \end{bmatrix}$$

where $\mathbf{H}_j(z^{-1}) = \sum_{i=1}^{j} H_i z^{-i}$. The predictions can be expressed in condensed form as:

$$\mathbf{y} = \mathbf{Gu} + \mathbf{Hv} + \mathbf{f}$$

Notice that if all initial conditions and future control moves are zero, the free response $\mathbf{f}$ and force response are also zero. If a unit step is applied to the first disturbance at time $t+1$; that is,

$$\triangle v(t+1) = [1, 0, \cdots, 0]^T, \triangle v(t+2) = 0, \cdots, \triangle v(t+N) = 0$$

the expected output sequence $[\hat{y}(t+2)^T, \hat{y}(t+3)^T, \ldots, \hat{y}(t+N)^T]^T$ is equal to the first column of matrix $\mathbf{H}$ or the first columns of matrices $H_1, \ldots, H_{N-1}$. That is, the first columns of matrix $\mathbf{H}$ can be interpreted as the step response of the plant when a unit step is applied to the first disturbance signal. Column i can be obtained in a similar manner by applying a unit step to the i disturbance. In general, matrix H_k could be obtained as follows

$$(H_k)_{i,j} = y_{i,j}(t+k+1)$$

where $(H_k)_{i,j}$ is the (i,j) element of matrix H_k and $y_{i,j}(t+k+1)$ is the i-output of the system when a unit step has been applied to the disturbance input j at time $t+1$. Notice that to do this test in practice external deterministic variables need to be manipulated, and this is not the usual case. However, they can be computed from the nominal model of the plant by simulation.

Notice that if matrix $\mathbf{A}(z^{-1})$ is diagonal, matrices $\mathbf{E}_j(z^{-1})$ and $\mathbf{F}_j(z^{-1})$ are also diagonal matrices and the problem is reduced to the recursion of n scalar Diophantine equations which are much simpler to program and require less computation. The computation of $\mathbf{G}_j(z^{-1})$, $\mathbf{H}_j(z^{-1})$ and $\mathbf{f}_j$ is also considerably simplified.

By making $\mathbf{f}' = \mathbf{Hv} + \mathbf{f}$, the prediction equation is now

$$\mathbf{y} = \mathbf{Gu} + \mathbf{f}'$$

which has the same shape as the prediction equation used for the case of zero external measured disturbances. The future control signal can now be found in the same way, but using as free response the response of the process due to initial conditions (including external disturbances) and future *known* disturbances.

6.2 Obtaining a Matrix Fraction Description

6.2.1 Transfer Matrix Representation

The transfer matrix is the most popular representation of multivariable processes. The reason for this is that transfer matrices can very easily be obtained by a frequency analysis or by applying pulses or steps to the plant, as in the case of the Reaction Curve method. For most plants in the process industry, any column of the plant transfer matrix can be obtained by applying a step to the corresponding input and measuring the static gain, time constant and equivalent delay time for each output. If the process is repeated for all the inputs, the full transfer matrix is obtained.

The input-output transfer matrix of the CARIMA multivariable model described by Equation (6.1) is given by the following $n \times m$ rational matrix:

$$\mathbf{T}(z^{-1}) = \mathbf{A}(z^{-1})^{-1}\mathbf{B}(z^{-1})z^{-1} \tag{6.24}$$

Given a rational matrix $\mathbf{T}(z^{-1})$, the problem consists of finding two polynomial matrices $\mathbf{A}(z^{-1})$ and $\mathbf{B}(z^{-1})$ so that Equation (6.24) holds. The simplest way of accomplishing this task is by making $\mathbf{A}(z^{-1})$ a diagonal matrix with its diagonal elements equal to the least common multipliers of the denominators of the corresponding row of $\mathbf{T}(z^{-1})$. Matrix $\mathbf{B}(z^{-1})$ is then equal to $\mathbf{B}(z^{-1}) = \mathbf{A}(z^{-1})\mathbf{T}(z^{-1})z$.

Matrices $\mathbf{A}(z^{-1})$ and $\mathbf{B}(z^{-1})$ obtained this way do not have to be left coprime in general. A left coprime representation can be obtained [80] as follows.

Find a right matrix fraction description $\mathbf{T}(z^{-1}) = \mathbf{N}_R(z^{-1})\mathbf{D}_R(z^{-1})^{-1}$ by making $\mathbf{D}_R(z^{-1})$ a diagonal matrix with its diagonal elements equal to the least common denominator of the corresponding column and form $\mathbf{N}_R(z^{-1})$ accordingly. Note that these polynomial matrices do not have to be right coprime in general.

Find a unimodular matrix $\mathbf{U}(z^{-1})$ such that

$$\begin{bmatrix} \mathbf{U}_{11} & \mathbf{U}_{12} \\ \mathbf{U}_{21} & \mathbf{U}_{22} \end{bmatrix} \begin{bmatrix} \mathbf{D}_R(z^{-1}) \\ \mathbf{N}_R(z^{-1}) \end{bmatrix} = \begin{bmatrix} \mathbf{R}(z^{-1}) \\ \mathbf{0} \end{bmatrix} \tag{6.25}$$

where $\mathbf{R}(z^{-1})$ is the greatest right common divisor of $\mathbf{D}_R(z^{-1})$ and $\mathbf{N}_R(z^{-1})$. That is, $\mathbf{R}(z^{-1})$ is a right divisor of $\mathbf{D}_R(z^{-1})$ and $\mathbf{N}_R(z^{-1})$ ($\mathbf{D}_R(z^{-1}) = \mathbf{D}_R'(z^{-1})\mathbf{R}(z^{-1})$, $\mathbf{N}_R(z^{-1}) = \mathbf{N}_R'(z^{-1})\mathbf{R}(z^{-1})$) and if there is another right divisor $\mathbf{R}'(z^{-1})$ then $\mathbf{R}(z^{-1}) = \mathbf{W}(z^{-1})\mathbf{R}'(z^{-1})$ where $\mathbf{W}(z^{-1})$ is a polynomial matrix.

The greatest right common divisor can be obtained by using the following algorithm (Goodwin and Sin [80]):

1. Form matrix

$$\mathbf{P}(z^{-1}) = \begin{bmatrix} \mathbf{D}_R(z^{-1}) \\ \mathbf{N}_R(z^{-1}) \end{bmatrix}$$

2. Make zero by elementary row transformation all the elements of the first column of $\mathbf{P}(z^{-1})$ below the main diagonal as follows: Choose the entry of the first column with smallest degree and interchange the corresponding rows to leave this element in position $(1,1)$ of the matrix (now $\tilde{\mathbf{P}}(z^{-1})$). Obtain $g_{i1}(z^{-1})$ and $r_{i1}(z^{-1})$ for all the elements of the first column such that $P_{i1}(z^{-1}) = \tilde{P}_{11}(z^{-1})g_{i1}(z^{-1}) + r_{i1}(z^{-1})$, with $\delta(r_{i1}(z^{-1})) < \delta(\tilde{P}_{i1}(z^{-1}))$. For all rows below the main diagonal subtract the first row multiplied by $g_{i1}(z^{-1})$, leaving $r_{i1}(z^{-1})$. Repeat the procedure until all the elements below the main diagonal are zero.

3. For the remaining columns use the same procedure described in step 2 to make zero all the elements below the main diagonal using the element (i,i) and at the same time reducing the order of elements on the right of the main diagonal as much as possible.

4. Apply the same elementary transformations to an identity matrix, the resulting unimodular matrix will be matrix $\mathbf{U}(z^{-1})$.

The submatrices $\mathbf{U}_{21}(z^{-1})$ and $\mathbf{U}_{22}(z^{-1})$ are left coprime, and $\mathbf{U}_{22}(z^{-1})$ is nonsingular and from (6.25) $\mathbf{N}_R(z^{-1})\mathbf{D}_R(z^{-1})^{-1} = -\mathbf{U}_{22}(z^{-1})^{-1}\mathbf{U}_{21}(z^{-1})$.

That is, $\mathbf{A}(z^{-1}) = \mathbf{U}_{22}(z^{-1})$ and $\mathbf{B}(z^{-1}) = -\mathbf{U}_{21}(z^{-1})$. Although $\mathbf{A}(z^{-1})$ and $\mathbf{B}(z^{-1})$ do not have to be left coprime for implementing a GPC, they will in general have higher degrees and may in some cases result in a less efficient algorithm.

Example

In order to illustrate how to obtain a matrix fraction description and how to apply GPC to a MIMO process given by its transfer matrix, consider the small signal model of a stirred tank reactor (Figure 6.1) described by the following transfer matrix (the time constants are expressed in minutes)

$$
\begin{bmatrix} Y_1(s) \\ Y_2(s) \end{bmatrix} = \begin{bmatrix} \dfrac{1}{1+0.7s} & \dfrac{5}{1+0.3s} \\[2mm] \dfrac{1}{1+0.5s} & \dfrac{2}{1+0.4s} \end{bmatrix} \begin{bmatrix} U_1(s) \\ U_2(s) \end{bmatrix}
$$

where the manipulated variables $U_1(s)$ and $U_2(s)$ are the feed flow rate and the flow of coolant in the jacket, respectively. The controlled variables $Y_1(s)$ and $Y_2(s)$ are the effluent concentration and the reactor temperature, respectively.

The discretized model for a sampling time of 0.03 minute is

$$
\begin{bmatrix} y_1(t) \\ y_2(t) \end{bmatrix} = \begin{bmatrix} \dfrac{0.0420z^{-1}}{1-0.9580z^{-1}} & \dfrac{0.4758z^{-1}}{1-0.9048z^{-1}} \\[3mm] \dfrac{0.0582z^{-1}}{1-0.9418z^{-1}} & \dfrac{0.1445z^{-1}}{1-0.9277z^{-1}} \end{bmatrix} \begin{bmatrix} u_1(t) \\ u_2(t) \end{bmatrix}
$$

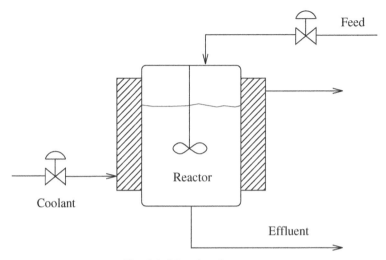

Fig. 6.1. Stirred tank reactor

A left matrix fraction description can be obtained by making matrix $\mathbf{A}(z^{-1})$ equal to a diagonal matrix with diagonal elements equal to the least common multiple of the denominators of the corresponding row of the transfer function, resulting in:

$$\mathbf{A}(z^{-1}) = \begin{bmatrix} 1 - 1.8629z^{-1} + 0.8669z^{-2} & 0 \\ 0 & 1 - 1.8695z^{-1} + 0.8737z^{-2} \end{bmatrix}$$

$$\mathbf{B}(z^{-1}) = \begin{bmatrix} 0.0420 - 0.0380z^{-1} & 0.4758 - 0.4559z^{-1} \\ 0.0582 - 0.0540z^{-1} & 0.1445 - 0.1361z^{-1} \end{bmatrix}$$

For a prediction horizon $N_2 = 3$, a control horizon $N_3 = 2$ and a control weight $\lambda = 0.05$, matrix $\mathbf{G}_{N_{123}}$ results in:

$$\mathbf{G}_{N_{123}} = \begin{bmatrix} 0.0420 & 0.4758 & 0 & 0 \\ 0.0582 & 0.1445 & 0 & 0 \\ 0.0821 & 0.9063 & 0.0420 & 0.4758 \\ 0.1131 & 0.2786 & 0.0582 & 0.1445 \\ 0.1206 & 1.2959 & 0.0821 & 0.9063 \\ 0.1647 & 0.4030 & 0.1131 & 0.2786 \end{bmatrix}$$

The evolution of the reactor temperature and of the effluent concentration obtained when the GPC is applied without prior knowledge of the references can be seen in Figure 6.2. The setpoints were increased by 0.5 and 0.3 at the beginning of the simulation. Once the variables reached the initial setpoint, a change in the setpoint of the effluent concentration from 0.5 to 0.4 was introduced.

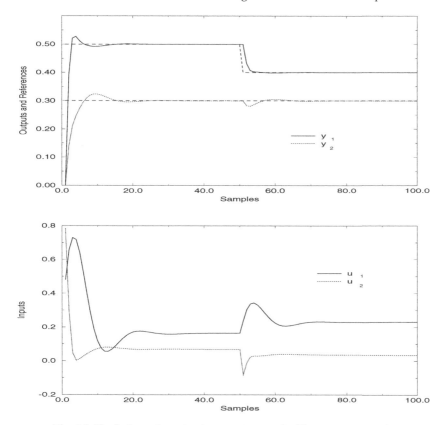

Fig. 6.2. Evolution of reactor temperature and effluent concentration

As can be seen, both variables reach their setpoint in a very short time exhibiting a very small overshoot. It can also be observed that the interactions are relatively small for the closed-loop system when the setpoint of one of the variables is changed. This is because the control action produced by the GPC in both variables acts simultaneously on both manipulated variables as soon as a change in the reference of any of them is detected. See [139] for a study about frequency response characteristics and interaction degree of MIMO GPC.

6.2.2 Parametric Identification

System identification can be defined as the process of obtaining a model for the behaviour of a plant based on plant input and output data. If a particular model structure is assumed, the identification problem is reduced to obtaining the parameters of the model. The usual way of obtaining the parameters of the model is by optimizing a function that measures how well the model, with a particular set of parameters, fits the existing input-output data. When process variables are perturbed by noise of a stochastic nature, the identification problem is usually interpreted as a parameter estimation problem. This problem has been extensively studied in literature for the case of processes which are linear on the parameters to be estimated and perturbed with a white noise. That is, processes that can be described by:

$$\mathbf{z}_k = \Theta \, \Phi_k + \mathbf{e}_k \tag{6.26}$$

where Θ is the vector of parameters to be estimated, Φ_k is a vector of past input and output measures, $\mathbf{z}_k$ is a vector of the latest output measures and $\mathbf{e}_k$ is a white noise.

A multivariable CARIMA model described by Equation (6.1) can easily be expressed as (6.26). Multiply Equation (6.1) by $\triangle$

$$\tilde{\mathbf{A}}(z^{-1})y(t) = \mathbf{B}(z^{-1}) \triangle u(t-1) + e(t)$$

which can be rewritten as

$$y(t) = \tilde{\mathbf{A}}'(z^{-1})y(t-1) + \mathbf{B}(z^{-1}) \triangle u(t-1) + e(t)$$

with

$$\tilde{\mathbf{A}}'(z^{-1}) = (I_{n \times n} - \tilde{\mathbf{A}}(z^{-1}))z$$
$$= -\tilde{A}_1 - \tilde{A}_2 z^{-2} - \cdots - \tilde{A}_{n_a} z^{-(n_a-1)} - \tilde{A}_{n_a+1} z^{-n_a}$$

This can be expressed as (6.26), by making

$$\Theta = \begin{bmatrix} \tilde{A}_1 \ \tilde{A}_2 \cdots \tilde{A}_{n_a} \ \tilde{A}_{n_a+1} \ B_0 \ B_1 \cdots B_{n_b} \end{bmatrix}$$

$$\Phi_k = \begin{bmatrix} -y(t-1) \\ -y(t-2) \\ \vdots \\ -y(t-na) \\ \triangle u(t-1) \\ \triangle u(t-2) \\ \vdots \\ \triangle u(t-n_b) \end{bmatrix}$$

The parameter can be identified by using a least squares identification algorithm [128],[215].

Notice that estimated parameters correspond to the coefficient matrices of polynomial matrices $\tilde{\mathbf{A}}(z^{-1})$ and $\mathbf{B}(z^{-1})$ which are used for the recursion of the Diophantine equation and for the prediction of forced and free responses.

Notice that if some knowledge about the structure of matrices $A(z^{-1})$ and $B(z^{-1})$ is available, the number of parameters to be identified can be reduced substantially, resulting in greater efficiency of the identification algorithms. For example, if matrix $A(z^{-1})$ is considered to be diagonal, only the parameters of the diagonal elements need to be identified and thus appear in Θ. The form of vectors Θ and Φ_k has to be changed accordingly.

6.3 State Space Formulation

Let us consider a multivariable process with n outputs and m inputs described by the following state space model:

$$x(t+1) = Mx(t) + N \triangle u(t) + Pv(t)$$
$$y(t) = Qx(t) + w(t) \tag{6.27}$$

where $x(t)$ is the state vector, $v(t)$ and $w(t)$ are the noises affecting the process and the output, respectively, and are assumed to be white stationary random processes with $E[v(t)] = 0$, $E[w(t)] = 0$, $E[v(t)v(t)^T] = \Gamma_v$, $E[w(t)w(t)^T] = \Gamma_w$, and $E[v(t)w(t)^T] = \Gamma_{vw}$.

The output of the model for instant $t+j$, assuming that the state at instant t and future control increments are known, can be computed by recursively applying Equation (6.27), resulting in:

$$y(t+j) = QM^j x(t) + \sum_{i=0}^{j-1} QM^{j-i-1} N \triangle u(t+i)$$

$$+ \sum_{i=0}^{j-1} QM^{j-i-1} Pv(t+i) + w(k+j)$$

Taking the expected value:

$$\hat{y}(t+j|t) = E[y(t+j)] = QM^j E[x(t)] + \sum_{i=0}^{j-1} QM^{j-i-1} N \triangle u(t+i)$$

$$+ \sum_{i=0}^{j-1} QM^{j-i-1} PE[v(t+i)] + E[w(k+j)]$$

As $E[v(t+i)] = 0$ and $E[w(t+j)] = 0$, the optimal j ahead prediction is given by:

$$\hat{y}(t+j|t) = QM^j E[x(t)] + \sum_{i=0}^{j-1} QM^{j-i-1} N \triangle u(t+i)$$

Let us now consider a set of N_2 j ahead predictions

$$\mathbf{y} = \begin{bmatrix} \hat{y}(t+1|t) \\ \hat{y}(t+2|t) \\ \vdots \\ \hat{y}(t+N_2|t) \end{bmatrix} = \begin{bmatrix} QME[x(t)] + QN \triangle u(t) \\ QM^2 E[x(t)] + \sum_{i=0}^{1} QM^{1-i} N \triangle u(t+i) \\ \vdots \\ QM^{N_2} E[x(t)] + \sum_{i=0}^{N_2-1} QM^{N_2-1-i} N \triangle u(t+i) \end{bmatrix}$$

which can be expressed as:

$$\mathbf{y} = \mathbf{F}\hat{x}(t) + \mathbf{H}\mathbf{u} \tag{6.28}$$

where $\hat{x}(t) = E[x(t)]$, $\mathbf{H}$ is a block lower triangular matrix with its non-null elements defined by $(\mathbf{H})_{ij} = QM^{i-j}N$ and matrix $\mathbf{F}$ is defined as:

$$\mathbf{F} = \begin{bmatrix} QM \\ QM^2 \\ \vdots \\ QM^{N_2} \end{bmatrix}$$

The prediction equation (6.28) requires an unbiased estimation of the state vector $x(t)$. If the state vector is not accessible, a Kalman filter [11] is required.

Let us now consider a set of j ahead predictions affecting the cost function: $\mathbf{y}_{N_{12}} = [\hat{y}(t+N_1|t)^T \cdots \hat{y}(t+N_2|t)^T]^T$ and the vector of N_3 future control moves $\mathbf{u}_{N_3} = [\triangle u(t)^T \cdots \triangle u(t+N_3-1)^T]^T$. Then

$$\mathbf{y}_{N_{12}} = \mathbf{F}_{N_{12}}\hat{x}(t) + \mathbf{H}_{N_{123}}\mathbf{u}_{N_3}$$

where matrices $\mathbf{F}_{N_{12}}$ and $\mathbf{H}_{N_{123}}$ are formed by the corresponding submatrices in $\mathbf{F}$ and $\mathbf{H}$ respectively. Equation (6.2) can be rewritten as:

$$J = (\mathbf{H}_{N_{123}}\mathbf{u}_{N_3} + \mathbf{F}_{N_{12}}\hat{x}(t) - \mathbf{w})^T \overline{R}(\mathbf{H}_{N_{123}}\mathbf{u}_{N_3} + \mathbf{F}_{N_{12}}\hat{x}(t) - \mathbf{w}) + \mathbf{u}_{N_3}^T \overline{Q}\mathbf{u}_{N_3}$$

If there are no constraints, the optimum can be expressed as:

$$\mathbf{u} = ((\mathbf{H}_{N_{123}}^T \overline{R}\mathbf{H}_{N_{123}}) + \overline{Q})^{-1}\mathbf{H}_{N_{123}}^T \overline{R}(\mathbf{w} - \mathbf{F}_{N_{12}}\hat{x}(t))$$

6.3.1 Matrix Fraction and State Space Equivalences

The output signal of processes described by Equations (6.27) and (6.1), with zero initial conditions, can be expressed as:

$$y(t) = Q(zI - M)^{-1}N \triangle u(t) + Q(zI - M)^{-1}Pv(t) + w(t)$$
$$y(t) = \tilde{\mathbf{A}}(z^{-1})^{-1}\mathbf{B}(z^{-1})z^{-1} \triangle u(t) + \tilde{\mathbf{A}}(z^{-1})^{-1}\mathbf{C}(z^{-1})e(t)$$

By comparing these equations, it is clear that both representations are equivalent if

$$Q(zI - M)^{-1}N = \tilde{\mathbf{A}}(z^{-1})^{-1}\mathbf{B}(z^{-1})z^{-1}$$
$$Q(zI - M)^{-1}Pv(t) + w(t) = \tilde{\mathbf{A}}(z^{-1})^{-1}\mathbf{C}(z^{-1})e(t)$$

This can be achieved by making $w(t) = 0$, $v(t) = e(t)$ and finding a left matrix fraction description of $Q(zI - M)^{-1}N$ and $Q(zI - M)^{-1}P$ with the same left matrix $\tilde{\mathbf{A}}(z^{-1})^{-1}$.

The state space description can be obtained from the matrix fraction description of Equation (6.1), used in the previous section, as follows

Consider the state vector $x(t) = [y(t)^T \cdots y(t - n_a)^T \; \triangle \, u(t - 1)^T \cdots \triangle \, u(t - n_b)^T e(t)^T \cdots e(t - n_c)^T]$ and the noise vector $v(t) = e(t + 1)$. Equation (6.1) can now be expressed as Equation (6.27) with:

$$
M =
\begin{bmatrix}
\tilde{A}'_1 & \cdots & \tilde{A}'_{n_a} & \tilde{A}'_{n_a+1} & B_1 & \cdots & B_{n_b-1} & B_{n_b} & C_1 & \cdots & C_{n_c-1} & C_{n_c} \\
I & \cdots & 0 & 0 & 0 & \cdots & 0 & 0 & 0 & \cdots & 0 & 0 \\
0 & \ddots & \vdots & \vdots & \vdots & \vdots & \vdots & \vdots & \vdots & \vdots & \vdots & \vdots \\
0 & \cdots & I & 0 & 0 & \cdots & 0 & 0 & 0 & \cdots & 0 & 0 \\
0 & \cdots & 0 & 0 & 0 & \cdots & 0 & 0 & 0 & \cdots & 0 & 0 \\
0 & \cdots & 0 & 0 & I & \cdots & 0 & 0 & 0 & \cdots & 0 & 0 \\
\vdots & \vdots & \vdots & \vdots & 0 & \ddots & \vdots & \vdots & \vdots & \vdots & \vdots & \vdots \\
0 & \cdots & 0 & 0 & 0 & \cdots & I & 0 & 0 & \cdots & 0 & 0 \\
0 & \cdots & 0 & 0 & 0 & \cdots & 0 & 0 & 0 & \cdots & 0 & 0 \\
0 & \cdots & 0 & 0 & 0 & \cdots & 0 & 0 & I & \cdots & 0 & 0 \\
\vdots & \vdots & \vdots & \vdots & \vdots & \vdots & \vdots & \vdots & 0 & \ddots & \vdots & \vdots \\
0 & \cdots & 0 & 0 & 0 & \cdots & 0 & 0 & 0 & \cdots & I & 0
\end{bmatrix}
$$

$$
N = \begin{bmatrix} B_0^T & 0 & \cdots & 0 \big| I & 0 & \cdots & 0 \big| 0 & \cdots & 0 & 0 \end{bmatrix}^T
$$

$$
P = \begin{bmatrix} I & 0 & \cdots & 0 \big| 0 & 0 & \cdots & 0 \big| I & 0 & \cdots & 0 \end{bmatrix}^T
$$

The measurement error vector $w(t)$ has to be made zero for both descriptions to coincide. If the colouring polynomial matrix is the identity matrix, the state vector is only composed of past inputs and outputs $x(t) = [y(t)^T \ldots y(t - n_a)^T \; \triangle \, u(t - 1)^T \ldots \triangle \, u(t - n_b)^T]$ and only the first two column blocks of matrices M, N and P and the first two row blocks of matrix M have to be considered.

Notice that no Kalman filter is needed to implement the GPC because the state vector is composed of past inputs and outputs. However, the description does not have to be minimal in terms of the state vector dimension. If there is a big difference in the degrees of the polynomials $(\tilde{A}(z^{-1}))_{ij}$ and $(B(z^{-1}))_{ij}$ it is better to consider only the past inputs and outputs that are really needed to compute future output signals. To do this, consider the i component of the output vector

$$
y_i(t + 1) = -\tilde{A}_{i1}(z^{-1})y_1(t) - \tilde{A}_{i2}(z^{-1})y_2(t) - \cdots - \tilde{A}_{in}(z^{-1})y_n(t)
$$

$$
+ B_{i1}(z^{-1}) \triangle u_1(t) + B_{i2}(z^{-1}) \triangle u_2(t) + \cdots + B_{im}(z^{-1}) \triangle u_m(t)
$$

$$
+ C_{i1}(z^{-1})e_1(t + 1) + C_{i2}(z^{-1})e_2(t + 1) + \cdots + C_{in}(z^{-1})e_n(t + 1)
$$

where $\tilde{A}_{ij}(z^{-1})$, $B_{ij}(z^{-1})$ and $C_{ij}(z^{-1})$ are the ij entries of polynomial matrices $\tilde{A}(z^{-1})$, $B(z^{-1})$ and $C(z^{-1})$, respectively.

The state vector can be defined as

$$
\begin{aligned}
x(t) = [&y_1(t) \cdots y_1(t - n_{y_1}), y_2(t) \cdots y_2(t - n_{y_2}), \cdots, y_n(t) \cdots y_n(t - n_{y_n}), \\
&\triangle u_1(t-1) \cdots \triangle u_1(t - n_{u_1}), \triangle u_2(t-1) \cdots \triangle u_2(t - n_{u_2}), \cdots, \\
&\triangle u_m(t-1) \cdots \triangle u_m(t - n_{u_m}), \\
&e_1(t) \cdots e_1(t - n_{e_1}), e_2(t) \cdots e_2(t - n_{e_2}), \cdots, e_n(t) \cdots e_n(t - n_{e_n})]^T
\end{aligned}
$$

where $n_{y_i} = \max_j \delta(\tilde{\mathbf{A}}_{ij}(z^{-1}))$, $n_{u_j} = \max_i \delta(\mathbf{B}_{ij}(z^{-1}))$ and $n_{e_j} = \max_j \delta(\mathbf{C}_{ij}(z^{-1}))$. Matrices M, N and P can be expressed as:

$$
M = \left[
\begin{array}{ccc|ccc|ccc}
Myy_{11} & \cdots & Myy_{1n} & Myu_{11} & \cdots & Myu_{1m} & Mye_{11} & \cdots & Mye_{1n} \\
\vdots & \vdots & \vdots & \vdots & \vdots & \vdots & \vdots & \vdots & \vdots \\
Myy_{n1} & \cdots & Myy_{nn} & Myu_{n1} & \cdots & Myu_{nm} & Mye_{n1} & \cdots & Mye_{nn} \\
0 & \cdots & 0 & Muu_{11} & \cdots & Muu_{1m} & 0 & \cdots & 0 \\
\vdots & \vdots & \vdots & \vdots & \vdots & \vdots & \vdots & \vdots & \vdots \\
0 & \cdots & 0 & Muu_{m1} & \cdots & Muu_{mm} & 0 & \cdots & 0 \\
0 & \cdots & 0 & 0 & \cdots & 0 & Mee_{11} & \cdots & Mee_{1n} \\
\vdots & \vdots & \vdots & \vdots & \vdots & \vdots & \vdots & \vdots & \vdots \\
0 & \cdots & 0 & 0 & \cdots & 0 & Mee_{n1} & \cdots & Mee_{nn}
\end{array}
\right]
$$

$$
N = \left[\; Ny_1^T \;\cdots\; Ny_n^T \;\big|\; Nu_1^T \;\cdots\; Nu_m^T \;\big|\; 0 \;\cdots\; 0 \right]^T
$$

$$
P = \left[\; Py_1^T \;\cdots\; Py_n^T \;\big|\; 0 \;\cdots\; 0 \;\big|\; 0 \cdots\; 0 \right]^T
$$

where the submatrices Myy_{ij}, Myu_{ij}, Mye_{ij}, Muu_{ij} and Ny_i have the following form:

$$
Myy_{ij} = \begin{bmatrix}
-\tilde{a}_{ij_1} & -\tilde{a}_{ij_2} & \cdots & \cdots & -\tilde{a}_{ij_{n_{y_i}}} \\
1 & 0 & \cdots & \cdots & 0 \\
0 & 1 & 0 & \cdots & 0 \\
\vdots & \ddots & \ddots & \ddots & \vdots \\
0 & \cdots & \cdots & 1 & 0
\end{bmatrix}
\qquad
Myu_{ij} = \begin{bmatrix}
b_{ij_1} & \cdots & b_{ij_{n_{u_j}}} \\
0 & \cdots & 0 \\
\vdots & \vdots & \vdots \\
0 & \cdots & 0
\end{bmatrix}
$$

$$
Mye_{ij} = \begin{bmatrix}
c_{ij_1} & \cdots & c_{ij_{n_{u_j}}} \\
0 & \cdots & 0 \\
\vdots & \vdots & \vdots \\
0 & \cdots & 0
\end{bmatrix}
\qquad
Muu_{ij} = \begin{bmatrix}
0 & 0 & 0 & \cdots & 0 \\
1 & 0 & 0 & \cdots & 0 \\
0 & 1 & 0 & \cdots & 0 \\
\vdots & \ddots & \ddots & \ddots & \vdots \\
0 & 0 & \cdots & 1 & 0
\end{bmatrix}
$$

$$
Ny_i = \begin{bmatrix}
b_{i1_0} & \cdots & b_{im_0} \\
0 & \cdots & 0 \\
\vdots & \vdots & \vdots \\
0 & \cdots & 0
\end{bmatrix}
$$

Matrix Mee_{ij} has the same form as matrix Muu_{ij}. Matrices Nu_j and Py_j have all elements zero except the j element of the first row which is 1. The noise vectors are $v(t) = e(t + 1)$ and $w(t) = 0$.

This state space description corresponds to the one used in [160] for the SISO case. Other state space descriptions have been proposed in literature in the MPC context. In [2] a state space description involving an artificial sampling interval equal to the prediction horizon multiplied by the sampling time is proposed. The vectors of predicted inputs and outputs over the control horizon are used as input and output signals. The GPC costing function, for the noise-free case, is then transformed into a one-step performance index. A state space description based on the step response of the plant has been proposed in [116]. Models based on the step response of the plant are widely used in industry because they are very intuitive and require less *a priori* information to identify the plant. The main disadvantages are that more parameters are needed and only stable processes can be modelled, although the description proposed in [116] allows for the modelling of processes containing integrators.

6.4 Case Study: Flight Control

This section shows an application of MPC to the control of climb rate/airspeed of an aircraft model. The model is taken from [37] and it corresponds to the longitudinal motion of a Boeing 747 airplane. The multivariable process is controlled using a predictive controller based on the state space model of the aircraft.

The autopilot will fly the airplane to the desired flight condition specified by the pilot. Two of the usual command outputs that must be controlled are airspeed, that is, velocity (or Mach number) with respect to air, and climb rate.

We will focus on the longitudinal motion of the aircraft, which can be controlled acting on elevator (e) and throttle (t). The aircraft longitudinal motion can be represented [37] by means of u (velocity in the longitudinal body axis, x), w (velocity in the y-axis), q (component of the angular velocity) and θ (angle of the x-axis with respect to the horizontal). Figure 6.3 shows the nomenclature for aircraft longitudinal motions.

The perturbation equations that model the dynamics of a 747 airplane cruising in level flight at an altitude of 40,000 ft and a velocity of 774 ft/sec (Mach number 0.80) can be written in continuous time as

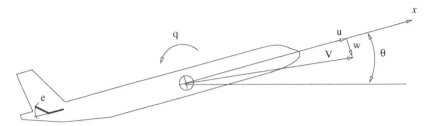

Fig. 6.3. Nomenclature for aircraft longitudinal motions

$$
\begin{bmatrix} \dot{u} \\ \dot{w} \\ \dot{q} \\ \dot{\theta} \end{bmatrix} = \begin{bmatrix} -0.003 & 0.039 & 0 & -0.322 \\ -0.065 & -0.319 & 7.74 & 0 \\ 0.020 & -0.101 & -0.429 & 0 \\ 0 & 0 & 1 & 0 \end{bmatrix} \begin{bmatrix} u - u_w \\ w - w_w \\ q \\ \theta \end{bmatrix}
$$
$$
+ \begin{bmatrix} 0.010 & 1 \\ -0.18 & -0.04 \\ -1.16 & 0.598 \\ 0 & 0 \end{bmatrix} \begin{bmatrix} e \\ t \end{bmatrix}
$$

where u_w and w_w are perturbations in wind velocity components. Velocities are given in ft/sec, the angles in crad and the angular velocity q in crad/sec.

The two outputs to be controlled are airspeed, $u - u_w$, and climb rate $\dot{h} = -w + u_0\theta$, with $u_0 = 774$ ft/sec. In order to use the state space predictive controller depicted in Section 6.3, the model must be converted to discrete time. If a sampling time of 0.1 second is used, the model turns to

$$
\begin{bmatrix} u(t+1) \\ w(t+1) \\ q(t+1) \\ \theta(t+1) \end{bmatrix} = \begin{bmatrix} 0.9996 & 0.0383 & 0.0131 & -0.0322 \\ -0.0056 & 0.9647 & 0.7446 & 0.0001 \\ 0.0020 & -0.0097 & 0.9543 & 0 \\ 0.0001 & -0.0005 & 0.0978 & 1 \end{bmatrix} \begin{bmatrix} u(t) - u_w(t) \\ w(t) - w_w(t) \\ q(t) \\ \theta(t) \end{bmatrix}
$$
$$
+ \begin{bmatrix} 0.0001 & 0.1002 \\ -0.0615 & 0.0183 \\ -0.1133 & 0.0586 \\ -0.0057 & 0.0029 \end{bmatrix} \begin{bmatrix} e(t) \\ t(t) \end{bmatrix}
$$

and the outputs are given by

$$
\begin{bmatrix} y_1(t) \\ y_2(t) \end{bmatrix} = \begin{bmatrix} 1 & 0 & 0 & 0 \\ 0 & -1 & 0 & 7.74 \end{bmatrix} \begin{bmatrix} u(t) - u_w(t) \\ w(t) - w_w(t) \\ q(t) \\ \theta(t) \end{bmatrix}
$$

The time evolution of the system with the predictive controller is shown in Figure 6.4. The figure shows the outputs evolution when the weighting

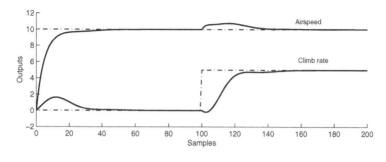

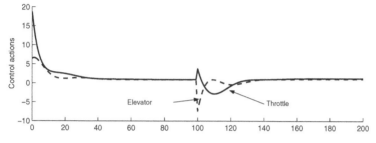

Fig. 6.4. Setpoint changes

matrices are $R = diag(5,5)$, $Q = diag(1,1)$, the control horizon is 10 and the prediction horizon is set to 30. A setpoint change from 0 to 10 ft/sec in airspeed is performed at the beginning of the simulation and a change from 0 to 5 ft/sec in climb rate is done 10 seconds later. It can be seen how the airplane responds to the commands and that some degree of interaction exists between the variables.

It can be observed that, since the process is multivariable, the climb rate is slightly affected by the change in airspeed and both control actions (elevator and throttle) act simultaneously to keep the outputs at the desired values. If, for instance, airspeed is to be given higher importance than the other output, the effect of climb rate changes on airspeed can be reduced by changing the error weighting matrix to $R = diag(10, 1)$. In this case, Figure 6.5 shows how the controller keeps the first output almost unaffected by a change in the second output setpoint.

6.5 Convolution Models Formulation

Step response and impulse response models can be easily extended to deal with multivariable processes. For a plant with n_u inputs each output j will reflect the effect of all the inputs in the following way

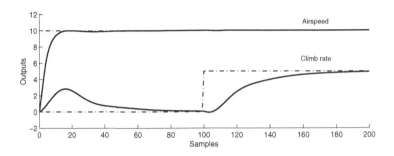

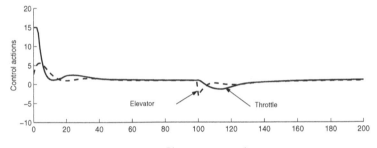

Fig. 6.5. Change in weights

$$y_j(t) = \sum_{k=1}^{n_u} \sum_{i=1}^{N_k} g_i^{kj} u^k(t - i)$$

where g_i^{kj} is the response of output j to a step (or an impulse if an impulse response model is being used) in input k.

The Dynamic Matrix Controller shown in Section 3.1 can be easily extended to the case of a process modelled by a multivariable step response model. The basic scheme of DMC already discussed extends to systems with multiple inputs and multiple outputs. The basic equations remain the same, except that the matrices and vectors become larger and appropriately partitioned.

Based upon model linearity, the superposition principle can be used to obtain the predicted outputs provoked by the system inputs. The vector of predicted outputs is now defined as

$$\hat{\mathbf{y}} = [y_1(t + 1 \mid t), \dots, y_1(t + p_1 \mid t), \dots, y_{ny}(t + 1 \mid t), \dots, y_{ny}(t + p_{ny} \mid t)]^T$$

the array of future control signals as

$$\mathbf{u} = [\triangle u_1(t), \dots, \triangle u_1(t + m_1 - 1), \dots, \triangle u_{nu}(t), \dots, \triangle u_{nu}(t + m_{nu} - 1)]^T$$

and the free response as

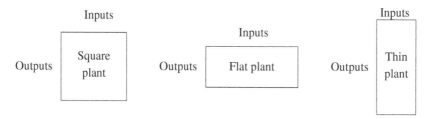

Fig. 6.6. Typical process transfer matrix shapes

$$\mathbf{f} = [f_1(t+1 \mid t), \ldots, f_1(t+p_1 \mid t), \ldots, f_{ny}(t+1 \mid t), \ldots, f_{ny}(t+p_{ny} \mid t)]^T$$

taking into account that the free response of output i depends on both the past values of y_i and the past values of all control signals.

With the vector defined earlier, the prediction equations are the same as (3.2) simply considering matrix $\mathbf{G}$ to be:

$$\mathbf{G} = \begin{bmatrix} G_{11} & G_{12} & \cdots & G_{1nu} \\ G_{21} & G_{22} & \cdots & G_{2nu} \\ \vdots & \vdots & \ddots & \vdots \\ G_{ny1} & G_{ny2} & \cdots & G_{nynu} \end{bmatrix}$$

Each matrix G_{ij} contains the coefficients of the ith step response corresponding to the jth input.

In case that an impulse response model is used, the extension to the multivariable case is exactly the same as that of DMC, so no more attention will be paid to the equations. However, some implementation issues of the resulting controller (Model Algorithmic Control) and the commercial product IDCOM-M (multivariable) will be addressed in this section.

The IDCOM-M algorithm [85] uses two separate objective functions, one for the outputs and if there are extra degrees of freedom one for the inputs. The degree of freedom available for the control depends on the plant structure. Figure 6.6 shows the shape of the process transfer matrix for three general cases.

The square plant case, which is rare in real situations, occurs when the plant has as many inputs as outputs and leads to a control problem with a unique solution. The flat plant case is more common (more inputs than outputs) and the extra degrees of freedom available can be employed in different objectives, such as moving the plant closer to an optimal operating point. In the last situation (thin plant case, where there are more outputs than inputs) it is not possible to meet all of the control objectives, and some specifications must be relaxed.

Thus, for flat plants, IDCOM-M incorporates the concept of *Ideal Resting Values* (IRV) for the inputs. In this case, in addition to the primary objective (minimize the output errors), the controller also tries to minimize the sum of squared deviations of the inputs from their respective IRVs, which may come from a steady-state optimizer (by default the IRV for a given input is set to its current measured value). So the strategy involves a two-step optimization problem that is solved using a quadratic programming approach: the primary problem involves the choice of the control sequence required to drive the controlled variables close to the setpoints, and the second involves optimizing the use of control effort in achieving the objective of the primary problem.

The input optimization makes the most effective use of available degrees of freedom without influencing the optimal output solution. Even when there are no excess inputs, the ideal resting values concept is of great interest when, for operational or economic reasons, there is a benefit in maintaining a manipulated variable at a specific steady-state value.

6.6 Case Study: Chemical Reactor

This section illustrates the application of a multivariable DMC. The chosen system is a chemical jacket reactor. The results have been obtained on a system simulation using the nonlinear differential equations which model its behaviour. The model used is taken from [7] and can be considered to be a very precise representation of this type of process.

6.6.1 Plant Description

The decomposition of a product A into a product B is produced in the reactor (see Figure 6.7). This reaction is exothermic and therefore the interior temperature must be controlled by means of cold water circulating through the jacket around the tank walls.

The variables which come into play are:

- A: feed product arriving at the reactor,
- B: product arising from the transformation of product A in the tank interior,
- C_{a0}: concentration of product A arriving at the reactor,
- T_{l0}: temperature of liquid containing product A,
- F_l : flow of liquid passing through the reactor (at the inlet it only contains product A and at the outlet it contains A and B),
- T_l: temperature of the liquid at the outlet of the reactor,
- C_b: concentration of product B at the outlet of the reactor and in the interior,
- C_a: concentration of A (the inequality $C_a < C_{a0}$ is always fulfilled and at stationary state $C_a + C_b = C_{a0}$),
- T_{c0}: temperature of coolant on entering the jacket,

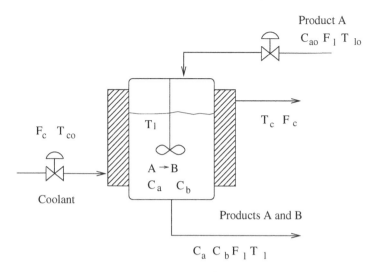

Fig. 6.7. Chemical jacket reactor

- T_c: temperature of coolant in the interior and at the outlet of the jacket, and
- F_c: coolant flow.

The concentrations are given by kmol/m³, the flows by m³/h and the temperatures in °C.

By applying the conservation laws of mass and energy the differential equations defining the dynamics of the system can be obtained. To do this, it is presumed that there is no liquid accumulated in the reactor, that the concentrations and temperature are homogeneous and that the energy losses to the exterior are insignificant.

The mass balance equations are as follows

$$\frac{d(V_l C_a)}{dt} = F_l C_{a0} - V_l k C_a - F_l C_a$$

$$\frac{d(V_l C_b)}{dt} = V_l k C_a - F_l C_b$$

and the energy balance equations are:

$$\frac{d(V_l \rho_l C_{pl} T_l)}{dt} = F_l \rho_l C_p l T_{l0} - F_l \rho_l C_p l T_l - Q + V_l k C_a H$$

$$\frac{d(V_c \rho_c C_{pc} T_c)}{dt} = F_c \rho_c C_p c (T_{c0} - T_c) + Q$$

Table 6.1 gives the meaning and nominal value of the parameters appearing in the equations.

Table 6.1. Process variables and values at operating point

Variable	Description	Value	Unit
k	Speed of reaction $k = \alpha e^{-E_a/R(272+T_l)}$		h^{-1}
α	Coefficient of speed of reaction	59.063	h^{-1}
R	Constant of ideal gas	8.314	kJ/kg kmol
E_a	Activation energy	2100	kJ/kmol
H	Enthalpy of reaction	2100	kJ/kmol
Q	Heat absorbed by coolant		kJ
U	Global heat transmission coefficient	4300	kJ/(h m^2 K)
ρ_l	Liquid density	800	kg/m^3
ρ_c	Coolant density	1000	kg/m^3
$C_p l$	Specific heat of liquid	3	kJ/(kg K)
$C_p c$	Specific heat of coolant	4.1868	kJ/(kg K)
S	Effective heat interchange surface	24	m^2
V_l	Tank volume	24	m^3
V_c	Jacket volume	8	m^3

The aim is to regulate the temperature in the tank interior (T_l) and the concentration at the reactor outlet of product B (C_b), the control variables being the flows of the liquid (F_l) and the cooling fluid (F_c). It is, therefore, a system with two inlets and two outlets.

6.6.2 Obtaining the Plant Model

The design of the controller calls for knowledge of the system dynamics to be controlled. To achieve this, step inputs are produced in the manipulated variables and the behaviour of the process variables is studied.

On the left-hand side of Figure 6.8 the response to a change in the feed flow from 25 to 26 m^3/h is shown. It can be seen that the concentrations present a fairly fast response of opposite sign. The temperatures, however, vary more slowly. The right-hand column of Figure 6.8 shows the effect of a step change of 1 m^3/h in the cooling flow. Due to thermic inertia the variation in temperatures is slow and dampened, presenting opposite in sign those corresponding to feeding and cooling fluids. A great deal of interaction is observed, therefore, between the circuits for feeding and cooling and the dynamics of the controlled variables are different depending on the control variable operating, to which is added slight effects of the nonminimum phase. All of this justifies the use of a multivariable controller instead of two monovariables ones.

Although the system is nonlinear it is possible to work with a model linearized about the operating point. The model is obtained from the response to steps shown in figure 6.8.

One has that

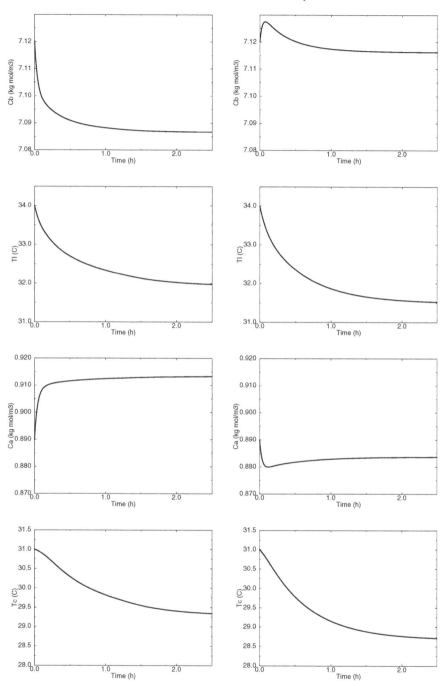

Fig. 6.8. System response to changes in the feed flow (left) and coolant flow (right)

$$y_1(t) = \sum_{i=1}^{N_{11}} g_i^{11} \, \triangle u_1(t) + \sum_{i=1}^{N_{12}} g_i^{12} \, \triangle u_2(t)$$

$$y_2(t) = \sum_{i=1}^{N_{21}} g_i^{21} \, \triangle u_1(t) + \sum_{i=1}^{N_{22}} g_i^{22} \, \triangle u_2(t)$$

where y_1 and y_2 correspond to the concentration of product B and the temperature in the interior of the reactor and u_1 and u_2 correspond to the flow of the liquid and the cooling fluid, respectively.

The sampling time is chosen as $T = 2.4$ minutes and the corresponding values of N_{ij} for this process are:

$$N_{11} = 40 \qquad N_{12} = 50 \qquad N_{21} = 55 \qquad N_{22} = 60$$

6.6.3 Control Law

In order to calculate the control law it is necessary to form matrix $\mathbf{G}$ and calculate the free response, as seen in Chapter 3.

As there are two inputs and two outputs, the free responses for C_b (f_1) and T_l (f_2) are given by:

$$f_1(t+k) = y_{m1}(t) + \sum_{i=1}^{N_{11}}(g_{k+i}^{11} - g_i^{11}) \, \triangle u_1(t-i) + \sum_{i=1}^{N_{12}}(g_{k+i}^{12} - g_i^{12}) \, \triangle u_2(t-i)$$

$$f_2(t+k) = y_{m2}(t) + \sum_{i=1}^{N_{21}}(g_{k+i}^{21} - g_i^{21}) \, \triangle u_1(t-i) + \sum_{i=1}^{N_{22}}(g_{k+i}^{22} - g_i^{22}) \, \triangle u_2(t-i)$$

Choosing the prediction and control horizons as $p = 5$, $m = 3$[1], matrix $\mathbf{G}$ is:

$$\mathbf{G} = \begin{bmatrix} 0 & 0 & 0 & 0 & 0 & 0 \\ -0.0145 & 0 & 0 & 0.0064 & 0 & 0 \\ -0.0201 & -0.0145 & 0 & 0.0074 & 0.0064 & 0 \\ -0.0228 & -0.0201 & -0.0145 & 0.0068 & 0.0074 & 0.0064 \\ -0.0244 & -0.0228 & -0.0201 & 0.0058 & 0.0068 & 0.0074 \\ 0 & 0 & 0 & 0 & 0 & 0 \\ -0.3073 & 0 & 0 & -0.3066 & 0 & 0 \\ -0.5282 & -0.3073 & 0 & -0.5449 & -0.3066 & 0 \\ -0.6946 & -0.5282 & -0.3073 & -0.7351 & -0.5449 & -0.3066 \\ -0.8247 & -0.6946 & -0.5282 & -0.8904 & -0.7351 & -0.5449 \end{bmatrix}$$

The control law is obtained from the minimization of the cost function where, since this is a multivariable process, the errors and control increments are weighted by matrices R and Q:

---

[1] Notice that better results can be obtained for bigger values of the horizon, although these small values have been used in this example for the sake of simplicity.

$$J = \sum_{j=1}^{p} \|\hat{y}(t+j \mid t) - w(t+j)\|_R^2 + \sum_{j=1}^{m} \| \triangle u(t+j-1)\|_Q^2$$

and R and Q are diagonal matrices of dimension $2p \times 2p$ and $2m \times 2m$, respectively. In this application the first m elements of R are taken equal to 1 and the second part is equal to 10 to compensate for the different range of values in temperature and concentration. The control weights are taken as 0.1 for both manipulated variables.

The solution is given by

$$\mathbf{u} = (\mathbf{G}^T R\mathbf{G} + Q)^{-1}\mathbf{G}^T \mathbf{R}(\mathbf{w} - \mathbf{f})$$

and the control increment at instant t is calculated multiplying the first row of $(\mathbf{G}^T \mathbf{RG} + \mathbf{Q})^{-1}\mathbf{G}^T\mathbf{R}$ by the difference between the reference trajectory and the free response

$$\triangle u(t) = \mathbf{l}(\mathbf{w} - \mathbf{f})$$

with

$$\mathbf{l} = [0 \; -0.1045 \; -0.1347 \; -0.1450 \; -0.1485 \; 0 \; -1.3695 \; -0.1112 \; -0.1579 \; 0.1381]$$

6.6.4 Simulation Results

In this section some results of applying the controller to a nonlinear model of the reactor are presented. Although the controller was designed using a linear model and the plant is nonlinear, the results obtained are satisfactory.

The charts on the left of Figure 6.9 show the behaviour of the process in the presence of changes in the composition reference (C_b). As can be observed, the output follows the reference by means of the contribution of the two manipulated variables F_l and F_c. It can also be seen that any change affects all the variables, such as the concentration of A (C_a) and the coolant temperature (T_c), and the other output T_l, which is slightly moved from its reference during the transient stage.

The response to a change in the temperature reference is drawn in the charts on the right of Figure 6.9. As can be seen, the temperature reference is followed satisfactorily but the concentration is affected and separated from its setpoint.

6.7 Dead Time Problems

Most plants in industry, especially in the process industry, exhibit input-output delays or dead times. That is, the effect of a change in the manipulated variable is not felt on the process output until the dead time has elapsed. Dead times are mainly caused by transport delays or sometimes as the result

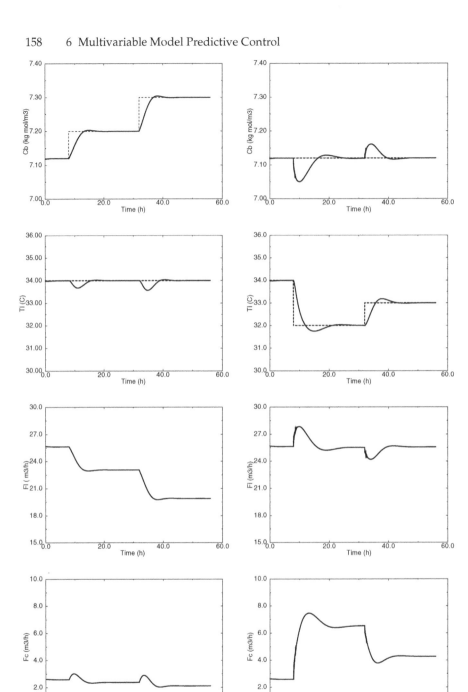

Fig. 6.9. Controller response to changes in concentration reference (left) and liquid temperature reference (right)

of processes with dynamics composed of multiple chained lags. The difficulties of controlling processes with significant dead time are well known and are due to the fact that a dead time produces a phase lag that deteriorates the phase margin. As a result, low gain controllers producing sluggish responses (which have to be added to the dead time of the process) have to be used in order to avoid high oscillations. There are different techniques to cope with delays. The most popular is, perhaps, the Smith predictor [193] which basically consists of getting the delay out of the closed loop by generating a prediction of the process output and designing a controller for the process minus the dead time. The error between process output and predictions is fed back to the controller to cope with plant and model mismatch.

Because of the predictive nature of model predictive controllers, time delays are inherently considered by them. Process input-output dead times are reflected in the polynomial matrix $\mathbf{B}(z^{-1})$. The dead time from the j-input to the i-output, expressed in sampling time units, is the maximum integer d_{ij} such that the entry $(\mathbf{B}(z^{-1}))_{ij}$ of polynomial matrix $\mathbf{B}(z^{-1})$ can be expressed as $(\mathbf{B}(z^{-1}))_{ij} = z^{-d_{ij}}(\mathbf{B}'(z^{-1}))_{ij}$. Let us define $d_{\min} = \min_{i,j} d_{ij}$, and $d_{\max} = \max_{i,j} d_{ij}$. Although process dead time is implicitly considered in the previous section by the first coefficient matrices of polynomial matrix $\mathbf{B}(z^{-1})$ being zero, the computation will not be efficient if precautions are not taken.

The natural extension of the dead time to multivariable processes is the interactor matrix [80] which represents the time delay structure of a multivariable process. The interactor matrix always exists if the transfer matrix $\mathbf{T}(z)$ is strictly proper with $\det(\mathbf{T}(z)) \neq 0$ for almost all z. It is defined as a polynomial matrix $\xi(z)$ such that

$$\det(\xi(z)) = z^k$$
$$\lim_{z \to \infty} \xi(z)\mathbf{T}(z) = K$$

where k is an integer and K is a nonsingular matrix. The interactor matrix can be made to have the following structure: $\xi(z) = M(z)D(z)$ where $D(z) = \mathrm{diag}(z^{d_1} \cdots z^{d_n})$ and $M(z)$ is a lower triangular matrix with the elements on the main diagonal equal to unity and the elements below the main diagonal either zero or divisible by z. The interactor matrix can be used to design precompensators as indicated in [201] by making the control signal $u(t) = \xi_r(z^{-1})z^{-d}v(t)$, with $\xi_r(z^{-1})$ equal to the right interactor matrix. The output vector is then equal to

$$y(t) = \mathbf{T}(z^{-1})u(t) = \mathbf{T}(z^{-1})\xi_r(z^{-1})z^{-d}v(t)$$
$$= [\mathbf{T}'(z^{-1})\xi_r(z^{-1})^{-1}]\xi_r(z^{-1})z^{-d}v(t) = \mathbf{T}'(z^{-1})z^{-d}v(t)$$

The process can now be interpreted as a process with a common delay d for all the variables. Notice that the precompensator consists of adding delays to the process. Model predictive control, as pointed out in [191], does not require the use of this type of pre- or postcompensation and the unwanted effects caused by adding extra delays at the input or output are avoided.

In most cases the interactor matrix will take a diagonal form, one corresponding to a single delay d_{min} for every output and the other with a delay d_i for each output. These two cases will be discussed in the following.

First consider the case where there is not much difference between d_{max} and d_{min} and a single delay d_{min} is associated to all output variables. The output of the process will not be affected by $\triangle u(t)$ until the time instant $t + d_{min} + 1$; the previous outputs will be a part of the free response and there is no point in considering them as part of the objective function. The lower limit of the prediction horizon N_1 can therefore be made equal to $d_{min} + 1$. Note that there is no point in making it smaller, and furthermore if it is made bigger, the first predictions, the ones predicted with greater certainty, will not be considered in the objective function. If the difference between d_{max} and d_{min} is not significant, and there is not much difference in the dynamics of the process variables, a common lower ($N_1 = d_{min} + 1$) and upper ($N_2 = N_1 + N - 1$) limit can be chosen for the objective function. Computation can be simplified by considering $\mathbf{B}(z^{-1}) = z^{-d_{min}}\mathbf{B}'(z^{-1})$ and computing the predictions as:

$$\hat{y}(t + N_1 + j|t) = \mathbf{E}_{N_1+j}(z^{-1})\mathbf{B}(z^{-1}) \triangle u(t + N_1 + j - 1) + \mathbf{F}_{N_1+j}(z^{-1})y(t)$$

$$= \mathbf{E}_{N_1+j}(z^{-1})\mathbf{B}'(z^{-1}) \triangle u(t + N_1 + j - 1 - d_{min}) + \mathbf{F}_{N_1+j}(z^{-1})y(t)$$

$$= \mathbf{E}_{N_1+j}(z^{-1})\mathbf{B}'(z^{-1}) \triangle u(t + j) + \mathbf{F}_{N_1+j}(z^{-1})y(t)$$

By making the polynomial matrix

$$\mathbf{E}_{N_1+j}(z^{-1})\mathbf{B}'(z^{-1}) = \mathbf{G}_{N1_j}(z^{-1}) + z^{-(j+1)}\mathbf{G}p_{N1_j}(z^{-1})$$

the prediction equation can now be written as:

$$\hat{y}(t + N_1 + j|t) = \mathbf{G}_{N1_j}(z^{-1}) \triangle u(t + j)$$

$$+\mathbf{G}p_{N1_j}(z^{-1}) \triangle u(t - 1) + \mathbf{F}_{N_1+j}(z^{-1})y(t) \qquad (6.29)$$

Notice that the last two terms of the right-hand side of Equation (6.29) depend on past values of the process output and process input and correspond to the free response of the process when the control signals are kept constant, while the first term depends only on the future values of the control signal and can be interpreted as the forced response, that is, the response obtained when the initial conditions are zero. Equation (6.29) can be rewritten as:

$$\hat{y}(t + N_1 + j|t) = \mathbf{G}_{N1_j}(z^{-1}) \triangle u(t + j) + \mathbf{f}_{N_1+j}$$

If there is a significant difference between d_{max} and d_{min}, there will be a lot of zero entries in the coefficient matrices of polynomial matrix $\mathbf{B}(z^{-1})$ resulting in low computational efficiency. Costing horizons should be defined independently in order to obtain higher efficiency.

Let us consider the polynomial matrix $\mathbf{A}(z^{-1})$ to be diagonal (this can easily be done from the process transfer matrix and, as shown previously, has many advantages). The minimum delay from the input variables to the i output variable d_i is given by: $d_i = \min_j d_{ij}$. The minimum meaningful value for the lower limit of the prediction horizon for output variable y_i is $N_{1_i} = d_i + 1$. The upper limit $N_{2_i} = N_{1_i} + N_i - 1$ will mainly be dictated by polynomial $A_{ii}(z^{-1})$. Let us define the pertinent set of optimal j ahead output predictions $\mathbf{y} = [\mathbf{y}_1^T\, \mathbf{y}_2^T \cdots \mathbf{y}_n^T]^T$ with

$$\mathbf{y}_i = [\hat{y}_i(t + N_{1_i}|t)\, \hat{y}_i(t + N_{1_i} + 1|t) \ldots \hat{y}_i(t + N_{2_i}|t)]^T$$

Notice that the set of optimal j ahead predictions for the i output variable can be computed by solving a one-dimension Diophantine equation

$$1 = E_{ik}(z^{-1})\tilde{A}_{ii}(z^{-1}) + z^{-k}F_{ik}(z^{-1})$$

with $\tilde{A}_{ii}(z^{-1}) = A_{ii}(z^{-1})\triangle$. The optimum prediction for the i component of the output variable vector is then given by

$$y_i(t + N_{1_i} + k|t) = \sum_{j=1}^{m} E_{ik}(z^{-1})B_{ij}(z^{-1})\triangle u_j(t + N_{1_i} + k - 1) + F_{ik}(z^{-1})y_i(t)$$

If we make $B_{ij}(z^{-1}) = z^{-d_i}B'_{ij}(z^{-1})$

$$y_i(t + N_{1_i} + k|t) = \sum_{j=1}^{m} E_{ik}(z^{-1})B'_{ij}(z^{-1})\triangle u_j(t + k) + F_{ik}(z^{-1})y_i(t)$$

which can be expressed as

$$y_i(t + N_{1_i} + k|t) = \sum_{j=1}^{m} G_{ij_k}(z^{-1})\triangle u_j(t + k) + \sum_{j=1}^{m} Gp_{ij_k}(z^{-1})\triangle u_j(t - 1)$$

$$+ F_{ik}(z^{-1})y_i(t)$$

where

$$E_{ik}(z^{-1})B'_{ij}(z^{-1}) = G_{ij_k}(z^{-1}) + z^{-(k+1)}Gp_{ij_k}(z^{-1})$$

Let us define $\mathbf{f}_i$ as the free response of $y_i(t)$:

$$\mathbf{f}_i = [f_i(t + N_{1_i}) \cdots f_i(t + N_{2_i})]^T$$

with

$$f_i(t + N_{1_i} + k) = \sum_{j=1}^{m} Gp_{ij_k}(z^{-1})\triangle u_j(t - 1) + F_{ik}(z^{-1})y_i(t)$$

The output prediction affecting the objective function can be expressed as

$$
\begin{bmatrix} \mathbf{y}_1 \\ \mathbf{y}_2 \\ \vdots \\ \mathbf{y}_n \end{bmatrix} = \begin{bmatrix} G_{11} & G_{12} & \cdots & G_{1m} \\ G_{21} & G_{22} & \cdots & G_{2m} \\ \vdots & \vdots & \ddots & \vdots \\ G_{n1} & G_{n2} & \cdots & G_{nm} \end{bmatrix} \begin{bmatrix} \mathbf{u}_1 \\ \mathbf{u}_2 \\ \vdots \\ \mathbf{u}_m \end{bmatrix} + \begin{bmatrix} \mathbf{f}_1 \\ \mathbf{f}_2 \\ \vdots \\ \mathbf{f}_n \end{bmatrix}
$$

with $\mathbf{u}_j = [\triangle u_j(t) \ \triangle u_j(t+1) \ldots \triangle u_j(t+Nu_j)]^T$ and $Nu_j = \max_i(N_i - d_{ij} - 1)$. The $N_i \times Nu_j$ block matrix G_{ij} has the following form

$$
G_{ij} = \begin{bmatrix} 0 & 0 & \cdots & 0 & 0 & \cdots & 0 \\ \vdots & \vdots & \vdots & \vdots & \vdots & \vdots & \vdots \\ 0 & 0 & \cdots & 0 & 0 & \cdots & 0 \\ g_{ij_0} & 0 & \cdots & 0 & 0 & \cdots & 0 \\ g_{ij_1} & g_{ij_0} & \ddots & 0 & 0 & \cdots & 0 \\ \cdots & \cdots & \ddots & \vdots & \vdots & \vdots & \vdots \\ g_{ij_l} & g_{ij_{l-1}} & \cdots & g_{ij_0} & 0 & \cdots & 0 \end{bmatrix}
$$

where the number of leading zero rows of matrix G_{ij} is $d_{ij} - N_{1_i}$ and the number of trailing zero columns is $Nu_j - N_i + d_{ij}$. Note that the dimension of matrix $\mathbf{G}$ is $(\sum_{i=1}^{i=n} N_i) \times (\sum_{j=1}^{m} \max_i(N_i - d_{ij} - 1))$, while for the single delay case it is $(N \times n) \times ((N - d_{\min} - 1) \times m)$ with $N \geq N_i$ and $d_{\min} \leq d_{ij}$ in general. The reduction of the matrix dimension, and hence the computation required depends on how the delay terms are structured.

In spite of the problems related to dead time management that may appear when using a CARIMA model to represent plant dynamics, worse problems appear when other types of models are used. In the case of convolution models, a delay of d sampling periods is represented by the inclusion of d zero elements in the model, that is, the first d elements (g_i for step response models or h_i for impulse response models) are zero. It means that a lot of zero elements must be stored, which can lead to ill-conditioned problems.

In the case of a state space model, delays have to be addressed by augmenting the state vector in such a way that it contains all the necessary past inputs. If the nondelayed process can be represented by a state vector $x(t)$, the new state vector is now:

$$
\bar{x}^T(t) = [x^T(t) \ u^T(t-1) \ u^T(t-2) \ldots u^T(t-d)]
$$

Note that the input vector at each sampling time contains all the plant inputs and therefore the new model is much bigger than the original. In order to show the size increase of delayed state space models, consider a very simple case: a first-order monovariable plant with a delay of the order of magnitude of the time constant. It is clear that if no delay exists the dimension of the state vector is 1. If the plant is sampled at an adequate sampling rate (one-tenth of the time constant, for instance), inputs that happened 10 sampling times before have influence on the current state, and therefore $u(t-1)$, $u(t-2)$, up to $u(t-10)$ are now part of the augmented vector, whose dimension is 11. Notice that this is a considerable increment for this extremely simple case and that the extrapolation to multivariable processes with different delays associated to the outputs can drive to very high-dimension models.

6.8 Case Study: Distillation Column

In order to illustrate the problem of controlling multivariable processes with different dead times for the output variables we are going to consider the control of a distillation column.

The model chosen corresponds to a heavy oil fractionator and is referred to in literature as the Shell Oil's heavy oil fractionator [10], [50]. The model was first described by Prett and Morari [167] and has been widely used to try different control strategies for distillation columns.

The process, shown in Figure 6.10, has three variables that have to be controlled: the top and side product compositions, which are measured by analyzers, and the bottom temperature. The manipulated variables are the top draw rate, the side draw rate and the bottom reflux duty. The feed provides all heat requirements for the column. Top and side product specifications are fixed by economic and operational goals and must be kept within 0.5 % of their setpoint at steady state. The bottom temperature must be controlled within limits fixed by operational constraints. The top endpoint must be maintained within the maximum and minimum values of -0.5 and 0.5. The manipulated variables are also constrained as follows: all draws must be within hard minimum and maximum bounds of -0.5 and 0.5. The bottom reflux duty is also constrained by -0.5 and 0.5. The maximum allowed slew rates for all manipulated variables is 0.05 per minute. The dynamics of the process can be described by the following

$$
\begin{bmatrix} Y_1(s) \\ Y_2(s) \\ Y_3(s) \end{bmatrix} = \begin{bmatrix} \dfrac{4.05e^{-27s}}{1+50s} & \dfrac{1.77e^{-28s}}{1+60s} & \dfrac{5.88e^{-27s}}{1+50s} \\[3mm] \dfrac{5.39e^{-18s}}{1+50s} & \dfrac{5.72e^{-14s}}{1+60s} & \dfrac{6.9e^{-15s}}{1+40s} \\[3mm] \dfrac{4.38e^{-20s}}{1+33s} & \dfrac{4.42e^{-22s}}{1+44s} & \dfrac{7.2}{1+19s} \end{bmatrix} \begin{bmatrix} U_1(s) \\ U_2(s) \\ U_3(s) \end{bmatrix}
$$

where $U_1(s)$, $U_2(s)$ and $U_3(s)$ correspond to the top draw, side draw and bottom reflux duties and $Y_1(s)$, $Y_2(s)$ and $Y_3(s)$ correspond to the top endpoint composition, side end point compositions and bottom reflux temperature, respectively.

Notice that the minimum dead time for the three output variables are 27, 14 and 0 minutes, respectively.

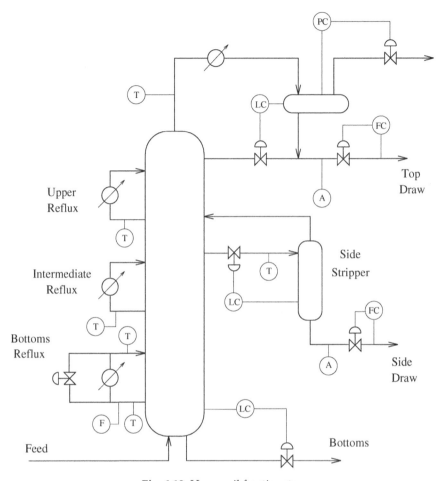

Fig. 6.10. Heavy oil fractionator

The discrete transfer matrix for a sampling time of 4 minutes is:

$$
\begin{bmatrix}
\dfrac{0.08(z^{-1}+2.88z^{-2})}{1-0.923z^{-1}}z^{-6} & \dfrac{0.114z^{-1}}{1-0.936z^{-1}}z^{-7} & \dfrac{0.116(z^{-1}+2.883z^{-2})}{1-0.923z^{-1}}z^{-6} \\[4mm]
\dfrac{0.211(z^{-1}+0.96z^{-2})}{1-0.923z^{-1}}z^{-4} & \dfrac{0.187(z^{-1}+0.967z^{-2})}{1-0.936z^{-1}}z^{-3} & \dfrac{0.17(z^{-1}+2.854z^{-2})}{1-0.905z^{-1}}z^{-3} \\[4mm]
\dfrac{0.5z^{-1}}{1-0.886z^{-1}}z^{-5} & \dfrac{0.196z^{-1}+0.955z^{-2}}{1-0.913z^{-1}}z^{-5} & \dfrac{1.367z^{-1}}{1-0.81z^{-1}}
\end{bmatrix}
$$

A left matrix fraction description can be obtained by making matrix $\mathbf{A}(z^{-1})$ equal to a diagonal matrix with diagonal elements equal to the least common multiple of the denominators of the corresponding row of the transfer function, resulting in:

$$A_{11}(z^{-1}) = 1 - 1.859z^{-1} + 0.8639z^{-2}$$
$$A_{22}(z^{-1}) = 1 - 2.764z^{-1} + 2.5463z^{-2} - 0.7819z^{-3}$$
$$A_{33}(z^{-1}) = 1 - 2.609z^{-1} + 2.2661z^{-2} - 0.6552z^{-3}$$
$$B_{11}(z^{-1}) = (0.08 + 0.155z^{-1} - 0.216z^{-2})z^{-6}$$
$$B_{12}(z^{-1}) = (0.114 - 0.105z^{-1})z^{-7}$$
$$B_{13}(z^{-1}) = (0.116 + 0.226z^{-1} - 0.313z^{-2})z^{-6}$$
$$B_{21}(z^{-1}) = (0.211 - 0.186z^{-1} - 0.194z^{-2} + 0.172z^{-3})z^{-4}$$
$$B_{22}(z^{-1}) = (0.187 - 0.161z^{-1} - 0.174z^{-2} + 0.151z^{-3})z^{-3}$$
$$B_{23}(z^{-1}) = (0.17 + 0.169z^{-1} - 0.755z^{-2} + 0.419z^{-3})z^{-4}$$
$$B_{31}(z^{-1}) = (0.5 - 0.8615z^{-1} + 0.369z^{-2})z^{-5}$$
$$B_{32}(z^{-1}) = (0.196 + 0.145z^{-1} - 1.77z^{-2} + 0.134z^{-3})z^{-5}$$
$$B_{33}(z^{-1}) = 1.367 - 2.459z^{-1} + 1.105z^{-2}$$

The minimum pure delay time for each of the output variables expressed in sampling time units are 6, 3 and 0, respectively. The results obtained when applying the multivariable GPC can be seen in Figure 6.11 for the case of a common prediction horizon of 30 and a control horizon of 5 for all the variables. The weighting matrices were chosen to be $Q = I$ and $R = 2\,I$. The reference trajectories are supposed to be equal to the actual setpoints: 0.5, 0.3 and 0.1, respectively. A change was produced in the setpoint of the top endpoint composition from 0.5 to 0.4 in the middle of the simulation.

The control increments needed could be, however, too big to be applied in reality and all the manipulated variables were saturated to the hard bounds described earlier. As can be seen, all the variables reach the setpoint quite rapidly and only the bottom reflux temperature exhibits a significant overshoot. The perturbations produced in the side endpoint composition and on the bottom reflux temperature due to the setpoint change of the top endpoint composition are quite small, indicating a low closed-loop interaction degree between the setpoint and controlled variables, in spite of the highly coupled open-loop dynamics.

The upper and intermediate reflux duties are considered to act as unmeasurable disturbances. The small-signal dynamic load model for the upper reflux duty is given by the following transfer functions:

$$\begin{bmatrix} Y_{p1}(s) \\ Y_{p2}(s) \\ Y_{p3}(s) \end{bmatrix} = \begin{bmatrix} \dfrac{1.44e^{-27s}}{1 + 40s} \\[2ex] \dfrac{1.83e^{-15s}}{1 + 20s} \\[2ex] \dfrac{1.26}{1 + 32s} \end{bmatrix} U(s)$$

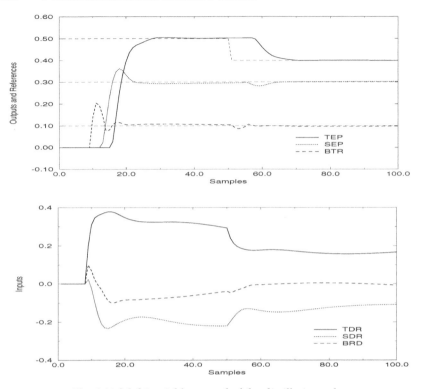

Fig. 6.11. Multivariable control of the distillation column

A step perturbation of 0.5 is introduced in the upper reflux duty, keeping all the setpoints at zero. The results obtained when applying the GPC with the previous weighting matrices are shown in Figure 6.12.

As can be seen, the perturbations are very rapidly cancelled in spite of the high load perturbations (the steady-state value for the load perturbation in the side endpoint composition is 0.91).

6.9 Multivariable MPC and Transmission Zeros

It is well known by control engineers that precautions have to be taken to avoid instability when controlling processes with *unstable* zeros[2], i.e. zeros outside the unit disk (OUD). Instability arises when, in order to achieve high performance, the controller contains an inverse of the process model. The OUD zeros are cancelled by OUD controller poles resulting in an internally unstable system. Model Predictive Control is one of those controllers which aims to achieve high performance, specially when the penalization of the

---

[2] Poles and not zeros are the cause of instability. However we are going to refer to *unstable* zeros as those zeros outside the unit circle.

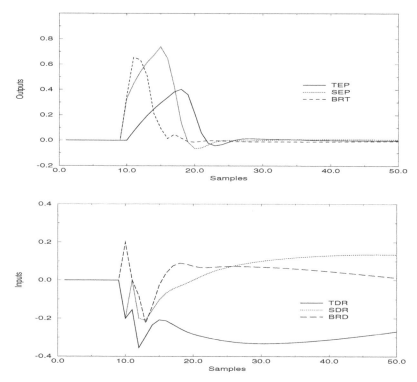

Fig. 6.12. Effect of a perturbation in the upper reflux duty

control effort is zero (aiming for perfect control whatever the cost in moving the manipulated variable). MPC can be considered an extension of minimum variance control (in fact, minimum variance control is a particular form of MPC when $N_1 = N_2 = d$, $N_u = 1$ and $\lambda = 0$). It is well known that minimum variance controllers cannot be applied to SISO nonminimum phase systems because an unstable closed loop is obtained. The reason is that nonminimum phase systems have OUD zeros which appear as unstable poles of the minimum variance controller. The idea of penalizing the control effort (extended minimum variance) came out to cope with this problem.

The instability problems arising when applying GPC to SISO systems with OUD zeros have been reported in literature. It was shown in [58] and [131] that nonminimum phase systems produce instability when $N_2 = 1$, $N_u = 1$. Instability in GPC has also been reported in [84], [123] and [194] when it is applied to nonminimum phase systems. A common practice to avoid this problem is increase the control weight as suggested in [27] and [59]. The way in which OUD transmission zeros affect the MPC behaviour is less understood in the Multiple Input Multiple Output (MIMO) case. In [141] the control limitations on closed-loop behaviour imposed by the process zeros are analyzed.

Extending the SISO concept of a zero to the MIMO case is not a trivial problem. In fact many definitions of MIMO zeros exist in literature. The definition that we are going to use throughout the text is that transmission zeros correspond to the poles of the inverse system. For square open-loop stable MIMO systems the transmission zeros are the complex values that makes the determinant of the system process matrix equal to zero. While the zeros of SISO processes can be detected by a simple inspection of the process transfer function, this cannot be done for the MIMO processes, where transmission zeros are not the zeros of the individual transfer functions. Transmission zeros of MIMO systems are, from that point of view, hidden dynamics. That is, it is not possible to detect transmission zeros of MIMO processes by a simple inspection of the transfer matrix.

Furthermore, in the case of step response models, widely used by industry, *unstable* transmission zeros are even more difficult to detect. A typical step response will contain around 50 terms. For a 5×5 MIMO system, the determinant of the transfer matrix will consist of a polynomial with degree equal to 250 which will have to be solved to determine its transmission zeros.

This section shows why the instability problem can be present for any MIMO system with OUD transmission zeros, when the control horizon has the same value as the prediction horizon and the control weight is equal to zero [76]. The section also indicates how can the problem be solved using an adequate predictive horizon, control horizon and the control weight as shown in [76].

In order to analyze how transmission zeros affect the closed-loop poles, consider a GPC controlling an (n-input $\times$ n-output) MIMO plant. The vector of future control moves, as seen, before can be computed as

$$\mathbf{u} = K(\mathbf{w} - \mathbf{f}) \tag{6.30}$$

with $K = (\mathbf{G}^T\mathbf{G} + \lambda\mathbf{I})^{-1}\mathbf{G}^T$.

The j component f_j of the free response $\mathbf{f}$ can be computed as $f_j = \mathbf{G}_{jp}(z^{-1})\Delta u(t-1) + \mathbf{F}_j(z^{-1})y(t)$. To compute the first control move $\Delta u(t)$, only the n first rows of K (6.30) are needed. Let us consider the matrices α_i ($n \times n$) composing the first n rows of matrix K in (6.30). Then

$$\Delta u(t) = \sum_{i=1}^{N_2} \alpha_i z^{-N_2+i} w(t+N_2)$$

$$- \sum_{i=1}^{N_2} \alpha_i \mathbf{G}_{ip}\Delta u(t-1) - \sum_{i=1}^{N_2} \alpha_i \mathbf{F}_i y(t) \tag{6.31}$$

Substituting $\Delta u(t)$ into the CARIMA model equation, the closed-loop relationship can be obtained

$$(\tilde{\mathbf{A}} + \mathbf{B} \sum_{i=1}^{N_2} \alpha_i \mathbf{G}_{ip} z^{-1} \mathbf{B}^{-1} \tilde{\mathbf{A}} + \mathbf{B} \sum_{i=1}^{N_2} \alpha_i \mathbf{F}_i z^{-1}) y(t)$$
$$= \mathbf{B} \sum_{i=1}^{N_2} \alpha_i z^{-N_2+i} w(t + N_2 - 1)$$
$$(\mathbf{I} + \mathbf{B} \sum_{i=1}^{N_2} \alpha_i \mathbf{G}_{ip} z^{-1} \mathbf{B}^{-1}) \xi(t) \qquad (6.32)$$

which shows that the closed-loop $\tilde{\mathbf{A}}_{CL}$ matrix is given by

$$\tilde{\mathbf{A}}_{CL} = \tilde{\mathbf{A}} + \mathbf{B} \sum_{i=1}^{N_2} \alpha_i \Gamma_i z^{-1} \mathbf{B}^{-1} \tilde{\mathbf{A}} + \mathbf{B} \sum_{i=1}^{N_2} \alpha_i \mathbf{F}_i z^{-1} \qquad (6.33)$$

When the control weight is zero ($\lambda = 0$), (6.30) can be written as

$$\mathbf{u} = (\mathbf{G}^T \mathbf{G})^{-1} \mathbf{G}^T (\mathbf{w} - \mathbf{f}) \qquad (6.34)$$

Also, if the prediction horizon and the control horizon have the same value, $N_u = N_2 - N_1 + 1$, $(\mathbf{G}^T \mathbf{G})^{-1} \mathbf{G}^T$ is given by $\mathbf{G}^{-1}$, whose first n rows are:

$$\left[\mathbf{G}_0^{-1}, 0, 0, \ldots, 0 \right] \qquad (6.35)$$

That is, $\alpha_1 = \mathbf{G}_0^{-1}$ and $\alpha_i = 0$ for $i \neq 1$. Consequently, using (6.33) and (6.35)

$$\tilde{\mathbf{A}}_{CL} = \tilde{\mathbf{A}} + \mathbf{B} \mathbf{G}_0^{-1} \mathbf{G}_{ip} z^{-1} \mathbf{B}^{-1} \tilde{\mathbf{A}} + \mathbf{B} \mathbf{G}_0^{-1} \mathbf{F}_1 z^{-1} \qquad (6.36)$$

Recalling from Section 6.1.1 that $\mathbf{F}_1 = z(\mathbf{I} - \tilde{\mathbf{A}})$, $\mathbf{E}_1 = \mathbf{I}$, $\mathbf{G}_1 = \mathbf{G}_0 + \mathbf{G}_{1p} z^{-1} \Rightarrow \mathbf{G}_{1p} z^{-1} = \mathbf{B} - \mathbf{G}_0$ and introducing them in (6.36):

$$\tilde{\mathbf{A}}_{CL} = \tilde{\mathbf{A}} + \mathbf{B} \mathbf{G}_0^{-1} (\mathbf{B} - \mathbf{G}_0) \mathbf{B}^{-1} \tilde{\mathbf{A}} + \mathbf{B} \mathbf{G}_0^{-1} z(\mathbf{I} - \tilde{\mathbf{A}}) z^{-1}$$
$$= \tilde{\mathbf{A}} + \mathbf{B} \mathbf{G}_0^{-1} \mathbf{B} \mathbf{B}^{-1} \tilde{\mathbf{A}} - \mathbf{B} \mathbf{G}_0^{-1} \mathbf{G}_0 \mathbf{B}^{-1} \tilde{\mathbf{A}} + \mathbf{B} \mathbf{G}_0^{-1} z(\mathbf{I} - \tilde{\mathbf{A}}) z^{-1}$$
$$= \mathbf{B} \mathbf{G}_0^{-1} \qquad (6.37)$$

The reference to output closed-loop dynamic is given by:

$$\mathbf{B}(z^{-1}) \mathbf{G}_0^{-1} y(t) = \mathbf{B}(z^{-1}) \mathbf{G}_0^{-1} w(t) \qquad (6.38)$$

which shows that the relationship between outputs and references is the identity matrix; consequently there will be internal instability due to the cancellation of the OUD zeros of the process with the OUD poles of the controller.

Using the polynomial matrices $\mathbf{R}(z^{-1})$, $\mathbf{S}(z^{-1})$ and $\mathbf{T}(z^{-1})$ as shown in Figure 6.13, Equation (6.31) can be written as

$$\mathbf{R} \Delta u(t) = -\mathbf{S} y(t) + \mathbf{T} w(t) \qquad (6.39)$$

The closed-loop equation from $e(t)$ to $u(t)$ is given by

$$\mathbf{R} \Delta u(t) = \mathbf{T} e(t) \qquad (6.40)$$

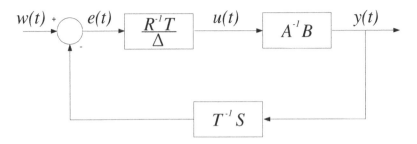

Fig. 6.13. Closed-loop block diagram

The polynomial matrix $\mathbf{R}(z^{-1})$ contains the poles of the controller; when $\lambda = 0$ and $N_u = N_2 - N_1 + 1$ this can be written as

$$\mathbf{R} = (\mathbf{I} + \sum_{i=1}^{N_2} \alpha_i \Gamma_i z^{-1}) = \mathbf{G}_0^{-1}\mathbf{B}(z^{-1}) \tag{6.41}$$

Consequently, matrix $\mathbf{B}(z^{-1})$ containing the OUD zero of plant model (2.4) is included in $\mathbf{R}$ and, therefore, the controller is unstable because some of its poles are outside the unit disc.

In the case of convolution models, which are frequently employed in industry in the MPC context, the OUD transmission zeros will appear as controller poles when $\lambda = 0$ and the prediction horizon and the number of control moves are equal.

Let us consider an (n-input $\times$ n-output) MIMO plant. When the control weight is zero ($\lambda = 0$), the future control move sequence can be,

$$\mathbf{u}_+ = (\mathbf{H}_1^T\mathbf{H}_1)^{-1}\mathbf{H}_1^T(\mathbf{w} - \mathbf{f}) \tag{6.42}$$

Recall that the prediction equations are $\mathbf{y} = \mathbf{H}_1\mathbf{u}_+ + \mathbf{H}_2\mathbf{u}_- + y(t)$. The free response $\mathbf{f}$ can be calculated as $\mathbf{f} = \mathbf{H}_2\mathbf{u}_- + y(t)$. By substituting (6.42) into the process model and considering equal control and cost horizons $N_u = N_2 - N_1 + 1$ (and $\mathbf{H}_1$ is non singular),

$$\mathbf{u}_+ = \mathbf{H}_1^{-1}(\mathbf{w} - \mathbf{H}_2\mathbf{u}_- - y(t)) = \mathbf{H}_1^{-1}(\mathbf{w} - \mathbf{y} + \mathbf{H}_1\mathbf{u}_+) \Rightarrow \mathbf{H}_1^{-1}\mathbf{w} = \mathbf{H}_1^{-1}\mathbf{y} \tag{6.43}$$

which shows that the relationship between outputs and references is the identity matrix and consequently it will be internally unstable due to cancellation of unstable transmission zeros of the process.

6.9.1 Simulation Example

As an example we are going to consider a nonlinear quadruple-tank process [100] shown in Figure 6.14. The target is to control the level of the two lower tanks with two pumps. The process inputs are v_1 and v_2 and the process outputs are y_1 and y_2. The model based on mass balances and Bernoulli's law is given by

$$\frac{dh_1}{dt} = -\frac{a_1}{A_1}\sqrt{2gh_1} + \frac{a_3}{A_1}\sqrt{2gh_3} + \frac{\gamma_1 k_1}{A_1}v_1$$

$$\frac{dh_2}{dt} = -\frac{a_2}{A_2}\sqrt{2gh_2} + \frac{a_4}{A_2}\sqrt{2gh_4} + \frac{\gamma_2 k_2}{A_2}v_2$$

$$\frac{dh_3}{dt} = -\frac{a_3}{A_3}\sqrt{2gh_3} + \frac{(1-\gamma_2)k_2}{A_3}v_2$$

$$\frac{dh_4}{dt} = -\frac{a_4}{A_4}\sqrt{2gh_4} + \frac{(1-\gamma_1)k_1}{A_4}v_1 \tag{6.44}$$

where A_i is the cross section of the tank, a_i is the cross section of the hole, h_i is the water level and g is the acceleration of gravity. The control signals are voltages (v_i) applied to pumps i and the corresponding flows are $k_i v_i$. The parameters $\gamma_1, \gamma_2 \in (0, 1)$ are how the valves are set. The flow to tank 1 is $\gamma_1 k_1 v_1$ and the flow to tank 4 is $(1 - \gamma_1)k_1 v_1$ and similarly for tanks 2 and 3. The measured level signals y_1 and y_2 are $k_c h_1$ and $k_c h_2$. The parameter values of the process are given in Table 6.2.

Table 6.2. Parameter values of the process

A_1, A_2	cm^2	28.000
A_2, A_4	cm^2	32.000
a_1, a_3	cm^2	0.071
a_2, a_4	cm^2	0.057
k_c	$\frac{V}{cm}$	0.500
g	$\frac{cm}{s^2}$	981.000

The nonlinear process is studied at two operating points: P_- at which the system has all its transmission zeros inside the unit disk and P_+ at which the process has *unstable* transmission zeros. The initial conditions of the chosen operating points are shown in Table 6.3. After linearizing the model (6.44) about the operating points P_- and P_+, the physical model gives the two following transfer matrices:

$$G_-(s) = \begin{bmatrix} \dfrac{2.6}{62s+1} & \dfrac{1.5}{(23s+1)(62s+1)} \\[3mm] \dfrac{1.4}{(30s+1)(90s+1)} & \dfrac{2.8}{90s+1} \end{bmatrix} \tag{6.45}$$

and

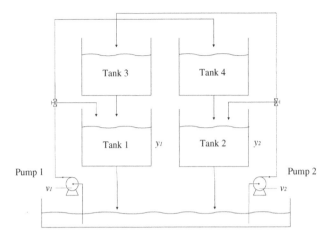

Fig. 6.14. Schematic diagram of the quadruple-tank process

Table 6.3. Initial conditions

		P_-	P_+
h_1^0, h_2^0	cm	$(12.4, 12.7)$	$(12.44, 13.17)$
h_3^0, h_4^0	cm	$(1.8, 1.4)$	$(4.73, 4.98)$
v_1^0, v_2^0	V	$(3.00, 3.00)$	$(3.15, 3.15)$
k_1, k_2	$\frac{cm^3}{Vs}$	$(3.33, 3.35)$	$(3.14, 3.29)$
γ_1, γ_2		$(0.70, 0.60)$	$(0.43, 0.34)$

$$G_+(s) = \begin{bmatrix} \dfrac{1.5}{63s+1} & \dfrac{2.5}{(39s+1)(63s+1)} \\[3mm] \dfrac{2.5}{(56s+1)(91s+1)} & \dfrac{1.6}{91s+1} \end{bmatrix} \qquad (6.46)$$

The discretization has been made with a sampling rate $T_s = 2s$. At first
sight, systems (6.45) and (6.46) are quite similar and the control engineer can
expect transmission zeros in both systems to be close; however, the two operating points P_- and P_+ have the transmission zeros showed in Table 6.4. The
presence of the OUD zeros is not a rare situation in multivariable processes.
In order to illustrate the problem, Figure 6.15 shows the performance of the

Table 6.4. Transmission zeros

	P_-	P_+
Zeros(s)	$(-0.0595, -0.0173)$	$(-0.0565, 0.0130)$
Zeros(z)	$(0.8878, 0.9660)$	$(0.8932, 1.0263)$

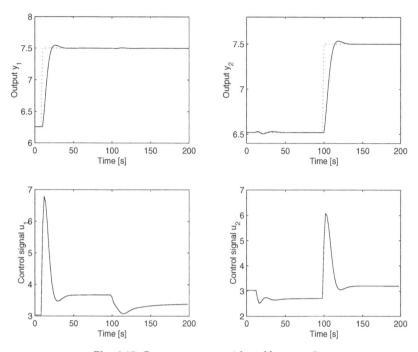

Fig. 6.15. Output process with *stable* zeros G_-

GPC at operating point P_-. The control level of tanks y_1 and y_2 have a good response without oscillations for changes in the level set points. That is, the GPC can control the four tanks without problems at this operating point.

Let us now consider that the operating point is changed to P_+, where the process linear model exhibits OUD transmission zeros. Figure 6.16 shows the GPC at operating point P_+, the controller has been tuned with the same parameters used in P_-: $N_1 = 1$, $N_2 = 40$, $N_u = 10$ and $\lambda = 0.1$ and a perfect linear model. It can be seen how a transmission zero outside the unit circle in (6.46) deteriorates the performance considerably, making the controller in Figure 6.16 unstable. A new tuning around the usual parameters of the Generalized Predictive Controller such as was used at operating point P_- cannot achieve a stable behaviour in MIMO systems with OUD zeros.

6.9.2 Tuning MPC for Processes with OUD Zeros

It would seem that the obvious way of avoiding the cancellation of the OUD zeros would be by increasing the control weight λ. However, it has been shown that the zero does not get cancelled exactly by a controller pole when $\lambda \neq 0$, however, poles close to the process zeros appear. By increasing λ the poles start shifting into the unit circle but the price that has to be paid is a

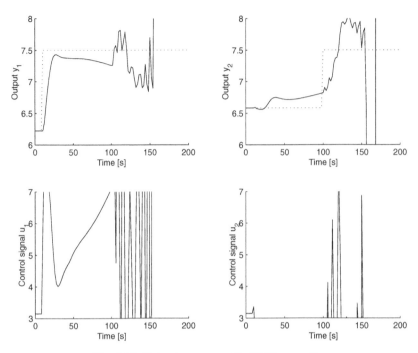

Fig. 6.16. Output process with OUD zeros G_+

very high λ and a very sluggish controller, as shown in [75], where it is also shown that more effective results are obtained for MIMO systems with OUD zeros by setting a big predictive horizon and a short control horizon.

An important characteristic of OUD transmission zeros of MIMO systems is that they are hidden. It is not possible to detect an OUD zero by a simple inspection. The poles of MIMO systems have a clear relationship with the pole of each transfer function but the zeros of multivariable processes are more subtle characteristics of the systems that are not easy to discern. This is even more difficult with convolution models extensively used in industry. Models that may look very simple (see Exercise 6.1) may have this hidden trap inside.

In order to achieve stable behaviour for MPC, the rule of thumb is to tune the GPC with a big predictive horizon more than twice the time constant of the OUD zero, the control horizon must be smaller than the predictive horizon, and the difference between the predictive and the control horizon helps avoid the situation shown in [74]. Control weight λ must be big enough to ensure soft control signals. Figure 6.17 shows a GPC with the process at operating point P_+ and the following tuning parameters: $N_1 = 1$, $N_2 = 78$, $N_u = 20$ and $\lambda = 25$. It can be seen how the new parameters achieve stable behaviour, though the performance is quite slow compared to the response obtained at operating point P_-.

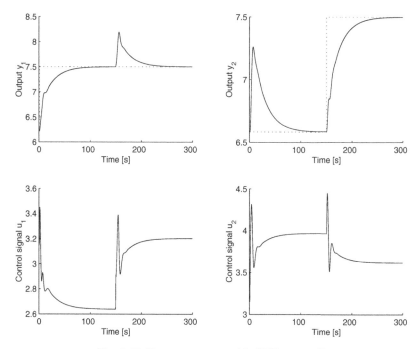

Fig. 6.17. Output process with OUD zeros G_+

Control engineers should be aware of the presence of OUD transmission zeros in MIMO plants. This kind of system deteriorates the performance of MPC, and instability problems may arise when MPC tries to achieve its performance by cancelling the OUD zeros with OUD poles. Increasing the control weight, as frequently proposed in literature, may help to avoid instability but the price to be paid is a sluggish controller. Using other stabilization techniques such a terminal weight or using a cost horizon much higher than the control horizon (which is somewhat similar to using a terminal weight) has proven to be more effective.

6.10 Exercises

6.1. Consider the stirred tank reactor (Figure 6.1) where the manipulated variables $U_1(s)$ and $U_2(s)$ are the feed flow rate and the flow of coolant in the jacket, respectively. The controlled variables $Y_1(s)$ and $Y_2(s)$ are the effluent concentration and the reactor temperature, respectively.

The continuous model is:

$$\begin{bmatrix} y_1(t) \\ y_2(t) \end{bmatrix} = \begin{bmatrix} \frac{1}{1+0.1s} & \frac{5}{1+s} \\ \frac{1}{1+0.5s} & \frac{2}{1+0.4s} \end{bmatrix} \begin{bmatrix} u_1(t) \\ u_2(t) \end{bmatrix}$$

1. Choose an appropriate sampling time (5 times smaller than the fastest time constant) and appropriate control horizons. Control the reactor with control weight $\lambda = 0$ and add a little noise. Use a long simulation time and comment on the results.
2. Compute the transmission zeros by solving $|G(s)| = 0$. Comment on the results obtained in the previous point.
3. Change λ and the rest of the GPC parameters and discuss the simulated results.

6.2. Obtain a step response model of the previous plant and apply a DMC with $\lambda = 0$. Comment on the results.

6.3. For the same plant: apply the tuning rules given at the end of Section 6.9.2 and comment on the results.

6.4. The continuous model of a 747 at landing configuration (weight of 564,000 lb and $u_0 = 221$ ft/sec) is given by:

$$
\begin{bmatrix} \dot{u} \\ \dot{w} \\ \dot{q} \\ \dot{\theta} \end{bmatrix} = \begin{bmatrix} -0.021 & 0.122 & 0 & -0.322 \\ -0.209 & -0.530 & 2.21 & 0 \\ 0.017 & -0.164 & -0.412 & 0 \\ 0 & 0 & 1 & 0 \end{bmatrix} \begin{bmatrix} u - u_w \\ w - w_w \\ q \\ \theta \end{bmatrix} + \begin{bmatrix} 0.010 & 1 \\ -0.064 & -0.44 \\ -0.378 & 0.544 \\ 0 & 0 \end{bmatrix} \begin{bmatrix} e \\ t \end{bmatrix}
$$

1. Obtain the discrete model for a sampling time of 0.1 second.
2. Use the state space controller to simulate the effect of a change in the airspeed setpoint from 0 to 10 ft/sec (without changes in the climb rate setpoint).
3. Use the state space controller to simulate the effect of a change in the climb-rate setpoint from 0 to 5 ft/sec without changes in the airspeed setpoint).
4. Compare the results with those obtained in Section 6.4.

6.5. Given the airplane model of Section 6.4:

1. Obtain the step response to both inputs and calculate the g_i^{kj} coefficients.
2. Use the companion software to implement a DMC and repeat the experiments done in Section 6.4.
3. Repeat the experiments when the setpoint for climb-rate is a ramp of 10 ft/sec per second from $t = 0$ to 60 samples and is kept constant until the end of the experiment. The airspeed setpoint remains 0 all the time.

7

Constrained Model Predictive Control

The control problem was formulated in the previous chapters considering all signals to possess an unlimited range. This is not very realistic because in practice all processes are subject to constraints. Actuators have a limited range of action and a limited slew rate, as is the case of control valves limited by a fully closed and fully open position and a maximum slew rate. Constructive or safety reasons, as well as sensor range, cause bounds in process variables, as in the case of levels in tanks, flows in pipes, and pressures in deposits. Furthermore, in practice, the operating points of plants are determined to satisfy economic goals and lie at the intersection of certain constraints. The control system normally operates close to the limits and constraint violations are likely to occur. The control system, especially for long-range predictive control, has to anticipate constraint violations and correct them in an appropriate way. Although input and output constraints are basically treated in the same way, as is shown in this chapter, the implications of the constraints differ. Output constraints are mainly due to safety reasons and must be controlled in advance because output variables are affected by process dynamics. Input (or manipulated) variables can always be kept in bound by the controller by clipping the control action to a value satisfying amplitude and slew rate constraints.

This chapter concentrates on how to implement generalized predictive controllers for processes with constrained input (amplitude and/or slew rate) and output signals.

7.1 Constraints and MPC

Recall that the MPC control actions were calculated by computing vector u of future control increments to minimize a quadratic objective function given by:

$$J(\mathbf{u}) = \frac{1}{2}\,\mathbf{u}^T\mathbf{H}\mathbf{u} + \mathbf{b}\mathbf{u} + \mathbf{f}_0 \qquad (7.1)$$

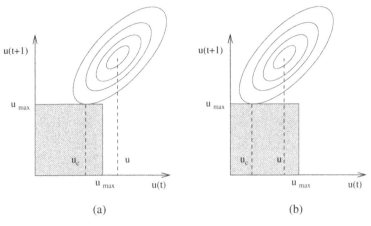

Fig. 7.1. Constraints on the control signal

The optimal solution of this problem is found by solving the linear equation:

$$\mathbf{Hu} = -\mathbf{b}$$

In practice, the normal way of using an MPC is to compute $u(t)$ as previously described and apply it to the process. If $u(t)$ violates the constraint it is saturated to its limits, by either the control program or the actuator. The case of $u(t + 1), \cdots, u(t + N)$ violating the constraints is not even considered, as in most cases these signals are not even computed.

This way of operating does not guarantee that the optimum will be obtained when constraints are violated. The main purpose of GPC, which is to apply the *best* possible control signal by minimizing expression (7.1), will not be achieved.

To illustrate this point, consider the cases of constraint violations shown in Figure 7.1 of an MPC problem with a control horizon of two. Figure 7.1(a) shows the case where $u(t) > u_{\max}$. In this case the normal way of operating would be to apply $u_{\max}$ to the process instead of u_c where the minimum of J is reached when restrictions are considered. In the case shown in figure 7.1(b), $u(t)$ does not violate the constraints, and it would be applied to the system instead of the optimum signal u_c that should be applied when constraints are taken into account.

Not considering constraints on manipulated variables to their full extent may result in higher values of the objective function and thus in poorer performance. However, manipulated variables can always be kept to their limits by either the control program or the actuator, and this is not the main reason for treating constraints in an appropriate way.

Violating the limits on the controlled variables may be more costly and dangerous as it could cause damage to equipment and losses in production. For example, in most batch reactors the quality of the production requires some of the variables to be kept within specified limits; violating these limits may create a bad-quality product and in some cases the loss of the whole batch. When the limits have been set because of safety reasons, the violation of these limits could cause damage to equipment, or spillage, or in most cases the activation of the emergency system which will normally produce an emergency stop of the process, losing, or delaying production, and a normally costly startup procedure.

Constraint violations on the output variables are not contemplated when the only way of handling constraints is by clipping the manipulated variables. One of the main advantages of MPC, its prediction capabilities, is not used to its full potential by this way of operating. Control systems, especially long-range predictive control, should anticipate constraint violations and correct them in an appropriate way.

The constraints acting on a process can originate from amplitude limits in the control signal, slew rate limits of the actuator and limits on the output signals, and can be described, respectively, by:

$$\underline{U} \leq u(t) \leq \overline{U} \quad \forall t$$
$$\underline{u} \leq u(t) - u(t-1) \leq \overline{u} \quad \forall t$$
$$\underline{y} \leq y(t) \leq \overline{y} \quad \forall t$$

For an m-input n-output process with constraints acting over a receding horizon N, these constraints can be expressed as:

$$\mathbf{1}\,\underline{U} \leq T\mathbf{u} + u(t-1)\,\mathbf{1} \leq \mathbf{1}\,\overline{U}$$
$$\mathbf{1}\,\underline{u} \leq \mathbf{u} \leq \mathbf{1}\,\overline{u}$$
$$\mathbf{1}\,\underline{y} \leq \mathbf{Gu} + \mathbf{f} \leq \mathbf{1}\,\overline{y}$$

where $\mathbf{1}$ is an $(N \times n) \times m$ matrix formed by N $m \times m$ identity matrices and T is a lower triangular block matrix whose nonnull block entries are $m \times m$ identity matrices. The constraints can be expressed in condensed form as:

$$\mathbf{R}\,\mathbf{u} \leq \mathbf{c}$$

with:

$$R = \begin{bmatrix} I_{N \times N} \\ -I_{N \times N} \\ T \\ -T \\ G \\ -G \end{bmatrix} \qquad \mathbf{c} = \begin{bmatrix} \mathbf{1}\,\overline{u} \\ -\mathbf{1}\,\underline{u} \\ \mathbf{1}\,\overline{U} - \mathbf{1}u(t-1) \\ -\mathbf{1}\,\underline{U} + \mathbf{1}u(t-1) \\ \mathbf{1}\,\overline{y} - \mathbf{f} \\ -\mathbf{1}\,\underline{y} + \mathbf{f} \end{bmatrix}$$

The constraints on the output variables of the type $\underline{y} \leq y(t) \leq \overline{y}$ are normally imposed because of safety reasons. Other types of constraint can be set on the process-controlled variables to force the response of the process to have certain characteristics, as shown in [114], and can also be expressed in a similar manner.

Band Constraints

Sometimes one wishes the controlled variables to follow a trajectory within a band. In the food industry, for example, it is very usual for some operations to require a temperature profile that has to be followed with a specified tolerance.

This type of requirement can be introduced in the control system by forcing the output of the system to be included in the band formed by the specified trajectory plus or minus the tolerance. That is:

$$\underline{y}(t) \leq y(t) \leq \overline{y}(t)$$

These constraints can be expressed in terms of the increments of the manipulated variables as follows:

$$\mathbf{Gu} \leq \overline{\mathbf{y}} - \mathbf{f}$$
$$\mathbf{Gu} \geq \underline{\mathbf{y}} - \mathbf{f}$$

Overshoot Constraints

In some processes overshoots are not desirable for different reasons. In the case of manipulators, for example, an overshoot may produce a collision with the workplace or with the piece it is trying to grasp.

Overshoot constraints have been treated in [114] and are very easy to implement. Every time a change is produced in the setpoint, which is considered to be kept constant for a sufficiently long period, the following constraints are added to the control system

$$y(t + j) \leq w(t) \text{ for } j = N_{o1} \cdots N_{o2}$$

where N_{o1} and N_{o2} define the horizon where the overshoot may occur (N_{o1} and N_{o2} can always be made equal to 1 and N if this is not known). These constraints can be expressed in terms of the increments of the manipulated variables as follows:

$$\mathbf{Gu} \leq \mathbf{1}w(t) - \mathbf{f}$$

Monotonic Behaviour

Some control systems tend to exhibit oscillations, known as kickback, on the controlled variables before they have gone over the setpoints. These oscillations are not desirable in general because, amongst other reasons, they may cause perturbations in other processes. Constraints can be added to the control system to avoid this type of behaviour by imposing a monotonic behaviour on the output variables. Each time a setpoint changes and is again considered to be kept constant for a sufficiently long period, new constraints with the following form are added to the control system:

$$y(t + j) \leq y(t + j + 1) \text{ if } y(t) < w(t)$$
$$y(t + j) \geq y(t + j + 1) \text{ if } y(t) > w(t)$$

These type of constraints can be expressed in terms of the manipulated variables as follows

$$\mathbf{Gu} + \mathbf{f} \leq \begin{bmatrix} \mathbf{0}^T \\ \mathbf{G'} \end{bmatrix} \mathbf{u} + \begin{bmatrix} y(t) \\ \mathbf{f'} \end{bmatrix}$$

where $\mathbf{G'}$ and $\mathbf{f'}$ result from clipping the last n rows (n is the number of output variables) of $\mathbf{G}$ and $\mathbf{f}$. These constraints can be expressed as

$$\begin{bmatrix} G_0 & 0 & \cdots & 0 \\ G_1 - G_0 & G_0 & \cdots & 0 \\ \vdots & \vdots & \ddots & \vdots \\ G_{N-1} - G_{N-2} & G_{N-2} - G_{N-3} & \cdots & G_0 \end{bmatrix} \mathbf{u} \leq \begin{bmatrix} y(t) - \mathbf{f}_1 \\ \mathbf{f}_1 - \mathbf{f}_2 \\ \vdots \\ \mathbf{f}_{N-1} - \mathbf{f}_N \end{bmatrix}$$

Nonminimum Phase Behaviour

Some processes exhibit a type of nonminimum phase behaviour. That is, when the process is excited by a step in its input the output variable tends to first move in the opposite direction prior to moving to the final position. This kind of behaviour may not be desirable in many cases.

Constraints can be added to the control system to avoid this type of behaviour. The constraints take the form

$$y(t + j) \geq y(t) \text{ if } \quad y(t) < w(t)$$
$$y(t + j) \leq y(t) \text{ if } \quad y(t) > w(t)$$

These constraints can be expressed in terms of the increments of the manipulated variables as follows:

$$\mathbf{Gu} \geq \mathbf{1}y(t) - \mathbf{f}$$

Actuator Nonlinearities

Most actuators in industry exhibit dead zones and other type of nonlinearities. Controllers are normally designed without taking into account actuator nonlinearities. Because of the predictive nature of MPC, actuator nonlinearities can be dealt with as suggested in [53].

Dead zones can be treated by imposing constraints on the controller in order to generate control signals outside the dead zone, say $(\underline{u}_d, \overline{u}_d)$ for a dead zone on the slew rate of actuators and $(\underline{U}_d, \overline{U}_d)$ for the dead zone on the amplitude of actuators. That is:

$$1\,\underline{U}_d \geq T\mathbf{u} + u(t-1)\,1 \geq 1\,\overline{U}_d$$
$$1\,\underline{u}_d \geq \mathbf{u} \geq 1\,\overline{u}_d$$

The feasible region generated by this type of constraint is nonconvex and the optimization problem is difficult to solve as pointed out in [53].

Terminal State Equality Constraints

These types of constraints appear when applying CRHPC [61] where the predicted output of the process is forced to follow the predicted reference during a number of sampling periods m after the costing horizon N_y. The terminal state constraints can be expressed as a set of equality constraints on the future control increments using the prediction equation for $\mathbf{y}_m = [y(t+N_y+1)^T \cdots y(t+N_y+m)^T]^T$:

$$\mathbf{y}_m = \mathbf{G}_m \mathbf{u} + \mathbf{f}_m$$

If the predicted response is forced to follow the future reference setpoint $\mathbf{w}_m$, the following equality constraint can be established:

$$\mathbf{G}_m \mathbf{u} = \mathbf{w}_m - \mathbf{f}_m$$

It will be seen later in this chapter that the introduction of this type of constraint simplifies the problem reducing the amount of computation required.

All constraints treated so far can be expressed as $\mathbf{R}\mathbf{u} \leq \mathbf{c}$ and $\mathbf{A}\mathbf{u} = \mathbf{a}$. The MPC problem when constraints are taken into account consists of minimizing Expression (7.1) subject to a set of linear constraints; that is, the optimization of a quadratic function with linear constraints, what is usually known as a quadratic programming problem (QP).

Terminal Set Constraints

In some cases, the final state (i.e. the state at the end of the predicting horizon) of the MPC problem is forced to belong to a terminal set. Terminal set constraints can be imposed by operational conditions or as a way of guaranteeing stability (see Section 9.5). The terminal set induces a set of constraints over the vector of control moves. Let us consider that the terminal region is defined by the polyhedron:

$$\mathbf{R}_T x(t+N) \leq \mathbf{r}_T \tag{7.2}$$

The vector of predicted states can be expressed as

$$\mathbf{x} = \mathbf{G}_u \mathbf{u} + \mathbf{F}_x x(t) \tag{7.3}$$

taking the last n rows ($n = dim(x(t))$) of (7.3) $x(t+N) = \mathbf{g}_{u_N} \mathbf{u} + \mathbf{f}_{x_N} x(t)$, where $\mathbf{g}_{u_N}$ and $\mathbf{f}_{x_N}$ are the last n rows of $\mathbf{G}_u$ and $\mathbf{F}_x$, respectively. Introducing $x(t+N)$ in (7.2)

$$\mathbf{R}_T \mathbf{R}_T (\mathbf{g}_{u_N} \mathbf{u} + \mathbf{f}_{x_N} x(t)) \leq \mathbf{r}_T \tag{7.4}$$

which shows that a polytopic terminal region induces a set of linear constraints in the vector of control moves.

7.1.1 Constraint General Form

All the constraints seen so far in this chapter, except for the dead zones of the actuators, can be described by

$$\mathbf{R} \mathbf{u} \leq \mathbf{r} + \mathbf{V} \mathbf{z}$$

where $\mathbf{z}$ is a vector composed of present and past signals. In the case of state space representation, $\mathbf{z}$ is $x(t)$, in the case of the CARIMA or CARMA models $\mathbf{z}$ is composed of the present output and finite series of past inputs and outputs (a way of representing the state). In the case of step or impulse response models, $\mathbf{z}$ is composed of the present output and a finite series of past inputs (another representation of the process state). In all cases, since the past output and output signals can be considered as a representation of the state, the constraints can be expressed by the general form:

$$\mathbf{R} \mathbf{u} \leq \mathbf{r} + \mathbf{V} x(t) \tag{7.5}$$

Notice that $\mathbf{R}$, $\mathbf{r}$, and $\mathbf{V}$ depend on process parameters and signal bounds and have to be recomputed only when they change (not very frequently). The right-hand side of the inequality constraints (7.5) depends on the process state that changes, in general, every sampling time, and it has to be recomputed accordingly.

7.1.2 Illustrative Examples

Constraints can be included in Generalized Predictive Control to improve performance, as demonstrated in [114]. Ordys and Grimble [161] have indicated how to analyze the influence of constraints on the stochastic characteristics of signals for a system controlled by a GPC algorithm. In order to illustrate how constraints can be used to improve the performance of different types of processes some simple illustrative examples are presented.

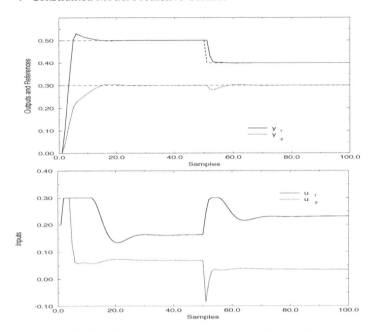

Fig. 7.2. Constraints in the manipulated variables

Input Constraints

In order to show the influence of constraints on the slew rate and on the amplitude of the manipulated variable, consider the reactor described in the previous chapter given by the following left fraction matrix description:

$$\mathbf{A}(z^{-1}) = \begin{bmatrix} 1 - 1.8629z^{-1} + 0.8669z^{-2} & 0 \\ 0 & 1 - 1.8695z^{-1} + 0.8737z^{-2} \end{bmatrix}$$

$$\mathbf{B}(z^{-1}) = \begin{bmatrix} 0.0420 - 0.0380z^{-1} & 0.4758 - 0.4559z^{-1} \\ 0.0582 - 0.0540z^{-1} & 0.1445 - 0.1361z^{-1} \end{bmatrix}$$

The constraints considered are maximum slew rate for the manipulated variables of 0.2 per sampling time and maximum value of -0.3 and 0.3.

The results obtained are shown in Figure 7.2. If we compare the results with the ones obtained in Chapter 6 (unconstrained manipulated variables) we can observe how the introduction of the constraints on the manipulated variables has produced a slower closed-loop response than was expected.

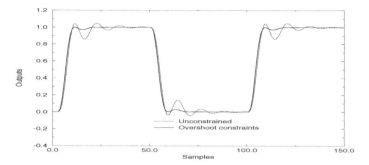

Fig. 7.3. Overshoot constraints

Overshoot Constraints

The system considered for this example corresponds to a discretized version of an oscillatory system $G(s) = 50/(s^2 + 25)$, taken from [114]. The discrete transfer function for a sampling time of 0.1 second is:

$$G(z^{-1}) = \frac{0.244835(z^{-1} + z^{-2})}{1 - 1.75516z^{-1} + z^{-2}}$$

The results obtained when applying an unconstrained GPC with a prediction horizon and a control horizon of 11 and a weighting factor of 50 are shown in Figure 7.3. As can be seen, the output shows a noticeable overshoot. The response obtained when overshoot constraints are taken into account is shown in the same figure. As can be seen, the overshoot has been eliminated.

Monotonic Behaviour

Although the overshoot has been eliminated from the process behaviour in the previous example by imposing the corresponding constraint, the system exhibits oscillations prior to reaching the setpoints (kickback). In order to avoid this type of behaviour, monotonic behaviour constraints were imposed. The results obtained for a prediction horizon and a control horizon of 11 and a weighting factor of 50 are shown in Figure 7.4. As can be seen, the oscillations have practically been eliminated. Notice that the prediction and control horizon used are quite large. The reason for this is that the oscillatory mode of the open-loop system has to be cancelled and a large number of control moves has to be considered to obtain a feasible solution.

Nonminimum Phase Process

In order to illustrate how constraints may be used to shape the closed loop behaviour consider the nonminimum phase system given by the following transfer function:

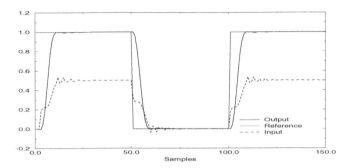

Fig. 7.4. Monotonic behaviour constraints

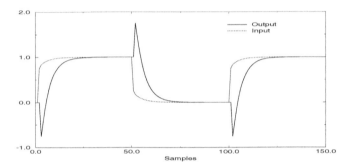

Fig. 7.5. Nonminimum phase behaviour

$$G(s) = \frac{1 - s}{1 + s}$$

If the system is sampled at 0.3 second, the discrete transfer function is given by

$$G(z^{-1}) = \frac{-1 + 1.2592z^{-1}}{1 - 0.7408z^{-1}}$$

The response obtained for step changes in the reference when a GPC with a prediction horizon of 30 a control horizon of 10 and a weighting factor of 0.1 is applied to the system is shown in figure 7.5. As can be seen, the closed-loop behaviour exhibits the typical nonminimum phase behaviour with an initial peak in the opposite direction to the setpoint change. The responses obtained when the inverse peaks are limited by 0.05 are shown in Figure 7.6. As can be seen, the system is slower but the peaks have been eliminated. Figure 7.6 also shows how the control signal generated grows slowly to avoid the inverse peaks.

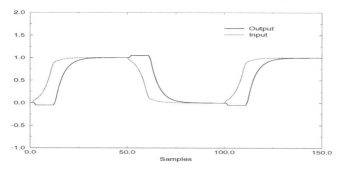

Fig. 7.6. Nonminimum phase constraints

7.2 Constraints and Optimization

The implementation of MPC in industry is not a trivial matter and is certainly a more difficult task than the commissioning of classical control schemes based on PID controllers. An MPC application is more costly, time consuming and requires personnel with a better training in control than implementing classical control schemes. For a start, a model of the process has to be found, and this requires a significant number of plant tests, which in most cases implies taking the plant away from its nominal operating conditions. The control equipment needs, in some cases, a more powerful computer and better instrumentation, and commercial MPC packages are expensive. Furthermore, control personnel need appropriate training for commissioning and using MPC.

In spite of these difficulties MPC has proven itself to be economically profitable by reducing operating costs or increasing production and is one of the most successful advanced control techniques used in industry. The reasons for this success depend on the particular application but are related to the abilities of MPC to optimize cost functions and treat constraints. The following reasons can be mentioned:

- optimization of operating conditions: MPC optimizes a cost function which can be formulated to minimize operating costs or any objective with economic implications.
- optimization of transitions: The MPC objective function can be formulated to optimize a function which measures the cost of taking the process from one operating point to another, with faster process start-up or commissioning times.
- minimization of error variance: An MPC can be formulated to minimize the variance of the output error. A smaller variance will produce economical benefits for the following reasons:
    - A smaller variance may increase the quality of the product as well as its uniformity.

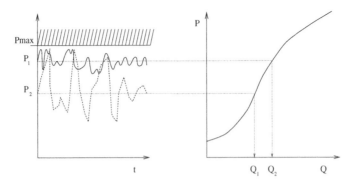

Fig. 7.7. Optimal operating point and constraints

- A smaller variance will allow the process to operate closer to the optimal operating conditions. As most processes are constrained, the optimal operating points usually lie at the intersection of some constraints. Processes operating with a smaller error variance can operate closer to the optimal operating points. Let us, as an example, consider a process where the production (flow) is related to the operating pressure as illustrated in Figure 7.7. Because of operating constraints, the pressure is limited to $p_{\max}$ which is the optimal operating point considering production. It is obvious that the process cannot operate at its limit because, due to perturbations, the process would be continuously violating the pressure limits and in most cases emergency systems would shut down the process. If the control system is able to keep the variance small, the setpoint can be established much closer to the optimal operating point as shown in Figure 7.7.
- The explicit handling of constraints may allow the process to operate closer to constraints and optimal operating conditions.
- The explicit considerations of constraints may reduce the number of constraint violations reducing the number of costly emergency shutdowns.

7.3 Revision of Main Quadratic Programming Algorithms

As was indicated in the previous section, the implementation of Generalized Predictive Controllers for processes with bounded signals requires the solution of a quadratic programming (QP) problem; that is, an optimization problem with a quadratic objective function and linear constraints. This section is dedicated to revising some of the main QP techniques. It is not intended to be an exhaustive description of all QP methods. There are other techniques, such as the ones based on decreasing ellipsoid volume methods that have been used in the GPC context [102] which are not described here.

7.3.1 The Active Set Methods

Equality Constraints

Although a set of inequality constraints is obtained when the GPC control problem is formulated, the first part of the section is dedicated to equality constraints because some of the methods for treating inequality constraints reduce the problem to an equality constraint problem and because in some cases, as in CRHPC [151], some equality constraints appear when the prediction of the future process output is forced to follow exactly the future reference.

The equality constrained QP problem can be stated as

$$\text{minimize} \quad \frac{1}{2}\mathbf{u}^T\mathbf{H}\mathbf{u} + \mathbf{b}^T\mathbf{u} + f_0$$
$$\text{subject to:} \quad \mathbf{A}\mathbf{u} = \mathbf{a}$$

where $\mathbf{A}$ is an $m \times n$ matrix and $\mathbf{a}$ is an m vector. It is assumed that $m < n$ and that $rank(\mathbf{A}) = m$.

A direct way of solving the problem is to use the constraints to express m of the $\mathbf{u}$ variables as a function of the remaining $n - m$ variables and then to substitute them in the objective function. The problem is reduced to minimizing a quadratic function of $n - m$ variables without constraints.

Usually a generalized elimination method is used instead of a direct elimination procedure. The idea is to express $\mathbf{u}$ as a function of a reduced set of $n - m$ variables: $\mathbf{u} = \mathbf{Y}\mathbf{a} + \mathbf{Z}\mathbf{v}$, where $\mathbf{Y}$ and $\mathbf{Z}$ are $n \times m$ and $n \times (n - m)$ matrices such that $\mathbf{A}\mathbf{Y} = I$, $\mathbf{A}\mathbf{Z} = 0$ and the matrix $[\mathbf{Y}\ \mathbf{Z}]$ has full rank. Notice that matrix $\mathbf{Y}$ can be interpreted as a generalized left inverse of $\mathbf{A}^T$ and that $\mathbf{Z}\mathbf{v}$ is the null column space of $\mathbf{A}^T$.

If this substitution is made, the equality constraints hold and the objective function

$$J(\mathbf{v}) = \frac{1}{2}[\mathbf{Y}\mathbf{a} + \mathbf{Z}\mathbf{v}]^T\mathbf{H}[\mathbf{Y}\mathbf{a} + \mathbf{Z}\mathbf{v}] + \mathbf{b}^T[\mathbf{Y}\mathbf{a} + \mathbf{Z}\mathbf{v}] + f_0$$
$$= \frac{1}{2}\mathbf{v}^T\mathbf{Z}^T\mathbf{H}\mathbf{Z}\mathbf{v} + [\mathbf{b}^T + \mathbf{a}^T\mathbf{Y}^T\mathbf{H}]\mathbf{Z}\mathbf{v} + [\frac{1}{2}\mathbf{a}^T\mathbf{Y}^T\mathbf{H} + \mathbf{b}^T]\mathbf{Y}\mathbf{a} + f_0$$

that is, an unconstrained QP problem of $n - m$ variables. If the matrix $\mathbf{Z}^T\mathbf{H}\mathbf{Z}$ is positive definite, there is only one global optimum point that can be found solving the linear set of equations

$$\mathbf{Z}^T\mathbf{H}\mathbf{Z}\mathbf{v} = -\mathbf{Z}^T(\mathbf{b} + \mathbf{H}\mathbf{Y}\mathbf{a})$$

Notice that if $\mathbf{u}^k$ is a point satisfying the constraints $\mathbf{A}\mathbf{u}^k = \mathbf{a}$, any other point $\mathbf{u}$ satisfying the constraints can be expressed as $\mathbf{u} = \mathbf{u}^k + \mathbf{Z}\mathbf{v}$. Thus the vector $\mathbf{Y}\mathbf{a}$ can be made equal to any point satisfying the constraints. Vector $\mathbf{v}$ can be expressed as the solution of the following linear equation

$$\mathbf{Z}^T \mathbf{H} \mathbf{Z} \mathbf{v} = -\mathbf{Z}^T \mathbf{g}(\mathbf{u}^k)$$

where $\mathbf{g}(\mathbf{u}^k) = \mathbf{H}\mathbf{u}^k + \mathbf{b}$ is the gradient of $J(\mathbf{u})$ at $\mathbf{u}^k$.

A general way of obtaining appropriate $\mathbf{Y}$ and $\mathbf{Z}$ matrices is to choose an $(n-m) \times n$ matrix $\mathbf{W}$ such that the matrix $\begin{bmatrix} \mathbf{A} \\ \mathbf{W} \end{bmatrix}$ is nonsingular. The inverse can then be expressed as:

$$\begin{bmatrix} \mathbf{A} \\ \mathbf{W} \end{bmatrix}^{-1} = [\mathbf{Y} \ \mathbf{Z}]$$

It then follows that $\mathbf{A}\mathbf{Y} = \mathbf{I}$ and $\mathbf{A}\mathbf{Z} = \mathbf{0}$.

If matrix $\mathbf{W}$ is chosen as $[\mathbf{0} \ \mathbf{I}]$, the method coincides with the direct elimination method. Another way of choosing $\mathbf{W}$ is related to the active set method that will be described later. The idea is to use inactive constraints $(\mathbf{a}_i)$ as the rows of $\mathbf{W}$. If an inactive constraint present in $\mathbf{W}$ becomes active (the rows of $\mathbf{R}$ where $\mathbf{r}_i \mathbf{u} = c_i$), the corresponding row of $\mathbf{W}$ is transferred to $\mathbf{A}$. When an active constraint becomes inactive, the corresponding row of $\mathbf{A}$ is transferred to $\mathbf{W}$. By doing this, the inverse of matrix $\begin{bmatrix} \mathbf{A} \\ \mathbf{W} \end{bmatrix}$ need not be recomputed to calculate $\mathbf{Y}$ and $\mathbf{Z}$.

Inequality Constraints

As shown at the beginning of this chapter, the GPC of processes with bounded signals results in a QP problem with linear inequality constraints.

The main idea of the active set method is to reduce the inequality constraint QP problem to a sequence of equality constraint QP problems that can be solved using the techniques described previously.

Consider a feasible point $\mathbf{u}^0$; that is, $\mathbf{R}\mathbf{u}^0 \leq \mathbf{c}$ and the set of active constraints (all the equality constraints and the rows of $\mathbf{R}$ where $\mathbf{r}_i \mathbf{u} = c_i$). Form matrix $\mathbf{A}$ and vector $\mathbf{a}$ by adding these rows $(\mathbf{r}_i)$ and corresponding limits (c_i) and the equality constraints.

The problem can now be solved with the method described previously. Suppose that $\mathbf{u}^1$ is the solution to the equally constrained QP problem. If $\mathbf{u}^1$ is feasible with respect to the inactive constraints, a test for optimality has to be performed to check if the global optimum has been found. This can be accomplished by verifying that the Lagrange multipliers for all equality constraints $\lambda_i \geq 0$. If this is not the case, the constraint with the most negative Lagrange multiplier is dropped from the active constraint set and the previous steps are repeated.

If point $\mathbf{u}^1$ is not feasible with respect to the inactive constraints, the nearest intersection from $\mathbf{u}^0$ of the line joining points $\mathbf{u}^0$ and $\mathbf{u}^1$ and the inactive constraints is computed. The corresponding constraint is added to the active set and the previous steps are repeated.

Notice that the method requires an initial feasible point. Procedures to find a feasible point will be described later in this chapter.

7.3.2 Feasible Direction Methods

The key idea of feasible direction methods is to improve the objective function by moving from a feasible point to an improved feasible point until the optimum is reached. Given a feasible point $\mathbf{u}^k$, an improving feasible direction $\mathbf{d}_k$ is determined such that by taking a sufficiently small step along $\mathbf{d}_k$ the new point will be feasible and will have a smaller value for the objective function.

There are various ways of generating feasible directions, one of the most popular ones in terms of simplicity is the gradient projection method of Rosen which is based on the following

Definition 7.1. An $n \times n$ matrix $\mathbf{P}$ is called a *projection matrix* if $\mathbf{P} = \mathbf{P}^T$ and $\mathbf{PP} = \mathbf{P}$.

Consider the problem of minimizing $J(\mathbf{u})$ subject to $\mathbf{Au} \leq \mathbf{a}$ and a feasible point $\mathbf{u}^k$ such that $\mathbf{A}_1\mathbf{u}^k = \mathbf{a}_1$ and $\mathbf{A}_2\mathbf{u}^k < \mathbf{a}_2$, where the matrices $\mathbf{A}_1$ and $\mathbf{A}_2$ and vectors $\mathbf{a}_1$ and $\mathbf{a}_2$ correspond to the active constraint and inactive constraint sets, respectively.

Lemma 7.1. *[15]: A nonzero direction $\mathbf{d}$ is an improving feasible direction if and only if $\mathbf{A}_1\mathbf{d} \leq 0$ and $\nabla J(\mathbf{u})^T\mathbf{d} < 0$.*

Lemma 7.2. *[15]: If $\mathbf{P}$ is a projection matrix such that $\mathbf{P}\nabla J(\mathbf{u}^k) \neq 0$ then $\mathbf{d} = -\mathbf{P}\nabla J(\mathbf{u}^k)$ is an improving direction of J at $\mathbf{u}^k$. Furthermore, if $\mathbf{A}_1$ has full rank and if $\mathbf{P}$ is of the form $\mathbf{P} = \mathbf{I} - \mathbf{A}_1^T(\mathbf{A}_1\mathbf{A}_1^T)^{-1}\mathbf{A}_1$, then $\mathbf{P}$ is an improving feasible direction.*

The proof is straightforward:

$$\nabla J(\mathbf{u})^T\mathbf{d} = -\nabla J(\mathbf{u})^T\mathbf{P}\nabla J(\mathbf{u}) = -\nabla J(\mathbf{u})^T\mathbf{P}^T\mathbf{P}\nabla J(\mathbf{u}) = -\left|\mathbf{P}\nabla J(\mathbf{u})\right|^2 < 0$$

That is, $\mathbf{d}$ is an improving direction. Moreover, $\mathbf{A}_1\mathbf{d} = -\mathbf{A}_1\mathbf{P}\nabla J(\mathbf{u}^k) = -\mathbf{A}_1(\mathbf{I} - \mathbf{A}_1^T(\mathbf{A}_1\mathbf{A}_1^T)^{-1}\mathbf{A}_1)\nabla J(\mathbf{u}^k) = 0$; that is, $\mathbf{A}_1\mathbf{d} = 0$ showing that $\mathbf{d}$ is a feasible direction.

Once a nonnull improving feasible direction $\mathbf{d}$ has been found, function $J(\mathbf{u})$ is minimized along $\mathbf{d}$. This can be done by making:

$$\mathbf{u}^{k+1} = \mathbf{u}^k + \lambda_k\mathbf{d}$$

The value of λ_k is computed as $\lambda_k = \min(\lambda^*, \lambda_{\max})$, where λ^* is the value of λ which minimizes $J(\mathbf{u}^k + \lambda\mathbf{d})$ and $\lambda_{\max}$ is the maximum value of λ such that $\mathbf{A}_2(\mathbf{u}^k + \lambda\mathbf{d}) \leq \mathbf{a}_2$.

Because $J(\mathbf{u})$ is a quadratic function, these values can easily be computed:

$$J(\mathbf{u}^k + \lambda \mathbf{d}) = \frac{1}{2}(\mathbf{u}^k + \lambda \mathbf{d})^T \mathbf{H}(\mathbf{u}^k + \lambda \mathbf{d}) + \mathbf{b}^T(\mathbf{u}^k + \lambda \mathbf{d}) + \mathbf{f}_0$$

$$= \frac{1}{2}\mathbf{d}^T \mathbf{H}\mathbf{d}\lambda^2 + (\mathbf{d}^T \mathbf{H}\mathbf{u}^k + \mathbf{b}^T \mathbf{d})\lambda + J(\mathbf{u}^k)$$

The optimum can be found for:

$$\lambda^* = -\frac{\mathbf{d}^T \mathbf{H}\mathbf{u}^k + \mathbf{b}^T \mathbf{d}}{\mathbf{d}^T \mathbf{H}\mathbf{d}} \tag{7.6}$$

The value of $\lambda_{\max}$ can be found as the minimum value of $\frac{c_j - \mathbf{a}_j^T \mathbf{u}^k}{\mathbf{a}_j^T \mathbf{d}}$ for all j such that $\mathbf{a}_j^T$ and c_j are the rows of the inactive constraint set and respective bound, and such that $\mathbf{a}_j^T \mathbf{d} > 0$.

The algorithm can be summarized as follows:

1. If the active constraint set is empty then let $\mathbf{P} = \mathbf{I}$; otherwise let $\mathbf{P} = \mathbf{I} - \mathbf{A}_1^T(\mathbf{A}_1\mathbf{A}_1^T)^{-1}\mathbf{A}_1$, where $\mathbf{A}_1$ corresponds to the matrix formed by the rows of $\mathbf{A}$ corresponding to active constraints.
2. Let $\mathbf{d}_k = -\mathbf{P}(\mathbf{H}\mathbf{u}^k + \mathbf{b})$.
3. If $\mathbf{d}_k \neq 0$ go to step 4; else:

   3.1 If the active constraint set is empty then STOP else:

   3.1.1 Let $\mathbf{w} = -(\mathbf{A}_1\mathbf{A}_1^T)^{-1}\mathbf{A}_1(\mathbf{H}\mathbf{u}^k + \mathbf{b})$

   3.1.2 If $\mathbf{w} \geq 0$ STOP, otherwise choose a negative component of $\mathbf{w}$, say w_j and remove the corresponding constraint from the active set. That is, remove row j from $\mathbf{A}_1$. Go to step 1.
4. Let $\lambda_k = \min(\lambda^*, \lambda\max)$ and $\mathbf{u}^{k+1} = \mathbf{u}^k + \lambda_k\mathbf{d}_k$. Replace k by $k+1$ and go to step 1.

7.3.3 Initial Feasible Point

Some of the QP algorithms discussed earlier start from a feasible point. If the bounds on the process only affect the control signals, a feasible point can very easily be obtained by making the control signal $u(k + j) = u(k - 1)$ (supposing that $u(k - 1)$ is in the feasible region and that $\underline{u} \leq 0 \leq \overline{u}$). This may not be a good starting point and may reflect in the efficiency of the optimization algorithm. A better way of obtaining a starting solution could be to use the feasible solution found in the previous iteration shifted one sampling time and adding a last term equal to zero. That is, if $\mathbf{u}^{k-1} = [\triangle u(k-1), \triangle u(k), \cdots, \triangle u(k+n-2), \triangle u(k+n-1)]$, the initial solution is made equal to $[\triangle u(k), \triangle u(k + 1), \cdots, \triangle u(k + n - 1), 0]$.

If the reference has changed at instant k, this may not be a good starting point and a better solution may be found by computing the unconstrained solution and clipping it to fit the constraints.

If more complex constraints are present, such as output constraints, the problem of finding an initial feasible solution cannot be solved as previously described because just clipping the input signal may not work and a procedure to find an interior point of a polytope has to be used. One of the simplest way of finding an initial solution is by using the following algorithm:

1. Fix any initial point u^0.
2. Let $r = Ru^0 - c$.
3. If $r \leq 0$ STOP. (u^0 is feasible).
4. $r_{max} = \max(r)$.
5. Solve the following augmented optimization problem using an active set:

$$\min_{u'} J'(u') = \min_{u'}[0\ 0\ \cdots\ 0\ 1]u'$$

$$R'u' \leq c$$

with

$$u' = \begin{bmatrix} u \\ z \end{bmatrix} \qquad R' = [R\ -1]$$

and the starting point

$$u' = \begin{bmatrix} u^0 \\ r_{max} \end{bmatrix}.$$

Notice that this starting point is feasible for the augmented problem.

6. If $J(u') \leq 0$ for the solution, a feasible solution has been found for the original problem; otherwise the original problem is unfeasible.

7.3.4 Pivoting Methods

Pivoting methods such as the Simplex have been widely used in linear programming because these algorithms are simple to program and they finish in a finite number of steps finding the optimum or indicating that no feasible solution exists. The minimization of a quadratic function subject to a set of linear constraints can be solved by pivoting methods and they can be applied to MPC as shown by Camacho [38]. One of the most popular pivoting algorithms is based on reducing the QP problem to a linear complementary problem.

The Linear Complementary Problem

Let q and M be a given m vector and a given $m \times m$ matrix respectively. The linear complementary problem (LCP) consists of finding two m vectors s and z so that:

$$s - Mz = q \quad sz \geq 0 \quad < s, z >= 0 \tag{7.7}$$

A solution (s, z) to this system is called a complementary basic feasible solution if for each pair of the complementary variables (s_i, z_i) one of them is basic for $i = 1, \cdots, m$, where s_i and z_i are the i entries of vectors s and z, respectively.

If q is nonnegative, a complementary feasible basic solution can be found by making $s = q$ and $z = 0$. If this is not the case, Lemke's algorithm [120] can be used. In this algorithm an artificial variable z_0 is introduced, leading to:

$$s - Mz - 1 z_0 = q, \quad s, z, z_0 \geq 0, \quad < s, z >= 0 \tag{7.8}$$

We obtain a starting solution to the above system by making $z_0 = \max(-q_i)$, $z = 0$, and $s = q + 1 z_0$. If by a sequence of pivoting, compatible with the system, the artificial variable z_0 is driven to zero, a solution to the linear complementary problem is obtained. An efficient way of finding a sequence of pivoting that converges to a solution in a finite number of steps under some mild assumptions on matrix M is by using Lemke's algorithm from the following tableau:

	s	z	z_0	
s	$I_{m \times m}$	$-M$	-1	q

Other advantages of using Lemke's algorithm are that the solution can be traced out as the parameter $z_0 \downarrow 0$, no special techniques are required to resolve degeneracy, and when applied to LCP generated by QP problems, the unconstrained solution of the QP problem can be used as the starting point, as will be shown later.

Transforming the GPC into an LCP

The constrained GPC problem can be transformed into an LCP problem as follows:

Make $u = 1 \underline{u} + x$. Constraints can then be expressed in condensed form as

$$x \geq 0 \tag{7.9}$$

$$R x \leq c$$

with

$$R = \begin{bmatrix} I_{N \times N} \\ T \\ -T \\ G \\ -G \end{bmatrix} \quad c = \begin{bmatrix} 1 \, (\overline{u} - \underline{u}) \\ 1 \, \overline{u} - T \, 1 \, \underline{u} - u(t-1) 1 \\ -1 \, \underline{u} + T \, 1 \, \underline{u} + u(t-1) 1 \\ 1 \, \overline{y} - f - G \, 1 \, \underline{u} \\ -1 \, \underline{y} + f + G \, 1 \, \underline{u} \end{bmatrix} \tag{7.10}$$

Equation (7.1) can be rewritten as

$$J = \frac{1}{2}\mathbf{x}^T \mathbf{H}\,\mathbf{x} + \mathbf{a}\,\mathbf{x} + f_1 \tag{7.11}$$

where $\mathbf{a} = \mathbf{b} + \underline{u}\,\mathbf{1}^T \mathbf{H}$ and $f_1 = f_0 + \underline{u}^2\,\mathbf{1}^T \mathbf{H}\,\mathbf{1} + \mathbf{b}\,\underline{u}$.

Denoting the Lagrangian multiplier vectors of the constraints $\mathbf{x} \geq 0$ and $R\mathbf{x} \leq \mathbf{c}$ by $\mathbf{v}$ and $\mathbf{v}_1$, respectively, and denoting the vector of slack variables by $\mathbf{v}_2$, the Karush-Kuhn-Tucker conditions (KKT) [15] can be written as

$$R\,\mathbf{x} + \mathbf{v}_2 = \mathbf{c} \tag{7.12}$$
$$-\mathbf{H}\,\mathbf{x} - R^T\mathbf{v} + \mathbf{v}_1 = \mathbf{a}$$
$$\mathbf{x}^T\mathbf{v}_1 = 0, \ \mathbf{v}^T\mathbf{v}_2 = 0$$
$$\mathbf{x}, \mathbf{v}, \mathbf{v}_1, \mathbf{v}_2 \geq 0$$

These expressions can be rewritten as:

$$\begin{bmatrix} \mathbf{I}_{m\times m} & \mathbf{0}_{m\times N} & \mathbf{0}_{m\times m} & R \\ \mathbf{0}_{N\times m} & \mathbf{I}_{N\times N} & -R^T & -\mathbf{H} \end{bmatrix} \begin{bmatrix} \mathbf{v}_2 \\ \mathbf{v}_1 \\ \mathbf{v} \\ \mathbf{x} \end{bmatrix} = \begin{bmatrix} \mathbf{c} \\ \mathbf{a} \end{bmatrix} \tag{7.13}$$

The Kuhn-Tucker conditions can be expressed as a linear complementary problem $\mathbf{s} - \mathbf{Mz} = q, \mathbf{s}^T\mathbf{z} = 0, \mathbf{s}, \mathbf{z} \geq 0$ with:

$$\mathbf{M} = \begin{bmatrix} 0 & -R \\ R^T & \mathbf{H} \end{bmatrix} \quad q = \begin{bmatrix} \mathbf{c} \\ \mathbf{a} \end{bmatrix} \quad \mathbf{s} = \begin{bmatrix} \mathbf{v}_2 \\ \mathbf{v}_1 \end{bmatrix} \quad \mathbf{z} = \begin{bmatrix} \mathbf{v} \\ \mathbf{x} \end{bmatrix} \tag{7.14}$$

This problem can be solved using Lemke's algorithm by forming the following tableau

	$\mathbf{v}_2$	$\mathbf{v}_1$	$\mathbf{v}$	$\mathbf{x}$	z_0	
$\mathbf{v}_2$	$\mathbf{I}_{m\times m}$	$\mathbf{0}_{m\times N}$	$\mathbf{0}_{m\times m}$	R	-1	$\mathbf{c}$
$\mathbf{v}_1$	$\mathbf{0}_{N\times m}$	$\mathbf{I}_{N\times N}$	$-R^T$	$-\mathbf{H}$	-1	$\mathbf{a}$

Although the algorithm will converge to the optimum solution in a finite number of steps as matrix $\mathbf{H}$ is positive definite [15], it needs a substantial amount of computation. One of the reasons for this is that the $\mathbf{x}$ variables in the starting solution of Lemke's algorithm are not part of the basis. That is, the algorithm starts from the solution $\mathbf{x} = 0$ which may be far away from the optimum solution. The efficiency of the algorithm can be increased by finding a better starting point.

If Equation (7.13) is multiplied by $\begin{bmatrix} \mathbf{I}_{m\times m} & R\,\mathbf{H}^{-1} \\ \mathbf{0}_{N\times m} & -\mathbf{H}^{-1} \end{bmatrix}$, we have:

$$\begin{bmatrix} \mathbf{I}_{m\times m} & R\mathbf{H}^{-1} & -R\mathbf{H}^{-1}R^T & \mathbf{0}_{m\times N} \\ \mathbf{0}_{N\times m} & -\mathbf{H}^{-1} & \mathbf{H}^{-1}R^T & \mathbf{I}_{N\times N} \end{bmatrix} \begin{bmatrix} \mathbf{v}_2 \\ \mathbf{v}_1 \\ \mathbf{v} \\ \mathbf{x} \end{bmatrix} = \begin{bmatrix} \mathbf{c} + R\mathbf{H}^{-1}\mathbf{a} \\ -\mathbf{H}^{-1}\mathbf{a} \end{bmatrix} \tag{7.15}$$

The vector on the right-hand side of Equation (7.15) corresponds to the vector of slack variables for the unconstrained solution and to the unconstrained solution, respectively. Furthermore, Equation (7.15) shows that if Lemke's algorithm is started from this point, all the $\mathbf{x}$ variables are in the basis. In most cases only a few constraints will be violated for the unconstrained solution of the GPC problem. Thus, the constrained solution will be close to the initial condition and the number of iterations required should decrease.

The algorithm can be described as follows:

1. Compute the unconstrained solution $\mathbf{x}_{min} = -\mathbf{H}^{-1}\mathbf{a}$.
2. Compute $\mathbf{v}_{2\,min} = \mathbf{c} - R\mathbf{x}_{min}$. If $\mathbf{x}_{min}$ and $\mathbf{v}_{2\,min}$ are non negative then stop with $u(t) = x_1 + \underline{u} + u(t-1)$.
3. Start Lemke's algorithm with $\mathbf{x}$ and $\mathbf{v}_2$ in the basis with the following

$$
\text{tableau:} \quad
\begin{array}{c|cccc|c}
 & \mathbf{v}_2 & \mathbf{x} & \mathbf{v} & \mathbf{v}_1 & z_0 \\
\hline
\mathbf{v}_2 & \mathbf{I}_{m\times m} & \mathbf{0}_{m\times N} & R\mathbf{H}^{-1}R^T & R\mathbf{H}^{-1} & -1 & \mathbf{v}_{2min} \\
\mathbf{x} & \mathbf{0}_{N\times m} & \mathbf{I}_{N\times N} & \mathbf{H}^{-1}R^T & -\mathbf{H}^{-1} & -1 & \mathbf{x}_{min}
\end{array}
$$

4. If x_1 is not in the first column of the tableau, make it zero. Otherwise give it the corresponding value.
5. $u(t) = u(t-1) + \underline{u} + x_1$.

7.4 Constraints Handling

In some cases, depending on the type of constraints acting on the process, some advantages can be obtained from the particular structure of the constraint matrix $\mathbf{R}$. This section deals with the way in which this special type of structure can be used to improve the efficiency of the QP algorithms.

7.4.1 Slew Rate Constraints

When only slew rate constraints are taken into account, the constraints can be expressed as:

$$
\begin{bmatrix} \mathbf{I} \\ -\mathbf{I} \end{bmatrix} \mathbf{u} \le \begin{bmatrix} \overline{\mathbf{u}} \\ -\underline{\mathbf{u}} \end{bmatrix}
$$

Active Set Methods

It can be seen that the active constraint matrix $\mathbf{A}$ can be expressed, after appropriate permutations to keep the active bounds on the m first variables, as

$$
\mathbf{A}^T = \mathbf{Y} = \begin{bmatrix} \mathbf{I} \\ \mathbf{0} \end{bmatrix} \begin{matrix} m \\ n-m \end{matrix} \qquad
\mathbf{Z} = \mathbf{V} = \begin{bmatrix} \mathbf{0} \\ \mathbf{I} \end{bmatrix} \begin{matrix} m \\ n-m \end{matrix}
$$

Matrix $\mathbf{H}$ can be partitioned, after reordering, as

$$\mathbf{H} \begin{bmatrix} \mathbf{H}_{11} \mathbf{H}_{12} \\ \mathbf{H}_{21} \mathbf{H}_{22} \end{bmatrix}$$

where $\mathbf{H}_{11}$ is an $m \times m$ matrix. The linear system to be solved is

$$\mathbf{Z}^T \mathbf{H} \mathbf{Z} \mathbf{v} = -\mathbf{Z}^T (\mathbf{b} + \mathbf{H} \mathbf{Y} \mathbf{a}) = \mathbf{H}_{22} \mathbf{v} = -\mathbf{b}_2 - \mathbf{H}_{22} \mathbf{a}_2$$

that is, no calculations are needed for the generalized elimination method other than reordering the variables and corresponding matrices.

Rosen's Gradient Projection Method

The active constraint matrix $\mathbf{A}_1$ will have m rows, corresponding to the m values of $\triangle u(k+j)$ which are bounded. Each of the rows of $\mathbf{A}_1$ will have all its elements equal to zero except element j which will be equal to 1 if the bound corresponds to the upper limit or -1 if it is bounded by the lower limit. The product $\mathbf{A}_1 \mathbf{A}_1^T$ is then:

$$(\mathbf{A}_1 \mathbf{A}_1^T)_{ij} = \sum_{l=1}^{N} a_{il} a_{jl} = \begin{cases} = 1 \text{ if } i = j \text{ and constraint } j \text{ is active} \\ = 0 \text{ otherwise} \end{cases}$$

That is, an $m \times m$ identity matrix. The projection matrix $\mathbf{P}$ is then:

$$\mathbf{P} = \mathbf{I} - \mathbf{A}_1^T \mathbf{A}_1$$

It can then easily be seen [194] that the projection matrix can now be expressed by:

$$p_{ij} = \begin{cases} = 0 \text{ when } i \neq j \text{ or one of the bounds on variable } \triangle u(k+j) \text{ is active} \\ = 1 \text{ when } i = j \text{ and neither bound on variable } \triangle u(k+j) \text{ is active} \end{cases}$$

The search direction is given by:

$$d_i = \begin{cases} = 0 \text{ when one of the bounds on variable } \triangle u(k+j) \text{ is active} \\ = -g_i \text{ when } i = j \text{ and neither bound on variable } \triangle u(k+j) \text{ is active} \end{cases}$$

The value of $\lambda_{\max}$ can easily be found by:

$$\lambda_{\max} = \min \left[\min_j \left(\frac{\overline{u} - \triangle u(k+j)}{d_j} \middle| d_j > 0 \right), \min_j \left(\frac{\underline{u} - \triangle u(k+j)}{d_j} \middle| d_j < 0 \right) \right]$$

The computation of vector $\mathbf{w}$, which is necessary to check the Kuhn-Tucker condition, can be written as $\mathbf{w} = -\mathbf{A}_1 \mathbf{g}$. The stopping criterion is also considerably simplified, and it can be stated as: for all active constraints j check that $g_j \leq 0$ if j corresponds to an upper bound otherwise check that $g_j \geq 0$.

7.4.2 Amplitude Constraints

When the only constraints present are the maximum and minimum value of the control signals $u(k + j)$, the constraints can be expressed as

$$1\underline{u} \leq \mathbf{T}u - 1u(k - 1)) \leq 1\overline{u}$$

or

$$1(\underline{u} - u(k - 1)) \leq \mathbf{T}u \leq 1(\overline{u} - u(k - 1))$$

where matrix $\mathbf{T}$ is a lower triangular matrix whose entries are ones, and 1 is a vector composed of ones the constraint matrix takes the form:

$$\mathbf{R} = \begin{bmatrix} \mathbf{T} \\ -\mathbf{T} \end{bmatrix}$$

Although some advantages can be gained from the particular shape of the constraint matrix, the GPC can be reformulated to reduce the case to the much simpler one seen in the previous section.

Recall from Chapter 4 that the optimal predictions for the process output can be expressed as: $\mathbf{y} = \mathbf{Gu} + \mathbf{f}$. The vector of future control increments is given by

$$\mathbf{u} = \begin{bmatrix} u(k) - u(k-1) \\ u(k+1) - u(k) \\ \vdots \\ u(k+N) - u(k+N-1) \end{bmatrix}$$

$$= \begin{bmatrix} 1 & 0 & 0 & \cdots & 0 \\ -1 & 1 & 0 & \cdots & 0 \\ 0 & -1 & 1 & \cdots & 0 \\ \vdots & & & \ddots & \vdots \\ 0 & 0 & 0 & \cdots & 1 \end{bmatrix} \begin{bmatrix} u(k) \\ u(k+1) \\ \vdots \\ u(k+N) \end{bmatrix} - \begin{bmatrix} u(k-1) \\ 0 \\ \vdots \\ 0 \end{bmatrix} = \mathbf{DU} - \mathbf{f}_1$$

If this substitution is made in the equation of future predictions we get

$$\mathbf{y} = \mathbf{G}(\mathbf{DU} - \mathbf{f}_1) + \mathbf{f}_0 = \mathbf{G}'\mathbf{U} + \mathbf{f}_2$$

where $\mathbf{G}'$ is a lower triangular matrix with all its diagonal elements equal to g_0 and its secondary diagonal elements are given by $g_i - g_{i-1}$. Vector $\mathbf{f}_2$ can be expressed as $(\mathbf{f}_2)_i = (\mathbf{f}_0)_i - g_i u(k - 1)$.

The objective function can now be expressed as a function of the future control actions $\mathbf{U}$:

$$J(\mathbf{U}) = (\mathbf{G}'\mathbf{U} + \mathbf{f}_2 - \mathbf{w})^T(\mathbf{G}'\mathbf{U} + \mathbf{f}_2 - \mathbf{w}) + \lambda(\mathbf{DU} - \mathbf{f}_1)^T(\mathbf{DU} - \mathbf{f}_1)$$
$$= \mathbf{U}^T(\mathbf{G}'^T\mathbf{G} + \mathbf{D}^T\mathbf{D})\mathbf{U} + 2[(\mathbf{f}_2 - \mathbf{w})^T\mathbf{G}' - \mathbf{f}_1^T\mathbf{D}]\mathbf{U}$$
$$+ (\mathbf{f}_2 - \mathbf{w})^T(\mathbf{f}_2 - \mathbf{w}) + \mathbf{f}_1^T\mathbf{f}_1$$

that is a quadratic form

$$J(\mathbf{U}) = \frac{1}{2}\mathbf{U}^T\mathbf{H}'\mathbf{U} + \mathbf{b}' + \mathbf{f}'$$

where

$$\mathbf{H}' = 2(\mathbf{G}'^T\mathbf{G}+\mathbf{D}^T\mathbf{D}), \ \ \mathbf{b}' = 2[(\mathbf{f}_2-\mathbf{w})^T\mathbf{G}'-\mathbf{f}_1^T\mathbf{D}], \ \ \mathbf{f}' = (\mathbf{f}_2-\mathbf{w})^T(\mathbf{f}_2-\mathbf{w})+\mathbf{f}_1^T\mathbf{f}_1$$

Notice that $\mathbf{f}_1^T\mathbf{f}_1 = u(k-1)^2$ and $\mathbf{D}^T\mathbf{D}$ is a tridiagonal matrix with the elements of the main diagonal equal to 2 and the elements of the other two accompanying subdiagonals equal to -1.

The problem has been reduced to optimizing a quadratic form with the constraint matrix $\mathbf{R} = [\mathbf{I} - \mathbf{I}]^T$ and the efficiency of the optimization procedure can be increased as shown in the previous section.

7.4.3 Output Constraints

When the only constraints present are the maximum and minimum value of the output signals $y(k+j)$, the constraints can be expressed as

$$\mathbf{y}_{\min} \leq \mathbf{G}\mathbf{u} + \mathbf{f} \leq \mathbf{y}_{\max}$$

which can also be expressed as:

$$\begin{bmatrix} \mathbf{G} \\ -\mathbf{G} \end{bmatrix} \mathbf{u} \leq \begin{bmatrix} \mathbf{y}_{max} - \mathbf{f} \\ \mathbf{y}_{min} - \mathbf{f} \end{bmatrix}$$

Notice that all the blocks of the constraint matrix are lower triangular blocks and some advantages may be gained from that as will be shown in the following.

7.4.4 Constraint Reduction

The computational requirements of the QP algorithms depend heavily on the number of constraints considered. Only those constraints which limit the feasible region of the space need to be taken into account. The efficiency of the algorithms can be increased if the superfluous constraints; that is, those constraints not limiting the feasible region, are eliminated. There are a number of algorithms for determining the minimum set of limiting constraints, or what is the same, the convex hull or polytope, corresponding to the feasible region of space. Although elimination of all superfluous constraints may reduce the amount of computation needed, the procedure itself requires a substantial amount of computation. The fact that in this case the constraint matrices are lower triangular can be used to detect nonlimiting constraints. Some constraints can easily be eliminated as follows.

By making $\mathbf{u} = 1\underline{u} + \mathbf{x}$, constraints on the slew rate and amplitude of actuators and on the process output signals can be expressed as

$$0 \leq \mathbf{x} \leq c_1 \tag{7.16}$$

$$c_3 \leq T\mathbf{x} \leq c_2$$

$$\frac{c_5}{g_0} \leq \frac{\mathbf{G}}{g_0}\mathbf{x} \leq \frac{c_4}{g_0}$$

with $\mathbf{c}^T = [c_1^T \ c_2^T \ c_3^T \ c_4^T \ c_5^T]$

Now consider the first row for each constraint in (7.16); that is, the constraint affecting only x_1,

$$x_1 \leq c_{11}, \quad x_1 \leq c_{21}, \quad x_1 \leq \frac{c_{41}}{g_0} \tag{7.17}$$

$$x_1 \geq 0, \quad x_1 \geq c_{31}, \quad x_1 \geq \frac{c_{51}}{g_0}$$

where c_{ij} is the j entry of vector c_i.

Notice that of the first three constraints in (7.17) only that with a smaller right-hand side has to be kept; the other two can be eliminated because they do not limit the feasible region. The same applies to the last three constraints. Thus, four constraints can be eliminated in this first step. The x_1 variable will be bounded by $l_1 \leq x_1 \leq r_1$, where r_1 is the smallest of all right-hand side terms of the first row of constraints in (7.17) and l_1 is the biggest of all right-hand side terms of the second row.

Let us now consider the constraints in (7.16) limiting x_1 and x_2; that is, the second row of each constraint block in (7.16). These constraints can be written as:

$$x_2 \leq c_{12}, \quad x_2 \leq c_{22} - x_1, \quad x_2 \leq \frac{c_{42}}{g_0} - \frac{g_1}{g_0}x_1 \tag{7.18}$$

$$x_2 \geq 0, \quad x_2 \geq c_{32} - x_1, \quad x_2 \geq \frac{c_{52}}{g_0} - \frac{g_1}{g_0}x_1$$

The right-hand sides of these constraints depend, in general, on x_1. As x_1 is bounded by $l_1 \leq x_1 \leq r_1$, the right-hand side of each constraint in (7.18) will be bounded by two limits. Consider, for example, the second of the first row of constraints in (7.18). The minimum of the right-hand side of this inequality is given by $\min(c_{22} - x_1) = c_{22} - \max(x_1) = c_{22} - r_1$. The same considerations can be applied to the maximum. The right-hand side will therefore be limited by $r_{22\,min} = c_{22} - r_1$ and $r_{22\,max} = c_{22} - l_1$, where the first subindex in r_{kij} refers to variable x_k, the second refers to constraint i and the last subindex indicates whether it is the minimum or maximum limit. Notice that $r_{23\,min}$ and $r_{23\,max}$ can be computed in a similar manner and that $r_{k1\,min} = r_{k1\,max}$.

A right boundary for x_2 can be defined by $r_2 = \min(r_{2j\,\max})$ for $j = 1, 2, 3$. Notice that variable x_2 must always be smaller than r_2, thus any constraint j having $r_2 < r_{2j\,\min}$ can be eliminated as it will not be limiting the feasible region.

A left boundary l_2 can be obtained for variable x_2 from the last three constraints in (7.18). For each of these constraints a minimum $l_{2j\,\min}$ and maximum $l_{2j\,\max}$ limit can be found in the same way. A left boundary for variable x_2 can now be given by $l_2 = \max(l_{2j\,\min})$ for $j = 1, 2, 3$. Constraint j can now be eliminated if $l_2 > l_{2j\,\max}$.

After this step, variables x_1 and x_2 will be bound by (l_1, r_1) and (l_2, r_2), respectively. As the constraint matrices are lower triangular, the same procedure can be applied to obtain the boundaries (and eliminate the superfluous restrictions) for x_3 and then recursively for the remaining variables.

Notice that constraints of the type $x_k \geq 0$ do not appear in the constraint matrix R in the algorithm described earlier. As the algorithm considers all variables to be positive, these constraints are implicitly taken into account. If any left bound l_i is positive, the constraint $x_i \geq 0$ can be eliminated. In order to do this, the following substitution can be made: $x_j = l(x_1, x_2, \cdots, x_{j-1}) + z_j$, where $l(x_1, x_2, \cdots, x_{j-1})$ is the right hand side of one of the remaining constraints of type $x_i \geq l(x_1, x_2, \cdots, x_{j-1})$. This constraint can now be substituted by $z_j \geq 0$. The constraint matrices have to be changed accordingly.

Notice that the procedure described does not guarantee a minimum number of constraints; further reductions could be achieved but it would require more computation and a more complex algorithm.

7.5 1-norm

Although quadratic programming algorithms are very efficient, the MPC problem can be solved by a much more efficient linear programming method if a 1-norm type of function is used.

The objective function is now

$$J(\mathbf{u}) = \sum_{j=N_1}^{N_2} \sum_{i=1}^{n} |y_i(t+j) - w_i(t+j)| + \lambda \sum_{j=1}^{N_u} \sum_{i=1}^{m} |\triangle u_i(t+j-1)| \quad (7.19)$$

where N_1 and N_2 define the costing horizon and N_u defines the control horizon. The absolute values of the output tracking error and the absolute values of the control increments are taken instead of the square of them, as is usual in GPC.

If a series of $\mu_i \geq 0$ and $\beta_i \geq 0$ such that

$$-\mu_i \leq (y_i(t+j) - w_i(t+j)) \leq \mu_i \quad i = 1, \ldots n \quad j = 1, \ldots N$$
$$-\beta_i \leq \Delta u_i(t+j-1) \leq \beta_i \quad i = 1, \ldots m \quad j = 1, \ldots N_u$$
$$0 \leq \sum_{i=1}^{n \times N} \mu_i + \lambda \sum_{i=1}^{m \times N_u} \beta_i \leq \gamma$$

then γ is an upper bound of $J(\mathbf{u})$. The problem is now reduced to minimizing the upper bound γ.

When constraints on the output variables $(\underline{\mathbf{y}}, \overline{\mathbf{y}})$, the manipulated variables $(\underline{\mathbf{U}}, \overline{\mathbf{U}})$, and the slew rate of the manipulated variables $(\underline{\mathbf{u}}, \overline{\mathbf{u}})$ are taken into account, the problem can be interpreted as an LP problem with:

$$\min_{\gamma, \mu, \beta, \mathbf{u}} \gamma$$
$$\text{subject to:} \quad \mu \geq G\mathbf{u} + \mathbf{f} - \mathbf{w}$$
$$\mu \geq -G\mathbf{u} - \mathbf{f} + \mathbf{w}$$
$$\overline{\mathbf{y}} \geq G\mathbf{u} + \mathbf{f}$$
$$-\underline{\mathbf{y}} \geq -G\mathbf{u} - \mathbf{f}$$
$$\beta \geq \mathbf{u}$$
$$\beta \geq -\mathbf{u}$$
$$\overline{\mathbf{u}} \geq \mathbf{u}$$
$$-\underline{\mathbf{u}} \geq \mathbf{u}$$
$$\overline{\mathbf{U}} \geq T\mathbf{u} + \mathbf{1}u(t-1)$$
$$-\underline{\mathbf{U}} \geq -T\mathbf{u} - \mathbf{1}u(t-1)$$
$$\gamma \geq \mathbf{1}^T \mu + \lambda \mathbf{1} \beta$$

The problem can be transformed into the form:

$$\min_{\mathbf{x}} \mathbf{c}^T \mathbf{x} \quad \text{subject to} \quad A\mathbf{x} \leq \mathbf{b}, \quad \mathbf{x} \geq 0$$

with

$$\mathbf{x} = \begin{bmatrix} \overline{\mathbf{u}} - \underline{\mathbf{u}} \\ \hline \mu \\ \hline \beta \\ \hline \gamma \end{bmatrix} \qquad \mathbf{c} = \begin{bmatrix} 0 \\ \hline 0 \\ \hline 0 \\ \hline 1 \end{bmatrix}$$

$$A = \begin{bmatrix} G & -I & 0 & 0 \\ -G & -I & 0 & 0 \\ \hline G & 0 & 0 & 0 \\ -G & 0 & 0 & 0 \\ \hline I & 0 & -I & 0 \\ -I & 0 & -I & 0 \\ \hline I & 0 & 0 & 0 \\ \hline T & 0 & 0 & 0 \\ -T & 0 & 0 & 0 \\ \hline 0 & \mathbf{1}^T & \mathbf{1}^T\lambda & -1 \end{bmatrix} \qquad \mathbf{b}_i = \begin{bmatrix} -G\underline{\mathbf{u}} - \mathbf{f} + \mathbf{w} \\ G\underline{\mathbf{u}} + \mathbf{f} - \mathbf{w} \\ \hline \overline{\mathbf{y}} - G\underline{\mathbf{u}} - \mathbf{f} \\ -\underline{\mathbf{y}} + G\underline{\mathbf{u}} + \mathbf{f} \\ \hline -\underline{\mathbf{u}} \\ \underline{\mathbf{u}} \\ \hline \overline{\mathbf{u}} - \underline{\mathbf{u}} \\ \hline \overline{\mathbf{U}} - T\underline{\mathbf{u}} - \mathbf{1}u(t-1) \\ -\underline{\mathbf{U}} + T\underline{\mathbf{u}} + \mathbf{1}u(t-1) \\ \hline 0 \end{bmatrix}$$

The number of variables involved in the linear programming problem is $2 \times m \times N_u + n \times N + 1$, while the number of constraints is $4 \times n \times N + 5 \times$

$m \times N_u + 1$. For a process with 5 input and 5 output variables with control horizon $N_u = 10$ and costing horizon $N = 30$, the number of variables for the LP problem is 251 and the number of constraints would be 851, which can be solved by any LP algorithm. As the number of constraints is higher than the number of decision variables, solving the dual LP problem should also be less computationally expensive. The number of constraints can be reduced because of the special form of the constraint matrix A by applying the constraint reduction algorithm.

7.6 Case Study: A Compressor

Compressed air is used in most industrial plants for different purposes. Air compressors supplying compressed air to the different processes of a plant can frequently be found in industry. The example studied in this section corresponds to a large air compressor (Figure 7.8) supplying air to a plant. The outlet pressure is controlled by manipulating the guide vanes of the compressor. A blow-off valve is installed to prevent a surge. When the blow-off valve is closed, the compressor is a single-input single-output process that can be controlled appropriately using standard control techniques. When the blow-off valve opens, the compressor is a multivariable process with two inputs and two outputs. The manipulated variables are the guide vane angle (u_1) and the position of the valve (u_2), and the controlled variables are the air pressure (y_1) and the airflow rate (y_2). The process model is given by the following transfer matrix [46]:

$$\begin{bmatrix} Y_1(s) \\ Y_2(s) \end{bmatrix} = \begin{bmatrix} \dfrac{0.1133e^{-0.715s}}{1 + 4.48s + 1.783s^2} & \dfrac{0.9222}{1 + 2.071s} \\ \dfrac{0.3378e^{-0.299s}}{1 + 1.09s + 0.361s^2} & \dfrac{-0.321e^{-0.94s}}{1 + 2.463s + 0.104s^2} \end{bmatrix} \begin{bmatrix} U_1(s) \\ U_2(s) \end{bmatrix}$$

The compressor can be controlled as shown in [46] by decoupling the process at zero frequency and using a PI controller for the first loop and a proportional controller for the second. These controllers were obtained with the help of the Inverse Nyquist Array (INA). The simulated closed-loop responses of the process to successive step changes in both references are shown in Figure 7.9(a). It can be seen that the responses are quite oscillatory. The evolution of the manipulated variables can be seen in Figure 7.9(b). As can be seen, the valve position exhibits high peaks for each change in the pressure setpoint.

A sampling time of 0.05 is chosen. The process can be approximated by the following discrete transfer matrix:

$$\begin{bmatrix} \dfrac{10^{-4}(0.7619z^{-1} + 0.7307z^{-2})}{1 - 1.8806z^{-1} + 0.8819z^{-2}} z^{-14} & \dfrac{0.022z^{-1}}{1 - 0.9761z^{-1}} \\ \dfrac{10^{-2}(0.1112z^{-1} + 0.1057z^{-2})}{1 - 1.8534z^{-1} + 0.8598z^{-2}} z^{-6} & \dfrac{10^{-2}(-0.2692z^{-1} - 0.1821z^{-2})}{1 - 1.2919z^{-1} + 0.306z^{-2}} z^{-19} \end{bmatrix}$$

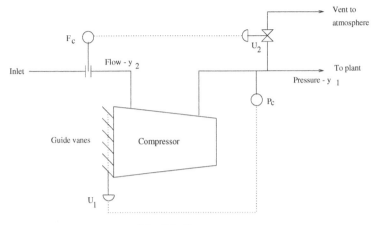

Fig. 7.8. Compressor

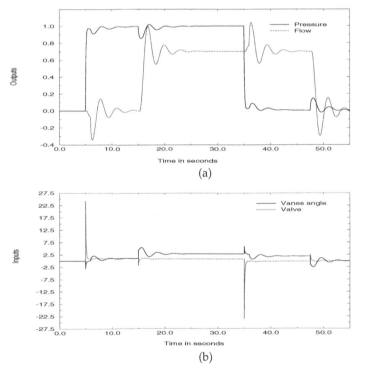

Fig. 7.9. Closed loop responses of compressor: INA controller

The process can be controlled with a multivariable GPC with the following design parameters: $N_1 = 20$, $N_2 = 23$, $N_3 = 3$, and $\lambda = 0.8$. The beginning of the costing horizon has been chosen as the maximum of the dead

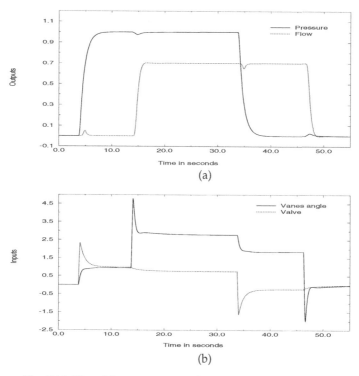

Fig. 7.10. Closed-loop responses of compressor: unconstrained GPC

times. The behaviour of the process can be seen in Figure 7.10(a). As can be seen, both controlled variables reach their setpoint rapidly and without oscillations. The perturbations caused in each of the controlled variables by a step change in the reference of the other variable are very small.

The evolution of the manipulated variables are shown in Figure 7.10(b). As can be seen, a high peak in the valve position is observed (although much smaller than the peaks observed when controlling the compressor with the INA controller).

To reduce the manipulated variable peak, a constrained GPC can be used. The manipulated variables are restricted to being in the interval $[-2.75, 2.75]$. The evolution of the controlled variables are shown in Figure 7.11(a). As can be seen, the response of the pressure to a step change in the reference is a bit slower than in the unconstrained case, but the manipulated variable is kept within the desired limits as shown in Figure 7.11(b).

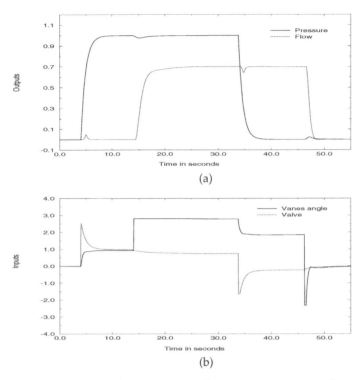

Fig. 7.11. Closed-loop responses of compressor: constrained GPC

7.7 Constraint Management

7.7.1 Feasibility

Sometimes, during the optimization stage, the region defined in the decision variables by the set of constraints is empty. In these conditions, the optimization algorithm cannot find any solution and the optimization problem is said to be infeasible. Unobtainable control objectives or perturbations that take the process away from the operating point may cause infeasibility. An optimization problem is feasible when the objective function is bounded and there are points in the space of decision variables that satisfy all constraints.

Infeasibility may appear in the steady-state regime and during transients. The infeasibility problems of the steady-state regime are usually caused by hyphenationun-obtainable unobtainable control objectives. This occurs, for example, when the setpoints cannot be reached because the manipulated variables are constrained. In general, if the manipulated variables are constrained to be in a hypercube, the reachable set-points are in a polytope in the controlled variable space whose vertices are defined by multiplying the vertex of the hypercube by the process DC gain matrix. These unfeasibilities can be easily handled during the design phase by eliminating these types of objectives.

Infeasibility can appear in the transitory regime, even when the imposed constraints seem to be reasonable. Constraints that do not cause problems in normal operation can cause problems under certain circumstances. A perturbation or large reference change may force a variable outside its limits so that it is impossible to introduce it into its permitted zone again using limited energy control signals. In these circumstances the constraints become temporarily incompatible. Infeasibility may also be produced when the operator redefines the operational variable limits while the process is in operation, as mentioned in [6]. If the variables are already outside the new limits, the problem will be unfeasible. These unfeasible solutions are more common in those cases where the optimum is close to the constraints and the system is subject to disturbances, taking the outlet to *forbidden areas*.

Feasibility is of great importance to constrained MPC not only because, as will be discussed in the following section, the stability proofs of constrained MPC strategies require feasibility, but also because if the optimization problem is not feasible the MPC will not work as it will not be able to compute the next control moves. Since unfeasibility is likely to arise in constrained MPC, some precautions have to be taken.

7.7.2 Techniques for Improving Feasibility

Constraint management methods try to recover feasibility by acting on the constraints according to varying criteria that depend on the type of limits imposed on process variables. The following types of limits can be considered:

- physical limits: These can never be exceeded because of the equipment construction itself and are usually associated to actuators.
- security limits: These limits should never be violated because their violation could be dangerous to process security or could induce a costly shutdown of the process by emergency equipment. These limits are usually associated to process-controlled variables.
- operational limits: These are fixed by the operators as bounds that should not be exceeded by the process variables to maintain appropriate operating conditions. They can be exceeded in certain circumstances.
- real limits: These are limits used by the control algorithm at each instance. They are provided by the constraint manager, who should calculate them in such a way that they never exceed the physical limits.

Possible solutions to this problem can be classified into the following groups:

1. **disconnection of the controller:** The easiest way of solving this type of problem is to pass the controller to a backup value or backup controller

when constraint incompatibilities arise and return to automatic operation when the admissibility of the solution is recovered. As can be understood, this method has serious disadvantages. Normally, when constraint incompatibility problems arise it is because the closed-loop system is at a critical stage where the operator usually has very little experience. Furthermore, if the constraints have to do with safety or economic aspects, then any decisions taken when constraint compatibility problems arise are usually critical because in these cases some of the control objectives cannot be satisfied. This method is usually used when constraint incompatibility problems are not frequent.

2. **constraint elimination:** Feasibility is analyzed at each sampling period and thus the elimination of constraints is temporary. The feasibility is checked periodically to be able to reinsert eliminated constraints. Notice that given a point in the decision variable space, the constraints that are violated can be computed easily, but optimization methods, in general, do not specify which constraints are causing infeasibility. When some of the constraints are dropped, the optimization algorithm has to be run again with the remaining constraints to check for feasibility.

   It is necessary to eliminate a group of constraints in those cases where the complete set of constraints imposed on the system is incompatible. Each time a constraint incompatibility problem arises, a set of inadmissible constraints is formed which is not taken into account in the optimization process. Various types can be distinguished in the constraint elimination methodology:

   - **indiscriminate elimination:** With this strategy all constraints are eliminated every time a feasible solution arises. This is not the best method to solve the problem, but it is the quickest. This method should not be used in cases where the constraints are directly related to safety.

   - **hierarchical elimination:** During the design stage, a priority is given to each constraint. Every time feasibility problems arise the controller eliminates, in an orderly manner, the lower priority constraints until the feasibility of the solution is reestablished. This is checked at every sampling period to reinsert those constraints that were temporarily dropped.

3. **constraints relaxation:** This method consists in temporarily relaxing the bounds (i.e., increasing their values) or changing *hard* constraints ($\mathbf{Ru} < \mathbf{a}$) to *soft* constraints ($\mathbf{Ru} < \mathbf{a} + \epsilon$), adding a term $\epsilon^T \mathbf{T} \epsilon$ to the cost function, so that any violation of the constraint is penalized. In the long run, the penalizing term in the objective function will take the auxiliary variable to zero.

4. **changing the constraint horizons:** Most of the constraint unfeasibility arises in the first part of the cost horizon, because sudden perturbations may take the process to an infeasible region. The main idea of this method is not to take into account the constraints during the first part

of the horizon. Some commercial MPCs use the concept of a constraint window.

7.8 Constrained MPC and Stability

Infinite horizon optimal controllers, such as the well-known Linear Quadratic Gaussian (LQG) optimal controller, are easy to implement and guarantee a stable closed-loop for linear processes under general assumptions. However, infinite horizon control problems can only be solved when all process variables are unconstrained. The main difficulty for using infinite horizons with processes with constrained variables comes from the fact that numerical methods, with a necessarily finite number of decision variables, have to be used to solve the optimization problem involved. The stability analysis of finite horizon controllers is a much more difficult task, especially if the variables are constrained giving rise to a nonlinear control law. Furthermore, no explicit functional description of the control law can be found, making the problem even more difficult. A breakthrough has been made in the last few years in this field. As pointed out by Morari [140], *the recent work has removed this technical and to some extent psychological barrier (people did not even try) and started widespread efforts to tackle extensions of this basic problem with the new tools.*

The basic idea of the new approaches is that infinite horizon cost functions can be shown to be monotonically decreasing, if there is a feasible solution, and thus can be interpreted as a Lyapunov function which guarantees stability. In order to find a numerical solution to the infinite costing horizon control problem, the number of decision variables has to be finite. Two basic approaches have been used for this: in the first one, the objective function is considered to be composed of two parts; one with a finite horizon and constrained, and the other with an infinite horizon and unconstrained. The second approach is essentially equivalent [61]; it consists of imposing terminal state constraints and using a finite control horizon.

The first type of approach has originated the following results obtained by Rawlings and Muske [177], who demonstrated asymptotic stability for processes described by

$$x(t+1) = Mx(t) + Nu(t)$$

with $u(t)$ generated by minimizing:

$$J_j = \sum_{j=t}^{\infty} (x(j)^t Rx(j) + u(j)^t Sx(j)) \text{ with } R, S > 0$$

and subject to

$$Du(i) \leq d \text{ for } i = t, t+1, \cdots t + Nu$$
$$Hx(i) \leq h \text{ for } i = t, t+1, \cdots \infty$$

for stabilizable pairs (M, N) with r unstable modes and $N_u \geq r$ if the minimization problem is feasible.

The central idea is that if the minimization problem is feasible at sampling time t, then J_t is finite and $J_{t+1} \leq J_t + x(t)^t Rx(t) + u(t)^t Su(t)$. The cost function J_t can then be interpreted as a monotonically decreasing Lyapunov function and asymptotic stability is therefore guaranteed. Notice that the problem at $t + 1$ is also feasible (for the noise-free case without external perturbations). Also note that the infinite and unconstrained part of the objective function can be solved by a Riccati equation and that a cost function depending on the state can be obtained. This cost function is introduced in the finite horizon optimization problem which is solved by numerical methods.

The second type of approach has been developed in the GPC context following Clarke and Scattolini [61] CRHPC. The main idea is to impose state terminal constraints or, in the input-output context, to force the predicted output to exactly follow the reference during a sufficiently large horizon m after the costing horizon. The problem can be stated as:

$$\min_{\mathbf{u}} \sum_{j=N_1}^{N_2} \|\hat{y}(t+j \mid t) - w(t+j)\|_R^2 + \sum_{j=1}^{N_u} \| \triangle u(t+j-1)\|_Q^2$$

$$\text{subject to} \quad \triangle u(t + N_u + j - 1) = 0$$
$$y(t + N_2 + m + j) = w(t + N_2 + m + 1)$$

Stability results for CRHPC have been obtained [61],[60] for the unconstrained case. Scokaert and Clarke [188] have demonstrated the stability property CRHPC in the presence of constraints. The principal idea is that if a feasible solution is found and the settling horizon N_y is large enough to cover the transient of the output variables, the cost function is monotonically decreasing (if there are no external disturbances and the process is noise-free) and can be interpreted as a Lyapunov function which will guarantee stability. It can also be shown that the problem will be feasible in the next iteration.

Stability results for constrained SGPC have also been obtained by Rossiter and Kouvaritakis [112], [186], who found that for any reference $w(t)$ which assumes a constant value w^* after a number (N) of sampling periods, if the constrained SGPC is feasible for sufficiently large values of the horizons (depending on N), the closed loop will be stable and the output will asymptotically go to w^*. The stability has been also demonstrated in [184] without imposing the terminal state conditions implicitly used in SGPC. The work is based on characterizing all stable predictions which are not necessarily of a finite impulse response type as in standard SGPC. This allows for more degrees of freedom and increases the feasibility of the problem.

All stability results require the feasibility of the control law. If no feasible solution is found, one can always use the unconstrained solution and clip it to the manipulated variable bounds, but this way of operating would not guarantee the stability of the closed loop (for the nominal plant). Note that input constraints can always be satisfied by saturating the control signals, but this is not the case of output or state constraints which are the real cause of infeasibility. Some suggestions have been made in the literature to cope with infeasibility.

Rawlings and Muske [177] proposed dropping the state constraints during the initial portion of the infinite horizon to make the problem feasible. Zheng and Morari [213] proposed changing the hard constraints on the state ($Hx(i) \leq h$) for soft constraints ($Hx(i) \leq h + \epsilon$ with $\epsilon \geq 0$) to ensure feasibility, adding the term $\epsilon^t Q \epsilon$ to the costing function to penalize constraint violation and thus obtain better performance. They also demonstrated that any stabilizable system can be asymptotically stabilized by the MPC with soft constraints and state feedback if N_u is chosen to be sufficiently large and that it stabilizes any open-loop stable system with output feedback (state vector computed by an observer). Muske *et al.* [146] have shown that an infinite horizon MPC with output feedback which does not enforce state constraints during the first stages produces a stable closed-loop when controlling open-loop stable systems and that it also stabilizes unstable processes provided that the initial process and observer states are inside the feasible region. Scokaert and Clarke [188] have proposed a way of removing constraints when no feasible solutions are found. Their idea is to increase the lower constraint horizon until a feasible solution has been found. They also suggest that another possible way of removing constraints would be to have them organized in a hierarchical way with the critical ones at one end and the less important ones at the other. This ordering may be used to drop the constraints when no feasible solution is found.

An idea is proposed in [82] to ensure the feasibility of constrained SGPC in the presence of bounded disturbances that could take the process away from the constrained region. The idea is to determine the minimum required control power to reject the worst perturbations in the future. To implement this idea, tighter constraints than the physical limits are imposed on the manipulated variables. The difference between the physical limits and the new constraints is the minimum control effort required to maintain the feasibility of the constrained optimization problem to guarantee stability.

Model predictive control schemes for nonlinear systems which guarantee stability have also been proposed in [135] and [136]. The main idea is to solve the constrained MPC in a finite horizon driving the state to zero or inside a region W where control is transferred to a linear stabilizing controller. The main problem with this idea is that region W is very difficult to compute and that, in general, the resulting optimization problem is nonconvex. This makes the optimization problem much more difficult to solve and the optimality of the solution cannot be assured. Fortunately, stability is guaranteed when the optimization problem is feasible and does not require optimality for this type of controller, although the performance may suffer by using local minima. These ideas have been extended by Chen and Allgöwer [48], who proposed a quasi-infinite constrained MPC. The idea of the terminal region and linear stabilizing controller is used, but only for computing the terminal cost. In quasi-infinite horizon nonlinear MPC the control signal is determined by solving the optimization problem online and the control is never transferred to the linear stabilizing controller even when inside the terminal region.

Another way to achieve closed-loop stability of nonlinear predictive control proposed by Yang and Polak [206] is by imposing *contraction* constraints on the state. The idea is to impose the following constraint $||x(t + 1)|| < \alpha ||x(t)||$ with $\alpha \in [0, 1)$. Stability is guaranteed if the optimization problem is feasible. The main advantage of the algorithm is that if it is feasible the closed loop is exponentially stable. Imposing the *contraction* constraints is, however, very restrictive for many control problems, and unfeasibility is encountered in many situations.

There have been many formulations to guarantee MPC stability. Most of the formulations have two key ingredients, a terminal state penalization and a terminal set where the final state is forced (see Section 9.5). The terminal set conditions can be translated into a set of constraints on the manipulated variables, as seen in Section 7.1. Stability is therefore linked to a constraint satisfaction problem.

7.9 Multiobjective MPC

All the MPC strategies analyzed previously are based on optimizing a single objective cost function, which is usually quadratic, in order to determine the future sequence of control moves that makes the process behave best. However, in many control problems the behaviour of the process cannot be measured by a single objective function, but most of the time, there are different, and sometimes conflicting, control objectives. The reasons for multiple control objectives are varied:

- Processes have to be operated differently when they are at different operating stages. For example, at the startup phase of the process, a minimum startup time may be desired, but once the process has reached the operating regime, a minimum variance of the controlled variables may be the primary control objective.
- Even if the process is working at a particular operating stage, the control objective may depend on the value of the variables. For example the control objective when the process is working at the nominal operating point may be to minimize the weighted sum of the square errors of the controlled variables with respect to their prescribed values. But if the value of one of the variables is too high, because of a sudden perturbation, for example, the main control objective may be to reduce the value of this variable as soon as possible.

Furthermore, in many cases, the control objective is not to optimize the sum of the squared errors but rather to keep some variables within specified bounds. Notice that this situation is different to the constraint control MPC, as the objective is to keep the variable there, although excursions of the variable outside this region, though not desirable, are permitted. In constrained MPC the variables should be kept within the prescribed region because of physical limitations, plant safety, or other considerations. Constraints which cannot be violated are referred to as *hard* constraints, while those which can are known as *soft* constraints. These types of objectives can be expressed by penalizing the amount by which the offending variable violates the limit. Consider, for example, the process with a controlled variable $y(t)$ where the control objective is to keep $y_l \leq y(t) \leq y_h$. The control objective may be formulated as

$$J = p(y(t+j) - y_h) \sum_{j=N_1}^{N_2} (y(t+j) - y_h)^2 + p(y_l - y(t+j)) \sum_{j=N_1}^{N_2} (y(t+j) - y_l)^2$$

where function p is a step function. That is, p takes the value 1 when its argument is greater than or equal to zero and the value zero when its argument is negative.

Notice that the objective function is no longer a quadratic function and QP algorithms cannot be used. The problem can be transformed into a QP problem by introducing slack variables $\epsilon_h(j)$ and $\epsilon_l(j)$. That is

$$y(t+j) \leq y_h + \epsilon_h(j)$$
$$y(t+j) \geq y_l - \epsilon_l(j)$$

The manipulated variable sequence is now determined by minimizing:

$$J = \sum_{j=N_1}^{N_2} \epsilon_h(j)^2 + \sum_{j=N_1}^{N_2} \epsilon_l(j)^2$$

subject to $\epsilon_l(j) \geq 0$ and $\epsilon_h(j) \geq 0$ and the rest of the constraints acting on the problem. Notice that the problem has been transformed into a QP problem with more constraints and decision variables.

Sometimes all control objectives can be summarized in a single objective function. Consider, for example, a process with a series of control objectives $J_1, J_2, ..., J_m$. Some of the control objectives may be to keep some of the controlled variables as close to their references as possible, while other control objectives may be related to keeping some of the variables within specified regions. Consider all objectives to have been transformed into minimizing a quadratic function J_i, subject to a set of linear constraints on the decision variables $R_i u \leq a_i$. The future control sequence can be determined by minimizing the following objective function

$$J = \sum_{i=1}^{m} \beta_i J_i$$

subject to: $\mathbf{R}_i \mathbf{u} \le \mathbf{a}_i$ for $i = 1, \cdots, m$

The importance of each objective can be modulated by appropriate setting of all β_i. This is, however, a nontrivial matter in general as it is very difficult to determine the set of weights which will represent the relative importance of the control objectives. Furthermore, practical control objectives are sometime qualitative, making the task of determining the weights even more difficult.

7.9.1 Priorization of Objectives

In some cases, the relative importance of the control objectives can be established by priorization. That is, the objectives of greater priority, for example, objectives related to security, must be accomplished before other objectives of lower priority are considered. Objectives can be prioritized by giving much higher values to the corresponding weights. However, this is a difficult task which is usually done by a trial-and-error method.

In [202] a way of introducing multiple prioritized objectives into the MPC framework is given. Consider a process with a series of m prioritized control objectives O_i. Suppose that objective O_i has a higher priority than objective O_{i+1} and that the objectives can be expressed as:

$$\mathbf{R}_i \mathbf{u} \le \mathbf{a}_i$$

The main idea consists of introducing integer variables L_i which take the value 1 when the corresponding control objective is met and zero otherwise. Objectives are expressed as

$$\mathbf{R}_i \mathbf{u} \le \mathbf{a}_i + K_i(1 - L_i) \tag{7.20}$$

where K_i is a conservative upper bound on $\mathbf{R}_i \mathbf{u} - \mathbf{a}_i$. If objective O_i is satisfied, $L_i = 1$ and the reformulated objective coincides with the original control objective. By introducing K_i, the reformulated objective (constraint) is always satisfied even when the corresponding control objective O_i is not met ($L_1 = 0$).

The priorization of objectives can be established by imposing the following constraints:

$$L_i - L_{i+1} \ge 0 \text{ for } i = 1, \cdots, m - 1$$

The problem is to maximize the number of satisfied control objectives:

$$\sum_{i=1}^{m} L_i$$

If the process model is linear, the problem can be solved with a Mixed Integer Linear Programming (MILP) algorithm. The number of integer variables can be reduced as indicated in [202]. The idea is to use the same variable L_i for constraints that cannot be violated at the same time, as is the case of upper and lower bounds on the same control or manipulated variable.

The set of constraints in (7.20) can be modified [202], to improve the degree of the constraint satisfaction of objectives that cannot be satisfied. Suppose that not all objectives can be satisfied at a particular instance. Suppose that objective O_f is the first objective that failed. In order to come as close as possible to satisfying this objective, a slack variable α satisfying the following set of constraints is introduced

$$\mathbf{R}_i \mathbf{u} \le \mathbf{a}_i + \alpha + K_i \left((i-1) + (1 - L_i) - \sum_{j=1}^{i-1} L_j \right) \tag{7.21}$$

and the objective function to be minimized is

$$J = -K_\alpha \sum_{i=1}^{m} L_i + f(\alpha) \tag{7.22}$$

where f is a penalty function of the slack variable α (positive and strictly increasing) and K_α is an upper bound on f. The optimization algorithm will try to maximize the number of satisfied objectives ($L_i = 1$) before attempting to reduce $f(\alpha)$ because the overall objective function can be made smaller by increasing the number of nonzero L_i variables than by reducing $f(\alpha)$. As all objectives O_i for $i < f$ are satisfied ($L_i = 1$), the constraints in (7.21) will also be satisfied. As O_f is the first objective that failed,

$$\sum_{i=1}^{i-1} L_i = f - 1 \text{ for } i \ge f$$

That is, the term multiplying K_i of constraint (7.21) is zero for $i = f$, while for $i > f$ this term is greater than one. This implies that all constraints in (7.21) will be satisfied for $i > f$. The only active constraint is:

$$\mathbf{R}_f \mathbf{u} \le \mathbf{a}_f + \alpha$$

That is, the optimization method will try to optimize the degree of satisfaction of the first objective that failed only after all higher priority objectives have been satisfied. Notice that $L_i = 0$ does not imply that objective O_i is not satisfied, it only indicates that the corresponding constraint has been relaxed.

If the process is linear and function f is linear, the problem of maximizing (7.22) can be solved by a MILP. If f is a quadratic function, the problem can be solved by a Mixed Integer Quadratic Programming (MIQP) algorithm. Although there are efficient algorithms to solve mixed integer programming

problems, the amount of computation required is much greater than that required for LP or QP problems. The number of objectives should be kept small
to implement the method in real time.

7.10 Exercises

7.1. Consider the system described by $y(t+1) = ay(t) + bu(t)$ with $-0.2 \leq \Delta u(t) \leq 0.2$, $-1 \leq u(t) \leq 1$, and $-3 \leq y(t) \leq 3$:

1. Formulate the MPC control problem (determining the constraint matrices) with $N = 3$ using a quadratic objective function.
2. Repeat the exercise considering a 1-norm type of objective function.
3. Simulate the previous problems and check that constraints are fulfilled.

7.2. Consider the system described by $y(t+1) = 1.75y(t) - y(t-1) + 0.25u(t) + 0.25u(t-1)$:

1. Formulate an MPC with $N = 2$ and a quadratic objective and imposing
   overshoot constraints.
2. Repeat the previous exercise with an ∞-norm imposing monotonic constraints.
3. Simulate the resulting controllers and comment on the results.
4. Increase the control horizon and control weighting to $N = 11$ and $\lambda = 50$
   and comment on the results.

7.3. For the system described by $x(t+1) = Ax(t) + Bu(t)$ with

$$A = \begin{bmatrix} 0 & 1 \\ 1 & 1 \end{bmatrix} \quad B = \begin{bmatrix} 1 \\ 1 \end{bmatrix}$$

formulate an MPC with $N = 3$ and a quadratic objective function and:

1. a terminal region described by $\|x(t+N)\|_{\infty} \leq 0.1$.
2. a terminal region described by $\|x(t+N)\|_{1} \leq 0.1$.
3. a terminal region described by $x(t+N)^T x(t+N) \leq 0.1$. Comment on
   the type of optimization problem encountered.
4. a terminal region described by $x(t+N)^T x(t+N) \leq x(t)^T x(t)$.

7.4. Repeat the experiments of Section 6.4 in the case that the airplane has
the following operational constraints:

1. $|u_1| \leq 10$ and $|u_2| \leq 5$.
2. $|\Delta u_1| \leq 10$ and $|\Delta u_2| \leq 5$.
3. both sets of constraints at the same time.

8

Robust Model Predictive Control

Mathematical models of real processes cannot contemplate every aspect of reality. Simplifying assumptions have to be made, especially when the models are going to be used for control purposes, where models with simple structures (linear in most cases) and sufficiently small size have to be used due to available control techniques and real-time considerations. Thus, mathematical models, especially control models, can only describe the dynamics of the process in an approximative way.

Most control design techniques need a control model of the plant with fixed structure and parameters (*nominal model*), which is used throughout the design. If the control model were an exact, rather than an approximate, description of the plant and there were no external disturbances, processes could be controlled by an open-loop controller. Feedback is necessary in process control because of the external perturbations and model inaccuracies in all real processes. The objective of robust control is to design controllers which preserve stability and performance in spite of the modelling inaccuracies or uncertainties. Although the use of feedback contemplates the inaccuracies of the model implicitly, the term of robust control is used in the literature to describe control systems that explicitly consider the discrepancies between the model and the real process [130].

There are different approaches for modelling uncertainties depending mainly on the type of technique used to design the controllers. The most extended techniques are frequency response uncertainties and transfer function parametric uncertainties. Most of the approaches assume that there is a family of models and that the plant can be exactly described by one of the models belonging to the family. That is, if the family of models is composed of linear models, the plant is also linear. The approach considered here is the one relevant to the key feature of MPC which is to predict future values of the output variables. The uncertainties can be defined about the prediction capability of the model. It will be shown that no assumptions have to be made regarding the linearity of the plant in spite of using a family of linear models for control purposes.

8.1 Process Models and Uncertainties

Two basic approaches are extensively used in the literature to describe modelling uncertainties: frequency response uncertainties and transfer function parametric uncertainties. Frequency uncertainties are usually described by a band around the nominal model frequency response. The plant frequency response is presumed to be included in the band. In the case of parametric uncertainties, each coefficient of the transfer function is presumed to be bounded by the uncertainties limit. The plant is presumed to have a transfer function, with parameters within the uncertainty set.

Both ways of modelling uncertainties consider that the exact model of the plant belongs to the family of models described by the uncertainty bounds. That is, the plant is linear with a frequency response within the uncertainty band for the first case and the plant is linear and of the same order as that of the family of models for the case of parametric uncertainties.

Control models in MPC are used to predict what is going to happen: future trajectories. The appropriate way of describing uncertainties in this context seems to be by a model or a family of models that, instead of generating a future trajectory, may generate a band of trajectories in which the process trajectory will be included when the same input is applied, in spite of the uncertainties. One should expect this band to be narrow when a good model of the process is available and the uncertainty level is low and to be wide otherwise.

The most general way of posing the problem in MPC is as follows: consider a process whose behaviour is dictated by the following equation

$$y(t+1) = f(y(t), \cdots, y(t-n_y), u(t), \cdots, u(t-n_u), z(t), \cdots, z(t-n_z), \psi) \quad (8.1)$$

where $y(t) \in \mathbf{Y}$ and $u(t) \in \mathbf{U}$ are n and m vectors of outputs and inputs, $\psi \in \mathbf{\Psi}$ is a vector of parameters possibly unknown, and $z(t) \in \mathbf{Z}$ is a vector of possibly random variables.

Now consider the model or family of models, for the process described by

$$\hat{y}(t + 1) = \hat{f}(y(t), \cdots, y(t - n_{n_a}), u(t), \cdots, u(t - n_{n_b}), \theta) \quad (8.2)$$

where $\hat{y}(t + 1)$ is the prediction of output vector for instant $t + 1$ generated by the model; $\hat{f}$ is a vector function, usually a simplification of f; n_{n_a} and n_{n_b} are the number of past outputs and inputs considered by the model; and $\theta \in \mathbf{\Theta}$ is a vector of uncertainties about the plant. Variables that although influencing plant dynamics are not considered in the model because of the necessary simplifications or for other reasons are represented by $z(t)$.

The dynamics of the plant in (8.1) are completely described by the family of models (8.2) if for any $y(t), \cdots, y(t - n_y) \in \mathbf{Y}$, $u(t), \cdots, u(t - n_u) \in \mathbf{U}$, $z(t), \cdots, z(t - n_z) \in \mathbf{Z}$ and $\psi \in \mathbf{\Psi}$, there is a vector of parameters $\theta_i \in \mathbf{\Theta}$ such that

$$f(y(t), \cdots, y(t - n_y), u(t), \cdots, u(t - n_u), z(t), \cdots, z(t - n_z), \psi)$$
$$= \hat{f}(y(t), \cdots, y(t - n_{n_a}), u(t), \cdots, u(t - n_{n_b}), \theta_i)$$

The way in which the uncertainties parameter θ and its domain Θ are defined mainly depends on the structures of f and $\hat{f}$ and on the degree of certainty about the model. In the following the most used model structures in MPC will be considered.

8.1.1 Truncated Impulse Response Uncertainties

For an m-input n-output MIMO stable plant the truncated impulse response is given by N real matrices $(n \times m)$ H_t. The (i, j) entry of H_t corresponds to the ith output of the plant when an impulse has been applied to input variable u_j.

The natural way of considering uncertainties is by supposing that the coefficients of the truncated impulse response, which can be measured experimentally, are not known exactly and are a function of the uncertainty parameters. Different types of functions can be used. The most general way will be by considering that the impulse response may be within a set defined by $(\underline{H}_t)_{ij} \leq (H_t)_{ij} \leq (\overline{H}_t)_{ij}$; that is, $(H_t)_{ij}(\Theta) = (Hm_t)_{ij} + \Theta_{t_{ij}}$, with Θ defined by $(Hm_t)_{ij} - (\overline{H}_t)_{ij} \leq \Theta_{t_{ij}} \leq (\underline{H}_t)_{ij} - (Hm_t)_{ij}$ and Hm_t is the *nominal* response. The dimension of the uncertainty parameter vector is $N \times (m \times n)$. For the case of $N = 40$ and a 5-input 5-output MIMO plant, the number of uncertainty parameters is 1000, which will normally be too high for the min-max problem involved.

This way of modelling does not take into account the possible structures of the uncertainties. When these are considered, the dimension of the uncertainty parameter set may be considerably reduced.

In [47] and [162] a linear function of the uncertainty parameters is suggested:

$$H_t = \sum_{j=1}^{q} G_{t_j} \theta_j$$

The idea is that the plant can be described by a linear combination of q known stable linear time-invariant plants with unknown weighting θ_j. This approach is suitable in the case when the plant is nonlinear and linear models are obtained at different operating regimes. It seems plausible that a linear combination of linearized models can describe the behaviour of the plant over a wider range of conditions than a single model.

As will be seen, considering the impulse response as a linear function of the uncertainty parameters is of great interest for solving the robust MPC problem. Furthermore, note that the more general description of uncertainties $(H_t)_{ij}(\Theta) = (\underline{H}_t)_{ij} + \Theta_{t_{ij}}$ can also be expressed this way by considering

$$H_t(\theta) = \sum_{i=1}^{n} \sum_{j=1}^{m} (\underline{H}_t)_{ij} + \Theta_{t_{ij}} H_{ij}$$

where H_{ij} is a matrix with entry (i, j) equal to one and the remaining entries are zero.

The predicted output can be computed as

$$y(t + j) = \sum_{i=1}^{N} (Hm_i + \theta_i)u(t + j - i)$$

while the predicted nominal response is

$$ym(t + j) = \sum_{i=1}^{N} Hm_i u(t + j - i)$$

The prediction band around the nominal response is then limited by:

$$\min_{\theta \in \Theta} \sum_{i=1}^{N} \theta_i u(t + j - i) \text{ and } \max_{\theta \in \Theta} \sum_{i=1}^{N} \theta_i u(t + j - i)$$

8.1.2 Matrix Fraction Description Uncertainties

Let us consider the following n-output m-input multivariable discrete-time model

$$\mathbf{A}(z^{-1})y(t) = \mathbf{B}(z^{-1})\, u(t - 1) \tag{8.3}$$

where $A(z^{-1})$ and $B(z^{-1})$ are polynomial matrices of appropriate dimensions.

Parametric uncertainties about the plant can be described by $(\underline{A}_k)_{ij} \le (A_k)_{ij} \le (\overline{A}_k)_{ij}$ and $(\underline{B}_k)_{ij} \le (B_k)_{ij} \le (\overline{B}_k)_{ij}$. That is, $(A_k)_{ij} = (\underline{A}_k)_{ij} + \Theta_{a_{k_{ij}}}$ $(B_k)_{ij} = (\underline{B}_k)_{ij} + \Theta_{b_{k_{ij}}}$.

The number of uncertainty parameters for this description is $n_a \times n \times n + (n_b + 1) \times n \times m$. Note that uncertainties about actuators and dead times will mainly reflect on coefficients of the polynomial matrix $B(z^{-1})$, while uncertainties about the time constants will mainly affect the polynomial matrix $A(z^{-1})$. Note that if the parameters of the polynomial matrices $\mathbf{A}(z^{-1})$ and $\mathbf{B}(z^{-1})$ have been obtained via identification, the covariance matrix indicates how big the uncertainty band for the coefficients is.

The most frequent case in industry is that each of the entries of the transfer matrix has been characterized by its static gain, time constant, and equivalent dead time. Bounds on the coefficients of matrices $\mathbf{A}(z^{-1})$ and $\mathbf{B}(z^{-1})$ can be obtained from bounds on the gain and time constants. Uncertainties about the dead time are, however, difficult to handle. If the uncertainty band about the dead time is higher than the sampling time used, it will translate into a change in the order of the polynomial or coefficients that can change from zero and to zero. If the uncertainty band about the dead time is smaller than the sampling time, the pure delay time of the discrete-time model does not have to be changed. The fractional delay time can be modelled by the first terms of a Padé expansion and the uncertainty bound of these coefficients can be calculated from the uncertainties of the dead time. In any case dead time uncertainty bounds tend to translate into a very high degree of uncertainty about the coefficients of the polynomial matrix $\mathbf{B}(z^{-1})$.

The prediction equations can be expressed in terms of the uncertainty parameters. Unfortunately, for the general case, the resulting expressions are too complicated and of little use because the involved min-max problem would be too difficult to solve in real time. If the uncertainties only affect polynomial matrix $\mathbf{B}(z^{-1})$, the prediction equation is an affine function of the uncertainty parameter and the resulting min-max problem is less computationally expensive, as will be shown later in the chapter. Uncertainties on $\mathbf{B}(z^{-1})$ can be given in various ways. The most general way is by considering uncertainties on the matrices $(B_i = Bn_i + \theta_i)$. If the plant can be described by a linear combination of q known linear time invariant plants with unknown weighting θ_j, polynomial matrix $\mathbf{B}(z^{-1})$ can be expressed as:

$$\mathbf{B}(z^{-1}) = \sum_{i=1}^{q} \theta_i \mathbf{P}_i(z^{-1})$$

The polynomial matrices $\mathbf{P}_i(z^{-1})$ are a function of the polynomial matrices $\mathbf{B}_i(z^{-1})$ and $\mathbf{A}_i(z^{-1})$ corresponding to the matrix fraction description of each plant. For the case of diagonal $\mathbf{A}_i(z^{-1})$ matrices, the polynomial matrices $\mathbf{P}_i(z^{-1})$ can be expressed as:

$$\mathbf{P}_i(z^{-1}) = \prod_{j=1,i\neq j}^{q} \mathbf{A}_j(z^{-1})\mathbf{B}_i(z^{-1})$$

Note that the general case of uncertainties on the coefficient parameters could have also been expressed this way but with a higher number of uncertainty parameters. Using prediction Equation (6.5)

$$\begin{aligned} y(t+j|t) &= \mathbf{F}_j(z^{-1})y(t) + \mathbf{E}_j(z^{-1})\mathbf{B}(z^{-1}) \triangle u(t+j-1) \\ &= \mathbf{F}_j(z^{-1})y(t) + \mathbf{E}_j(z^{-1})(\textstyle\sum_{i=1}^{q} \theta_i \mathbf{P}_i(z^{-1})) \triangle u(t+j-1) \end{aligned}$$

that is, an affine function in θ_i.

8.1.3 Global Uncertainties

The key idea of this way of modelling the uncertainties is to assume that all modelling errors are globalized in a vector of parameters, such that the plant can be described by the following family of models

$$\hat{y}(t+1) = \hat{f}(y(t), \cdots, y(t-n_{n_a}), u(t), \cdots, u(t-n_{n_b})) + \theta(t)$$

with $\dim(\theta(t))=n$.

Notice that global uncertainties can be related to other types of uncertainties. For the impulse response model, the output at instant $t+j$ with parametric and temporal uncertainties description is given by:

$$\hat{y}(t+j) = \sum_{i=1}^{N}(Hm_i + \theta_i)u(t+j-i)$$

$$\hat{y}(t+j) = \sum_{i=1}^{N}(Hm_i)u(t+j-i) + \theta(t+j)$$

Therefore $\theta(t+j) = \sum_{i=0}^{N}\theta_i u(t+j-i)$ and the limits for the i component $(\underline{\theta}_i, \bar{\theta}_i)$ of vector $\theta(t+j)$ when $u(t)$ is bounded (in practice always) are given by:

$$\underline{\theta}_i = \min_{u(\cdot)\in\mathbf{U},\theta_i\in\Theta}\sum_{i=0}^{N}\theta_{t_i}u(t+j-i)$$

$$\bar{\theta}_i = \max_{u(\cdot)\in\mathbf{U},\theta_i\in\Theta}\sum_{i=0}^{N}\theta_{t_i}u(t+j-i)$$

The number of uncertainty parameters is reduced from $N \times (m \times n)$ to n but the approach is more conservative because the limits of the uncertainty parameter domain have to be increased to contemplate the worst global situation. The way of defining the uncertainties is, however, much more intuitive and directly reflects how good the j step ahead prediction model is.

For the left matrix fraction description, the uncertainty model is defined by

$$\tilde{\mathbf{A}}(z^{-1})y(t) = \mathbf{B}(z^{-1})\,\triangle\,u(t-1) + \theta(t) \tag{8.4}$$

with $\theta(t) \in \Theta$ defined by $\underline{e}(t) \le \theta(t) \le \bar{e}(t)$.

Notice that with this type of uncertainty one does not have to presume the model to be linear, as is the case of parametric uncertainty or frequency uncertainty modelling. Here it is only assumed that the process can be approximated by a linear model in the sense that all trajectories will be included in bands that depend on $\underline{\theta}(t)$ and $\bar{\theta}(t)$. It may be argued that this type of global uncertainties are more disturbances than uncertainties because they seem to work as external perturbations. However, the only assumption made is that they are bounded; in fact $\theta(t)$ may be a function of past inputs and outputs. If the process variables are bounded, the global uncertainties can also be bounded.

The model given by Expression (8.4) is an extension of the integrated error concept used in CARIMA models. Because of this, the uncertainty band will grow with time. To illustrate this point consider the system described by the first-order difference equation

$$y(t+1) = ay(t) + bu(t) + \theta(t+1) \text{ with } \underline{\theta} \le \theta(t) \le \overline{\theta}$$

that is, a model without integrated uncertainties. Let us suppose that the past inputs and outputs and future inputs are zero, thus producing a zero nominal trajectory. The output of the uncertain system is given by

$$y(t+1) = \theta(t+1)$$
$$y(t+2) = a\,\theta(t+1) + \theta(t+2)$$

$$\vdots$$

$$y(t+N) = \sum_{j=0}^{N-1} a^j \theta(t + N - j)$$

The upper bound will grow as $|a|^{(j-1)}\overline{\theta}$ and the lower bound as $|a|^{(j-1)}\underline{\theta}$. The band will stabilize for stable systems to a maximum value of $\overline{\theta}/(1 - |a|)$ and $\underline{\theta}/(1 - |a|)$. This type of model will not incorporate the possible drift in the process caused by external perturbations.

For the case of integrated uncertainties defined by the following model

$$y(t+1) = ay(t) + bu(t) + \frac{\theta(t)}{\triangle}$$

the increment of the output is given by

$$\triangle y(t+k) = \sum_{j=0}^{k-1} a^j \theta(t+k-j)$$

and

$$y(t+N) = \sum_{k=1}^{N} \triangle y(t+j) = \sum_{k=1}^{N} \sum_{j=0}^{k-1} a^j \theta(t+k-j)$$

indicating that the uncertainty band will grow continuously. The rate of growth of the uncertainty band stabilizes to $\overline{\theta}/(1 - |a|)$ and $\underline{\theta}/(1 - |a|)$ after the transient caused by process dynamics.

In order to generate the j step ahead prediction for the output vector, let us consider the Bezout identity:

$$I = \mathbf{E}_j(z^{-1})\tilde{\mathbf{A}}(z^{-1}) + \mathbf{F}_j(z^{-1})z^{-j} \qquad (8.5)$$

Using Equations (8.4) and (8.5) we get

$$y(t+j) = \mathbf{F}_j(z^{-1})y(t) + \mathbf{E}_j(z^{-1})\mathbf{B}(z^{-1})\triangle u(t+j-1) + \mathbf{E}_j(z^{-1})\theta(t+j) \quad (8.6)$$

Notice that the prediction will be included in a band around the nominal prediction $ym(t+j) = \mathbf{F}_j(z^{-1})y(t) + \mathbf{E}_j(z^{-1})\mathbf{B}(z^{-1})\triangle u(t+j-1)$ delimited by

$$ym(t+j) + \min_{\theta(\cdot)\in\Theta} \mathbf{E}_j(z^{-1})\theta(t+j) \le y(t+j) \le ym(t+j) + \max_{\theta(\cdot)\in\Theta} \mathbf{E}_j(z^{-1})\theta(t+j)$$

Because of the recursive way in which polynomial $\mathbf{E}_j(z^{-1})$ can be obtained, when $\bar{e}(t)$ and $\underline{e}(t)$ are independent of t, the band can also be obtained recursively by increasing the limits obtained for $y(t+j-1)$ by

$$\max_{\theta(t+j)\in\Theta} E_{j,j-1}\theta(t+1) \quad \text{and} \quad \min_{\theta(t+j)\in\Theta} E_{j,j-1}\theta(t+1)$$

where $\mathbf{E}_j(z^{-1}) = \mathbf{E}_{j-1}(z^{-1}) + E_{j,j-1}z^{-(j-1)}$.

Consider the set of j ahead optimal predictions $\mathbf{y}$ for $j = 1, \cdots, N$, which can be written in condensed form as

$$\mathbf{y} = \mathbf{G}_u\mathbf{u} + \mathbf{G}_\theta\theta + \mathbf{f} \quad (8.7)$$

where $\theta = [\theta(t+1), \theta(t+2), \cdots, \theta(t+N)]^T$ is the vector of future uncertainties, $\mathbf{y}$ is the vector of predicted outputs, $\mathbf{f}$ is the free response, that is, the response due to past outputs (up to time t) and past inputs and vector $\mathbf{u}$ corresponds to the present and future values of the control signal.

8.2 Objective Functions

The objective of predictive control is to compute the future control sequence $u(t)$, $u(t+1)$, ..., $u(t+N_u)$ in such a way that the optimal j step ahead predictions $y(t+j \mid t)$ are driven close to $w(t+j)$ for the prediction horizon. The way in which the system will approach the desired trajectories will be indicated by a function J which depends on the present and future control signals and uncertainties. The usual way of operating, when considering a stochastic type of uncertainty, is to minimize function J for the most expected situation; that is, supposing that the future trajectories are going to be the future expected trajectories. When bounded uncertainties are considered explicitly, bounds on the predictive trajectories can be calculated and it would seem that a more robust control would be obtained if the controller tried to minimize the objective function for the worst situation; that is, by solving

$$\min_{\mathbf{u}\in U} \max_{\theta\in\Theta} J(\mathbf{u}, \theta)$$

The function to be minimized is the maximum of a norm that measures how well the process output follows the reference trajectories. Different types of norms can be used for this purpose.

8.2.1 Quadratic Cost Function

Let us consider a finite horizon quadratic criterion

$$J(N_1, N_2, N_u) = \sum_{j=N_1}^{N_2} [\hat{y}(t + j \mid t) - w(t + j)]^2 + \sum_{j=1}^{N_u} \lambda[\triangle u(t + j - 1)]^2 \quad (8.8)$$

If the prediction Equation (8.7) is used, Equation (8.8) can now be written as

$$\begin{aligned} J &= (\mathbf{G}_u\mathbf{u} + \mathbf{G}_\theta\theta + \mathbf{f} - \mathbf{w})^T(\mathbf{G}_u\mathbf{u} + \mathbf{G}_\theta\theta + \mathbf{f} - \mathbf{w}) + \lambda\mathbf{u}^T\mathbf{u} \quad (8.9) \\ &= \mathbf{u}^T M_{uu}\mathbf{u} + M_u\mathbf{u} + M + M_\theta\theta + \theta^t M_{\theta\theta}\theta + \theta^t M_{eu}\mathbf{u} \end{aligned}$$

where $\mathbf{w}$ is a vector containing the future reference sequences $\mathbf{w} = [w(t + N_1), \cdots, w(t + N_2)]^T$

The function $J(\mathbf{u}, \theta)$ can be expressed as a quadratic function of θ for each value of $\mathbf{u}$

$$J(\mathbf{u}, \theta) = \theta^t M_{\theta\theta}\theta + M'_e(\mathbf{u})\theta + M'\mathbf{u} \quad (8.10)$$

with $M'_\theta = M_\theta + \mathbf{u}^t M_\theta$ and $M' = M + M_u\mathbf{u} + \mathbf{u}^t M_{uu}\mathbf{u}$.

Let us define:

$$Jm(\mathbf{u}) = \max_{\theta \in \Theta} J(\mathbf{u}, \theta)$$

Matrix $M_{\theta\theta} = \mathbf{G}_\theta^t \mathbf{G}_\theta$ is a positive definite matrix because G_θ is a lower triangular matrix having all the elements on the principal diagonal equal to one. Since matrix $M_{\theta\theta}$ is positive definite, the function is strictly convex ([15] theorem 3.3.8) and the maximum of J will be reached in one vertex of the polytope Θ ([15] theorem 3.4.6).

For a given $\mathbf{u}$ the maximization problem is solved by determining which of the $2^{(N \times n)}$ vertices of the polytope Θ produces the maximum value of $J(\mathbf{u}, \theta)$.

It can easily be seen that function $Jm(\mathbf{u})$ is a piecewise quadratic function of $\mathbf{u}$. Let us divide the $\mathbf{u}$ domain $\mathbf{U}$ in different regions $\mathbf{U}_p$ such that $\mathbf{u} \in \mathbf{U}_p$ if the maximum of $J(\mathbf{u}, \theta)$ is attained for the polytope vertex θ_p. For the region $\mathbf{U}_p$ the function $Jm(\mathbf{u})$ is defined by

$$Jm(\mathbf{u}) = \mathbf{u}^t M_{uu}\mathbf{u} + M_u^*(\theta_p)\mathbf{u} + M^*\theta_p$$

with $M_u^* = M_u + \theta_p^t M_u$ and $M^* = M + M_\theta\theta_p + \theta_p^t M_{\theta\theta}\theta_p$.

Matrix M_{uu}, which is the Hessian matrix of function $Jm(\mathbf{u})$, can be assured to be positive definite by choosing a value of $\lambda > 0$. This implies that the function is convex ([15] theorem 3.3.8) and that there are no local optimal solutions different from the global optimal solution ([15] theorem 3.4.2).

One of the main problems of nonlinear programming algorithms, local minima, is avoided, and any nonlinear programming method can be used to minimize function $Jm(\mathbf{u})$. However, and because the evaluation of $Jm(\mathbf{u})$

implies finding the minimum at one vertex of the polytope Θ, the computation time can be prohibitive for real-time applications with long costing and control horizons. The problem gets even more complex when the uncertainties on the parameters of the transfer function are considered. The amount of computation required can be reduced considerably if other types of objective functions are used, as will be shown in the following sections.

8.2.2 ∞-∞ norm

Campo and Morari [47], showed that by using an ∞-∞ type of norm the min-max problem involved can be reduced to a linear programming problem that requires less computation and can be solved with standard algorithms. Although the algorithm proposed by Campo and Morari was developed for processes described by the truncated impulse response, it can easily be extended to the left matrix fraction descriptions used throughout the text.

The objective function is now described as

$$J(\mathbf{u}, \theta) = \max_{j=1\cdots N} \|\hat{y}(t+j|t) - w(t)\|_\infty = \max_{j=1\cdots N} \max_{i=1\cdots n} |\hat{y}_i(t+j|t) - w_i(t)| \quad (8.11)$$

Note that this objective function will result in an MPC which minimizes the maximum error between any of the process outputs and the reference trajectory for the worst situation of the uncertainties; the control effort required to do so is not taken into account.

By making use of the prediction equation $\mathbf{y} = G_u\mathbf{u} + G_\theta\theta + \mathbf{f}$ and defining $g(\mathbf{u}, \theta) = (\mathbf{y} - \mathbf{w})$, the control problem can be expressed as

$$\min_{\mathbf{u} \in U} \max_{\theta \in \Theta} \max_{i=1\cdots n \times N} |g_i(\mathbf{u}, \theta)|$$

Define $\mu^*(\mathbf{u})$ as

$$\mu^*(\mathbf{u}) = \max_{\theta \in \Theta} \max_{i=1\cdots n \times N} |g_i(\mathbf{u}, \theta)|$$

If there is any positive real value μ satisfying $-\mu \leq g_i(\mathbf{u}, \theta) \leq \mu$, $\forall \theta \in \Theta$ and for $i = 1 \cdots n \times N$ it is clear that μ is an upper bound of $\mu^*(\mathbf{u})$. The problem can now be transformed into finding the smallest upper bound μ and some $\mathbf{u} \in U$ for all $\theta \in \Theta$. When constraints on the controlled variables $(\underline{\mathbf{y}}, \overline{\mathbf{y}})$ are taken into account, the problem can be expressed as

$$\min_{\mu, \mathbf{u}} \mu$$

subject to: $\left. \begin{array}{c} -\mu \leq g_i(\mathbf{u}, \theta) \leq \mu \\ \underline{\mathbf{y}}_i - \mathbf{w}_i \leq g_i(\mathbf{u}, \theta) \leq \overline{\mathbf{y}}_i - \mathbf{w}_i \end{array} \right\} \begin{array}{l} \text{for } i = 1, \cdots, n \times N \\ \forall \theta \in \Theta \end{array}$

The control problem has been transformed into an optimization problem with an objective function which is linear in the decision variables $(\mu, \mathbf{u})$ and with an infinite (continuous) number of constraints. If $g(\mathbf{u}, \theta)$ is an affine function of θ, $\forall \mathbf{u} \in \mathbf{U}$, the maximum and minimum of $g(\mathbf{u}, \theta)$ can be obtained at one of the extreme points of Θ [47]. Let us call $\mathcal{E}$ the set formed by the $2^{n \times N}$ vertices of Θ. If constraints are satisfied for every point of $\mathcal{E}$ they will also be satisfied for every point of Θ. Thus the infinite, and continuous, constraints can be replaced by a finite number of constraints.

When the global uncertainty model is used and constraints on the manipulated variables $(\underline{\mathbf{U}}, \overline{\mathbf{U}})$ and on the slew rate of the manipulated variables $(\underline{\mathbf{u}}, \overline{\mathbf{u}})$ are also taken into account, the problem can be stated as

$$\min_{\mu, \mathbf{u}} \mu$$

subject to

$$\left.\begin{array}{c} 1\mu \geq G_u \mathbf{u} + G_\theta \theta + \mathbf{f} - \mathbf{w} \\ 1\mu \geq -G_u \mathbf{u} - G_\theta \theta - \mathbf{f} + \mathbf{w} \\ \overline{\mathbf{y}} \geq G_u \mathbf{u} + G_\theta \theta + \mathbf{f} \\ -\underline{\mathbf{y}} \geq -G_u \mathbf{u} - G_\theta \theta - \mathbf{f} \end{array}\right\} \forall \theta \in \mathcal{E}$$

$$\overline{\mathbf{u}} \geq \mathbf{u}$$
$$-\underline{\mathbf{u}} \geq -\mathbf{u}$$
$$\overline{\mathbf{U}} \geq T\mathbf{u} + 1u(t-1)$$
$$-\underline{\mathbf{U}} \geq -T\mathbf{u} - 1u(t-1)$$

where 1 is an $(N \times n) \times m$ matrix formed by N $m \times m$ identity matrices and T is a lower triangular block matrix whose non null block entries are $m \times m$ identity matrices. The problem can be transformed into the usual form

$$\min_{\mathbf{x}} \mathbf{c}^t \mathbf{x} \quad \text{subject to} \quad A\mathbf{x} \leq \mathbf{b}, \quad \mathbf{x} \geq 0$$

with

$$\mathbf{x} = \begin{bmatrix} \mathbf{u} - \underline{\mathbf{u}} \\ \mu \end{bmatrix} \quad \mathbf{c}^t = [\overbrace{0, \cdots, 0}^{m \times N_u}, 1] \quad A^t = [A_1^t, \cdots, A_{2^N}^t, A_u^t] \quad \mathbf{b}^t = [\mathbf{b}_1^t, \cdots, \mathbf{b}_{2^N}^t, \mathbf{b}_u^t]$$

The block matrices have the form:

$$A_i = \begin{bmatrix} G_u & -1 \\ -G_u & -1 \\ G_u & 0 \\ -G_u & 0 \end{bmatrix} \quad A_u = \begin{bmatrix} I & 0 \\ T & 0 \\ -T & 0 \end{bmatrix} \quad \mathbf{b}_i = \begin{bmatrix} -G_u\underline{\mathbf{u}} - G_\theta\theta_i - \mathbf{f} + \mathbf{w} \\ G_u\underline{\mathbf{u}} + G_\theta\theta_i + \mathbf{f} - \mathbf{w} \\ \overline{\mathbf{y}} - G_u\underline{\mathbf{u}} - G_\theta\theta_i - \mathbf{f} \\ -\underline{\mathbf{y}} + G_u\underline{\mathbf{u}} + G_\theta\theta_i - \mathbf{f} \end{bmatrix}$$

$$\mathbf{b}_u = \begin{bmatrix} \overline{\mathbf{u}} - \underline{\mathbf{u}} \\ \overline{\mathbf{U}} + T\mathbf{u} + 1u(t-1) \\ -\underline{\mathbf{U}} - T\mathbf{u} - 1u(t-1) \end{bmatrix}$$

where θ_i is the ith vertex of $\mathcal{E}$.

The number of variables involved in the linear programming problem is $m \times N_u + 1$ while the number of constraints is $4 \times n \times N \times 2^{n \times N} + 3m \times N_u$. As the number of constraints is much higher than the number of decision variables, solving the dual LP problem should be less computationally expensive, as pointed out by Campo and Morari [47].

The number of constraints can, however, be dramatically reduced because of the special form of matrix A. Consider the jth row for each of the constraint blocks $A_i x \leq b_i$. As $A_1 = A_2 = \cdots = A_{2N}$, the only constraint limiting the feasible region will be the one with the smallest value on the jth element of vector b_i. Therefore, all the other $(2^N - 1)$ constraints can be eliminated and the number of constraints can be reduced to $4 \times n \times N + 3m \times N_u$. Notice that any uncertainty model giving rise to an affine function $g(\mathbf{u}, \theta)$ can be transformed into an LP problem as shown by Campo and Morari [47]. The truncated impulse response uncertainty model or uncertainties in the $B(z^{-1})$ polynomial matrix produce an affine function $g(\mathbf{u}, \theta)$. However, the constraint reduction mechanism described earlier cannot be applied and the number of constraints would be very high.

8.2.3 1-norm

Although the type of ∞-∞ norm used earlier seems to be appropriate in terms of robustness, the norm is only concerned with the maximum deviation and the rest of the behaviour is not taken explicitly into account. Other types of norms are more appropriate for measuring the performance. Allwright [5] has shown that this method can be extended to the 1-norm for processes described by their truncated impulse response. The derivation for the left matrix representation is also straightforward

The objective function is

$$J(\mathbf{u}, \theta) = \sum_{j=N_1}^{N_2} \sum_{i=1}^{n} |y_i(t+j \mid t, \theta) - w_i(t+j)| + \lambda \sum_{j=1}^{N_u} \sum_{i=1}^{m} |\triangle u_i(t+j-1)| \quad (8.12)$$

where N_1 and N_2 define the prediction horizon and N_u defines the control horizon. If a series of $\mu_i \geq 0$ and $\beta_i \geq 0$ such that for all $\theta \in \Theta$,

$$-\mu_i \leq (y_i(t+j) - w_i(t+j)) \leq \mu_i$$
$$-\beta_i \leq \triangle u_i(t+j-1) \leq \beta_i$$
$$0 \leq \sum_{i=1}^{n \times N} \mu_i + \lambda \sum_{i=1}^{m \times N_u} \beta_i \leq \gamma$$

then γ is an upper bound of

$$\mu^*(\mathbf{u}) = \max_{\theta \in \mathcal{E}} \sum_{j=1}^{n} \sum_{i=1}^{} |y_i(t+j, \theta) - w_i(t+j)| + \lambda \sum_{j=1}^{N_u} \sum_{i=1}^{m} |\triangle u_i(t+j-1)|$$

The problem is reduced to minimizing the upper bound γ.

When the global uncertainty model is used and constraints on the output variables, the manipulated variables $(\underline{U}, \overline{U})$, and the slew rate of the manipulated variables $(\underline{u}, \overline{u})$ are taken into account, the problem can be interpreted as an LP problem with:

$$\min_{\gamma,\mu,\beta,u} \gamma$$

subject to

$$\left.\begin{array}{c} \mu \geq G_u u + G_\theta \theta + f + w \\ \mu \geq -G_u u - G_\theta \theta - f + w \\ \overline{y} \geq G_u u + G_\theta \theta + f \\ -\underline{y} \geq -G_u u - G_\theta \theta - f \end{array}\right\} \quad \forall \theta \in \mathcal{E}$$

$$\begin{array}{rcl} \beta & \geq & u \\ \beta & \geq & -u \\ \overline{u} & \geq & u \\ -\underline{u} & \geq & u \\ \overline{U} & \geq & T\,u + \mathbf{1}u(t-1) \\ -\underline{U} & \geq & -T u - \mathbf{1}u(t-1) \\ \gamma & \geq & \mathbf{1}^t \mu + \lambda \mathbf{1}\beta \end{array}$$

The problem can be transformed into the usual form

$$\min_{x} c^t x \text{ subject to } Ax \leq b, \ x \geq 0$$

with

$$x = \begin{bmatrix} \underline{u} - u \\ \hline \mu \\ \hline \beta \\ \hline \gamma \end{bmatrix} \qquad c^t = [\overbrace{0, \cdots, 0}^{m \times N_u}, \overbrace{0, \cdots, 0}^{n \times N}, \overbrace{0, \cdots, 0}^{m \times N_u}, 1]$$

$$A^t = [A_1^t, \cdots, A_{2N}^t, A_u^t] \quad b^t = [b_1^t, \cdots, b_{2N}^t, b_u^t]$$

where the block matrices take the following form

$$A_i = \begin{bmatrix} G_u & -I & 0 & 0 \\ -G_u & -I & 0 & 0 \\ G_u & 0 & 0 & 0 \\ -G_u & 0 & 0 & 0 \end{bmatrix} \quad b_i = \begin{bmatrix} -G_u \underline{u} - G_\theta \theta_i - f + w \\ G_u \underline{u} + G_\theta \theta_i + f - w \\ \overline{y} - G_u \underline{u} - G_\theta \theta_i - f \\ -\underline{y} + G_u \underline{u} + G_\theta \theta_i + f \end{bmatrix}$$

$$A_u = \begin{bmatrix} I & 0 & 0 & 0 \\ I & 0 & -I & 0 \\ -I & 0 & -I & 0 \\ T & 0 & 0 & \\ -T & 0 & 0 & 0 \\ 0 & \underbrace{1,\cdots,1}_{n \times N} & \underbrace{1,\cdots,1}_{m \times N_u} & -1 \end{bmatrix} \quad b_u = \begin{bmatrix} \overline{u} - \underline{u} \\ \hline -\underline{u} \\ \hline \underline{u} \\ \hline \overline{U} - T\underline{u} - \mathbf{1}u(t-1) \\ \hline -\underline{U} + T\underline{u} + \mathbf{1}u(t-1) \\ \hline 0 \end{bmatrix}$$

and θ_i is the ith vertex of $\mathcal{E}$. The number of variables involved in the linear programming problem is $2 \times m \times N_u + n \times N + 1$, while the number of constraints is $4 \times n \times N \times 2^{n \times N} + 5 \times m \times N_u + 1$. As the number of constraints is much higher than the number of decision variables, solving the dual LP problem should also be less computationally expensive than the primal problem.

The number of constraints can be reduced considerably as in the ∞-∞ norm case because of the special form of the constraint matrix A. Consider the jth row for each constraint block $A_i \mathbf{x} \le \mathbf{b}_i$. As $A_1 = A_2 = \cdots = A_{2N}$; the only constraint limiting the feasible region will be the one with the smallest value on the jth element of vector $\mathbf{b}_i$. Therefore, all the other $(2^N - 1)$ constraints can be eliminated. The number of constraints can be reduced to $4 \times n \times N + 5m \times N_u + 1$.

8.3 Robustness by Imposing Constraints

A way of guaranteeing robustness in MPC is to impose that stability conditions are satisfied for all possible realizations of the uncertainties. The stability conditions (see Section 9.5) were summarized in [137]. The key ingredients of the stabilizing MPC are a terminal set and a terminal cost. The terminal state (i.e. the state at the end of the prediction horizon) is forced to reach a terminal set that contains the steady state. An associated cost denoted as terminal cost, which is added to the cost function, is associated to the terminal state.

The robust MPC consists of finding a vector of future control moves such that it minimizes an objective function (including a terminal cost satisfying the stability conditions [137]) and forces the final state to reach the terminal region for all possible value of the uncertainties; that is

$$\min_{\mathbf{u}} \ J(x(t), \mathbf{u}) \tag{8.13}$$

$$\text{subject to} \quad \left. \begin{array}{l} \mathbf{Ru} \le \mathbf{r} + \mathbf{V}x(t) \\ x(t+N) \in \mathbf{\Omega}_T \end{array} \right\} \forall \theta \in \mathbf{\Theta}$$

where the terminal set $\mathbf{\Omega}_T$ is usually defined by a polytope $\mathbf{\Omega}_T \triangleq \{x : \mathbf{R}_T x \le \mathbf{r}_T\}$. The inequalities $\mathbf{Ru} \le \mathbf{r} + \mathbf{V}x(t)$ contain the operating constraints. If there are operating constraints on the process output and/or state, vector $\mathbf{r}$ is an affine function of the uncertainties θ; i.e., $\mathbf{r} = \mathbf{r}_0 + \mathbf{R}_\theta \theta$. The vector of predicted state can be written as:

$$\mathbf{x} = \mathbf{G}_u \mathbf{u} + \mathbf{G}_\theta \theta + \mathbf{f}_x x(t) \tag{8.14}$$

Taking the rows corresponding to $x(t+N)$ and substituting them into the inequality defining the terminal region

$$\mathbf{R}_T(\mathbf{g}_{u_N} \mathbf{u} + \mathbf{g}_{\theta_N} \theta + \mathbf{f}_{x_N} x(t)) \le \mathbf{r}_T \tag{8.15}$$

where $\mathbf{g}_{u_N}$, $\mathbf{g}_{\theta_N}$, and $\mathbf{f}_{x_N}$ are the last n rows of $\mathbf{G}_u$, $\mathbf{G}_\theta$, and $\mathbf{F}_x$ respectively, with $n = dim(x)$. The left-hand side of Inequalities (8.15) are affine functions of the uncertainty vector θ. Problem (8.13) results in a QP or LP problem (depending on the type of the objective function) with an infinite number of constraints. As in the previous cases, because the constraints are affine expressions of the uncertainties, if the inequalities hold for all extreme points (vertices) of Θ they also hold for all points inside Θ; that is, the infinite constraints can be replaced by a finite number (although normally very high) of constraints and the problem is solvable. The problem can be expressed as

$$\min_{\mathbf{u}} \ J(x(t), \mathbf{u}) \tag{8.16}$$

$$\text{subject to} \quad \left. \begin{array}{l} \mathbf{Ru} \le \mathbf{r}_0 + \mathbf{R}_\theta \theta_i + \mathbf{V}x(t) \\ \mathbf{R}_T(\mathbf{g}_{u_N}\mathbf{u} + \mathbf{g}_{\theta_N}\theta_i + \mathbf{f}_{x_N}x(t)) \le \mathbf{r}_T \end{array} \right\} \forall \theta_i \in \varepsilon$$

where ε is the finite set of extreme points (vertices) of Θ.

8.4 Constraint Handling

As seen in the previous sections, when uncertainties are additive, the robust MPC consists of solving at each sampling step a problem of the form

$$\min_{\mathbf{u}} \ J(x(t), \mathbf{u}) \tag{8.17}$$

$$\text{subject to} \ \ \mathbf{R}_u \mathbf{u} \le \mathbf{r} + \mathbf{R}_\theta \theta_i + \mathbf{R}_x x(t) \ \ \forall \theta_i \in \varepsilon \tag{8.18}$$

where function $J(x(t), \mathbf{u})$ can be a linear, quadratic, or piecewise quadratic objective function. The number of constraints in (8.18) is equal to the number of rows of $\mathbf{R}_u$ times the number of vertices of the polytope defining the uncertainties along the prediction horizon. The number of constraints can be reduced dramatically as indicated in Sections 8.2.2 and 8.2.3. Consider the j rows of all the constraints in (8.18)

$$\mathbf{r}_{u_j} \mathbf{u} \le \mathbf{r}_j + \mathbf{r}_{\theta_j} \theta_i + \mathbf{r}_{x_j} x(t) \ \ \forall \theta_i \in \varepsilon \tag{8.19}$$

where $\mathbf{r}_{u_j}$, $\mathbf{r}_{\theta_j}$, and $\mathbf{r}_{x_j}$ are the j rows of matrices $\mathbf{R}_u$, $\mathbf{R}_\theta$, and $\mathbf{R}_x x(t)$, respectively. The scalar $\mathbf{r}_j$ is the j entry of vector $\mathbf{r}$. Let us define $m_j \triangleq \min_{\theta_i \in \epsilon} \mathbf{r}_{\theta_j} \theta_i$. Now consider the inequality:

$$\mathbf{r}_{u_j} \mathbf{u} \le \mathbf{r}_j + m_j + \mathbf{r}_{x_j} x(t) \tag{8.20}$$

It is easy to see that if Inequality (8.20) is satisfied, all constraints in (8.19) will also be satisfied. Furthermore, if any constraints in (8.19) is not satisfied then constraint (8.20) will not be satisfied. Problem (8.17) can be expressed with a considerably smaller number of constraints:

$$\min_{\mathbf{u}} \; J(x(t), \mathbf{u}) \tag{8.21}$$

$$\text{subject to} \;\; \mathbf{R}_u \mathbf{u} \le \mathbf{r} + \mathbf{m} + \mathbf{R}_x x(t) \tag{8.22}$$

where $\mathbf{m}$ is a vector with its j entry equal to $\min_{\theta_i \in \epsilon} \mathbf{r}_{\theta_j} \theta_i$. Notice that these quantities are constant and can be computed offline.

8.5 Illustrative Examples

8.5.1 Bounds on the Output

The setpoint of many processes in industry is determined by an optimization program to satisfy economic objectives. As a result, the optimal setpoint is usually on the intersection of some constraints. This is, for example, the normal situation when maximizing the throughput, which normally results in operating the process at extreme conditions as near as possible to the safety or quality constraints. Consideration of uncertainties may be of great interest for this type of situation. If an MPC that takes into account the constraints is used, the MPC will solve the problem, keeping the expected values of the output signals within the feasible region, but, because of external perturbations or uncertainties, this does not guarantee that the output is going to be bound. When uncertainties are taken into account, the MPC will minimize the objective function for the worst situation and keep the value of the variables within the constraint region for all possible cases of uncertainties.

To illustrate this point, consider the process described by the following difference equation

$$y(t+1) = -1.4y(t) + 0.42y(t-1) + 0.1u(t) + 0.2u(t-1) + \frac{\theta(t+1)}{\triangle}$$

with $-0.03 \le \theta(t) \le 0.03$, $y(t) \le 1$, and $-1 \le \triangle u(t) \le 1$. A 1-norm MPC is applied with a weighting factor of 0.2, and predictions and control horizon of 3 and 1, respectively. The setpoint is set at the output constraint value. The uncertainties are randomly generated within the uncertainty set with a uniform distribution. The results obtained are shown in Figure 8.1(a). Note that the output signal violates the constraint because the MPC only checked the constraints for the expected values.

The results obtained when applying a min-max algorithm are shown in Figure 8.1(b). As can be seen, the constraints are always satisfied because the MPC checked all possible values of the uncertainties.

8.5.2 Uncertainties in the Gain

The next example is the frequently found case of uncertainties in the gain. Consider a second-order system described by the following difference equation

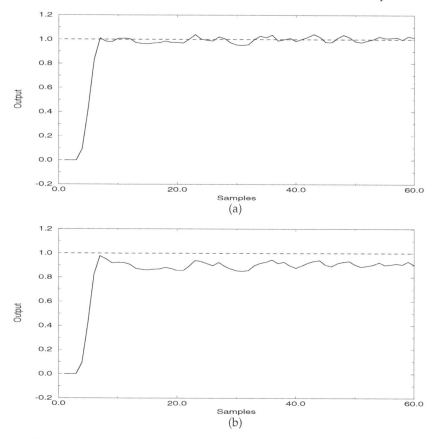

Fig. 8.1. (a) Output bound violation and (b) output with min-max algorithm

$$y(t+1) = 1.97036y(t) - 0.98019y(t-1) + 0.049627\, K\, (u(t) + 0.99335u(t-1))$$

where $0.5 \leq K \leq 2$. That is, the process static gain can be anything from half to twice the nominal value. A quadratic norm is used with a weighting factor of 0.1 for the control increments, a control horizon of 1, and a prediction horizon of 10. The control increments were constrained between -1 and 1. Figure 8.2(a) shows the results obtained by applying a constrained GPC for three different values of the process gain (nominal, maximum and minimum). As can be seen, the results obtained by the GPC deteriorate when the gain takes the maximum value giving rise to an oscillatory behaviour.

The results obtained when applying a min-max GPC for the same cases are shown in Figure 8.2b. The min-max problem was solved in this case by using a gradient algorithm in the control increments space. For each point visited in this space the value of K maximizing the objective function had to be determined. This was done by computing the objective function for the extreme points of the uncertainty polytope (two points in this case). The

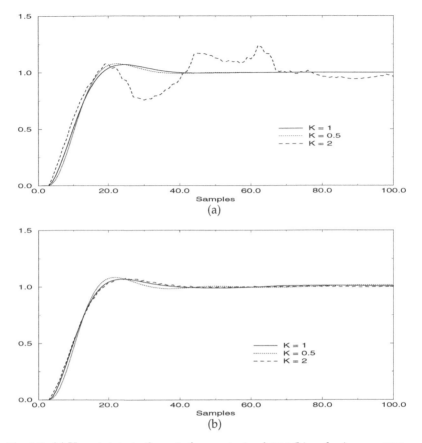

Fig. 8.2. (a) Uncertainty in the gain for constrained GPC (b) and min-max GPC

responses of the min-max GPC, which takes into account the worst case, are acceptable for all situations as can be seen in Figure 8.2(b).

A simulation study was carried out with 600 cases varying the process gain uniformly in the parameter uncertainty set from the minimum to maximum value. The bands limiting the output for the constrained GPC and the min-max constrained GPC are shown in Figure 8.3. As can be seen, the uncertainty band for the min-max constrained GPC is much smaller than the one obtained for the constrained GPC.

8.6 Robust MPC and Linear Matrix Inequalities

Linear matrix inequalities (LMI) are becoming very popular in control and have also been used in the MPC context.

A linear matrix inequality is an expression of the form

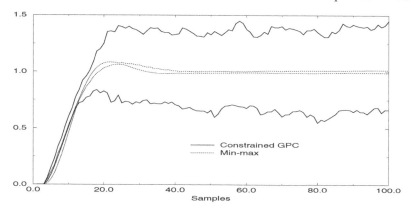

Fig. 8.3. Output limiting bands

$$F(x) = F_0 + \sum_{i=1}^{m} x_i F_i > 0 \qquad (8.23)$$

where F_i are symmetrical real $n \times n$ matrices, x_i are variables and $F(x) > 0$, means that $F(x)$ is positive definite. The three main LMI problems are:

1. the feasibility problem: determining variables $x_1, x_2, ..., x_m$ so that Inequality (8.23) holds.
2. the linear programming problem: finding the optimum of $\sum_{i=1}^{m} c_i x_i$ subject to $F(x) > 0$.
3. the generalized eigenvalue minimization problem: finding the minimum λ such that: $\lambda A(x) - B(x) > 0$, $A(x) > 0$, $B(x) > 0$

Many problems can be expressed as LMI problems [35], even inequality expressions that are not affine in the variables. This is the case of quadratic inequalities, frequently used in control, which can be transformed into an LMI form using Schur complements: Let $Q(x)$, $R(x)$, and $S(x)$ depend affinely on x and be $Q(x)$, $R(x)$ symmetrical. Then the LMI problem

$$\begin{bmatrix} Q(x) & S(x) \\ S(x)^T & R(x) \end{bmatrix} > 0$$

is equivalent to

$$R(x) > 0, \quad Q(x) - S(x)R(x)^{-1}S(x)^T > 0$$
$$\text{and } Q(x) > 0, \quad R(x) - S(x)^T Q(x)^{-1} S(x) > 0$$

There are efficient algorithms to solve LMI problems which have been applied to solve control problems such as robust stability, robust pole placement, optimal LQG, and robust MPC. In this last context, Kothare *et al.* [109] proposed a robust constrained model predictive control as follows

Consider the linear time-varying system:

$$x(k+1) = A(k)x(k) + B(k)x(k)$$
$$y(k) = Cx(k)$$
$$[A(k)B(k)] \in \Omega \tag{8.24}$$
$$R_x x(k) \le a_x$$
$$R_u u(k) \le a_u$$

and the following cost function

$$J(k) = \sum_{i=0}^{\infty} \left(\hat{x}(k+i|k)^T Q_1 \hat{x}(k+i|k) + u(k+i)^T R u(k+i) \right) \tag{8.25}$$

The robust optimization problem is stated as

$$\min_{u(k+i|k),i\ge0} \quad \max_{[A(k+i),B(k+i)]\in\Omega,i\ge0} \quad J(k) \tag{8.26}$$

Define a quadratic function $V(x) = x^T P x$ with $P > 0$ such that $V(x)$ is an upper bound on $J(k)$:

$$\max_{[A(k+i),B(k+i)]\in\Omega,i\ge0} \quad J(k) \le V(x(k|k)) \tag{8.27}$$

The problem is solved by finding a linear feedback control law $u(k + i|k) = Fx(k + i|k)$ such that $V(x(k|k))$ is minimized. Suppose that there are no constraints in the state and inputs and that the model uncertainties are defined as follows

$$\Omega = Co\{[A_1, B_1], [A_2, B_2], \cdots, [A_L, B_L]\} \tag{8.28}$$

where Co denotes the convex hull defined by vertices $[A_i, B_i]$. That is, any plant $[A, B] \in \Omega$ can be expressed as

$$[A, B] = \sum_{i=1}^{L} \lambda_i [A_i, B_i]$$

with $\lambda_i \ge 0$ and $\sum_{i=1}^{L} \lambda_i = 1$

In these conditions, the robust MPC can be transformed into the following LMI problem

$$\min_{\gamma,Q,Y} \gamma \tag{8.29}$$

subject to

$$\begin{bmatrix} 1 & x(k|k)^T \\ x(k|k) & Q \end{bmatrix} \ge 0 \tag{8.30}$$

$$\begin{bmatrix} Q & QA_j^T + Y^TB_j^T & QQ_1^{1/2} & Y^TR^{1/2} \\ A_jQ + B_jY & Q & 0 & 0 \\ Q_1^{1/2}Q & 0 & \gamma I & 0 \\ R^{1/2}Y & 0 & 0 & \gamma I \end{bmatrix} \geq 0, \quad j = 1, \cdots, L \quad (8.31)$$

Once this LMI problem is solved, the feedback gain can be obtained by:

$$F = YQ^{-1}$$

Kothare *et al.* [109] demonstrated that constraints on the state and manipulated variables and other types of uncertainties can also be formulated and solved as LMI problems.

The main drawbacks of this method are:

- Although LMI algorithms are supposed to be numerically efficient, they are not as efficient as specialized LP or QP algorithms.
- The manipulated variables are computed as a linear feedback of the state vector satisfying constraints, but when constraints are present, the optimum does not have to be linear.
- Feasibility problems are more difficult to treat as the physical meaning of constraints is somehow lost when transforming them into the LMI format.

8.7 Closed-Loop Predictions

When solving min-max MPC in the presence of output or state constraints, the solution has to fulfill the constraints for all possible realisations of the uncertainties. That is, the uncertainty bounds of the evolution of the trajectories have to be included in the polytopes defined by constraints along the prediction horizon. In many cases, the uncertainty bounds of the predictions grow to such an extent with the prediction horizon that a feasible solution cannot be found. Furthermore, even if a feasible solution can be found, it tends to be too conservative, giving rise to an overcautious controller. This excessive conservatism is caused by the open-loop nature of the predictions in conventional MPC, which does not take into account the moving horizon principle and the real closed-loop nature of MPC. That is, the prediction is made at time t for the whole control horizon in an open-loop mode without taking into account that in future sampling times the controller will have more information about the process.

Another way of illustrating the closed-loop prediction idea is by considering control in the presence of uncertainties as a game problem (min-max algorithms are used extensively in game problems). The robust control problem can be considered as a game, say chess, with two players. The first player (control action) will try to make a move that minimises a cost function (the lower the cost function the better for the first player), the second player (uncertainty) will then make a move trying to maximise the cost function. When the first player has to make his second move, the moves made at the first stage by both players are known and a more informed decision can be made. That is, the problem can be posed as

$$\min_{u(t)}[\max_{\theta(t)}[\min_{u(t+1)}[\max_{\theta(t+1)} \cdots \min_{u(t+N-1)}[\max_{\theta(t+N-1)} J(.)]]\cdots] \tag{8.32}$$

instead of the conventional open-loop formulation:

$$\min_{u(t),u(t+1),\cdots,u(t+N-1)} \max_{\theta(t),\theta(t+1),\cdots,\theta(t+N-1)} J(.) \tag{8.33}$$

The control action at time t should be computed supposing the receding horizon principle is going to be used and that control action at future time instants will be computed minimizing the cost function and, at the same time, the uncertainties will do their best to maximize the cost function. It is easy to see that because of the receding horizon principle, this turns into a recursive strategy that leads in fact to an infinite horizon unless a winning (losing) position is found (checkmate or terminal region in MPC) for all possible realizations of uncertainties in a finite number of moves. If the prediction horizon is N and the real closed-loop situation is considered, at time $t+1$ the computation will be made with a horizon of N and so on, leading to a min-max problem of infinite dimension.

As in most games, the real closed-loop strategy can be approximated by considering that the control horizon will diminish at each sampling instant and by introducing a function that evaluates (estimates) the merit of the final position reached.

8.7.1 An Illustrative Example

Consider a process given by

$$y(t+1) = ay(t) + bu(t) + \theta(t) \tag{8.34}$$

with $a = 0,9$, $b = 1$, $|u(t)| \leq 10$, $|y(t)| \leq 2$, and bounded uncertainties $|\theta(t)| \leq 1$. Let us suppose that the problem is to maintain $y(t)$ as close to zero as possible and bounded by $|y(t+j)| \leq 2$ for all possible values of uncertainties in the following N steps of the control horizon. If $y_n(t+j)$ are the nominal predictions, that is, the predictions made when no uncertainties are present, the output is given by:

$$y(t+j) = y_n(t+j) + \sum_{i=1}^{j} a^{j-i}\theta(t+i-1)$$

Given any possible combination of the control moves $(u(t),\ldots u(t+N-1))$, it is always possible to find a combination of the uncertainties $(\theta(t),\ldots\theta(t+N-1))$ that will make the process violate the constraint

$|y(t)| \leq 2$. Notice that if the uncertainties take one of the extreme values $\theta(t+j) = 1$ or $\theta(t+j) = -1$ and $\theta(t+j) = sign(y_n(t+j))$ is chosen then:

$$|y(t+j)| = |y_n(t+j)| + |(1 + a + \ldots + a^{j-i})| \geq 2 \text{ for } j > 2$$

That is, there is no sequence of the control moves that guarantees that process variables will be within bounds for all possible realizations of uncertainties.

However, if the manipulated variable is chosen as $u(t+j) = -ay(t+j)/b$, the prediction equations are now:

$$y(t+1) = ay(t) + b(-ay(t)/b) + \theta(t) = \theta(t)$$

$$\vdots$$

$$y(t+j) = \theta(t+j)$$

Then $|y(t+j)| = |\theta(t+j-1)| \leq 1 \leq 2$; that is, the constraints are fulfilled with this simple control law for all possible values of the uncertainties. The difference is that $u(t+j)$ is now computed with $\theta(t), \ldots \theta(t+j-1)$ known while in the previous case, $u(t+j)$ was computed with no knowledge of $\theta(t) \ldots \theta(t+j-1)$.

8.7.2 Increasing the Number of Decision Variables

The previous example has shown that the traditional (open-loop) prediction strategy used in min-max MPC results in an infeasible problem. The reason is that a single control profile cannot handle all possible future uncertainties. The example also shows that a simple linear controller can find a feasible solution to the problem by using feedback. This is the key issue: the open-loop MPC tries to find a solution to the control problem $(u(t), u(t+1), \cdots, u(t+N-1))$ with the information available at sampling time t but the reality is that because of the receding control strategy, at time $t+1$ the information about the process state (and therefore the uncertainties) at time $t+1$ will be available. By using the open-loop prediction, the future control moves are computed as

$$[u(t), u(t+1), \cdots, u(t+N-1)] = [f_0(x(t)), f_1(x(t)), \cdots, f_{N-1}(x(t))]$$

that is, $u(t+j)$ is computed as a function of the state at time t, while in the second case, a control law is given $(u(t+j) = -ay(t+j)/b$, in the example) by a function of the state at $t+j$:

$$[u(t), u(t+1), \cdots, u(t+N-1)] = [g_0(x(t)), g_1(x(t+1)), \cdots, g_{N-1}(x(t+N-1))]$$

In an MPC framework, this will translate into optimizing over the possible control laws: the decision variables now are not $u(t+j)$ but all the possible functions $g_{t+j}(x(t+j))$. The optimizer will have to search in the space of all possible functions of $x(t+j)$. This is a much harder problem to solve.

Another approach to the problem is to consider different variables for each possible realization of the perturbations (uncertainties) as proposed in [189]. Suppose that the realization of the perturbation and uncertainties are known. This would be the ideal situation from a control point of view: no uncertainty in the model or disturbances. The process could be controlled in an open loop manner applying a previously computed control law optimizing some operational criteria. Suppose that we compute the optimum for every possible realization of the perturbations. We would have for each particular realization of the perturbations, the initial state and the possible realization of the reference (notice that if no future references are known they can be considered uncertainties)

$$[u(t), \ldots, u(t+N-1)] = f(x(t), \theta(t+1), \ldots, \theta(t+N), r(t+1), \cdots, r(t+N))$$
(8.35)

Notice that $u(t)$ can be different for each realization of the uncertainties. However, we would like to have a $u(t)$ which depends only on state $x(t)$. If this $u(t)$ is applied to the process, the next possible states will be given by

$$x(t+1) = f(x(t), u(t), \theta(t)) \quad \text{with} \quad \theta(t) \in \Theta \quad (8.36)$$

where $\theta(t)$ is the vector of all possible uncertainties (including the reference if necessary) at time t. Let us consider for the sake of simplicity that $\theta(t)$ is either θ^- or θ^+ and that $t = 0$. Then we will have two next possible states

$$x_1^+ = f(x_0, u_0, \theta^+)$$
$$x_1^- = f(x_0, u_0, \theta^-)$$

In either of these two states we would like to apply a control law which depends only on the state. That is, we have two more variables, u_1^+ and u_1^-, associated to each possible realization of the uncertainty. We now have the following set of possible states for the next time instant:

$$x_2^{++} = f(x_1^+, u_1^+, \theta^+)$$
$$x_2^{+-} = f(x_1^+, u_1^+, \theta^-)$$
$$x_2^{-+} = f(x_1^-, u_1^-, \theta^+)$$
$$x_2^{--} = f(x_1^-, u_1^-, \theta^-)$$

We can now associate the following decision variables to the next step: $u_2^{++}, u_2^{+-}, u_2^{-+}, u_2^{--}$. If the process uncertainties can take only two possible values (or when these two values are the only two values which are relevant to the max problem), the number of decision variables added at each sampling instant j is 2^j. In general, at each sampling instant, the number of decision variables added is m^j, where m is the number of possible uncertainties to be considered at sampling time j. The number of decision variables for the min problem is $\sum_{j=1}^{N} m^j$.

In a multivariable case with four states and only one uncertainty parameter for each state and two possible values of interest for each uncertainty parameter, the number of possible realizations of the uncertainties at the extreme points is $m = 16$. In this case, if the control horizon is $N = 10$, the number of decision variables for the minimization problem would be 7.3×10^{10}. By using causality arguments the number of decision variables decreases but the problem gets more complex because additional constraints have to be added. The method is regarded as impractical except for very small problems.

8.7.3 Dynamic Programming Approach

Another approach to closed-loop MPC proposed in the literature is based on Dynamic Programming (see Appendix B). The idea of Dynamic Programming is intimately related to closed-loop control. Let us consider a system described by $x(t+1) = f(x(t), u(t), \theta(t))$ and the cost function defined as:

$$J_t(x(t), \mathbf{u}, \theta) \triangleq \sum_{j=0}^{N-1} L(x(t+j), u(t+j)) + F(x(t+N))$$

Define $\bar{J}_t(x(t), \mathbf{u}) \triangleq \max_{\theta \in \Theta} J_t(x(t), \mathbf{u}, \theta)$. Suppose we want to optimize the closed-loop nested problem:

$$\min_{u(t)} [\max_{\theta(t)} [\min_{u(t+1)} [\max_{\theta(t+1)} \cdots \min_{u(t+N-1)} [\max_{\theta(t+N-1)} J(\cdot)]] \cdots] \qquad (8.37)$$

The key idea in Dynamic Programming is to solve Problem (8.37) from the inner bracket outward, that is, first the inner most problem:

$$J^*_{t+N-1}(x(t+N-1)) \triangleq \min_{u(t+N-1)} \bar{J}_{t+N-1}(x(t+N-1), u(t+N-1)) \quad (8.38)$$

with $\bar{J}_{t+N-1}(x(t+N-1), u(t+N-1)) \triangleq \max_{\theta(t+N-1)} L(x(t+N-1), u(t+N-1)) + F(x(t+N))$.

Notice that $F(x(t+N))$ measures the merit of the last position, and this is the way to avoid entering an infinite loop.

Suppose we are able to solve Problem (8.38) explicitly, i.e., determining $J^*_{t+N-1}(x(t+N-1))$ as a function of $x(t+N-1)$. This is the cost of going from $x(t+N-1)$ to the end. At the next stage we would encounter the following problem:

$$J^*_{t+N-2}(x(t+N-2)) \triangleq \min_{u(t+N-2)} \overline{J}_{t+N-2}(x(t+N-2), u(t+N-2)) \quad (8.39)$$

with

$$\overline{J}_{t+N-2}(x(t+N-2), u(t+N-2)) \triangleq \max_{\theta(t+N-2)} L(x(t+N-2), u(t+N-2))$$
$$+ J^*_{t+N-1}(f(x(t+N-2), u(t+N-2), \theta(t+N-2)))$$

Again, if Problem (8.39) could be solved explicitly, we would obtain $J^*_{t+N-2}(x(t+N-2))$ and so forth until we arrive at

$$J^*_t(x(t)) \triangleq \min_{u(t)} \overline{J}_t(x(t), u(t)) \quad (8.40)$$

with

$$\overline{J}_t(x(t), u(t)) \triangleq \max_{\theta(t)} L(x(t), u(t)) + J^*_{t+1}(f(x(t), u(t), \theta(t)))$$

The closed-loop min-max MPC control move $u^*(t)$ for a particular value of $x(t)$ is the minimum of $\overline{J}_t(x(t), u(t))$. The key factor in Dynamic Programming is finding the functions $J^*_{t+j}(x(t+j))$. If we include constraints in the min-max MPC, each of the steps taken earlier can be described as

$$J^*_t(x(t)) \triangleq \min_{u(t)} \overline{J}_t(x(t), u(t)) \quad (8.41)$$

$$\text{s.t. } \mathbf{R}_x x(t) + \mathbf{R}_u u(t) \leq \mathbf{r}$$
$$f(x(t), u(t), \theta(t)) \in \mathcal{X}(t+1)$$
$$\text{with } \theta(t) \in \Theta$$
$$J_t(x(t), u(t)) \triangleq \max_{\theta(t) \in \Theta} L(x(t), u(t)) + J^*_{t+1}(x(t+1)) \quad (8.42)$$

Notice that constraints are not taken into account in the optimization Problem (8.42) because keeping the process within constraints is the mission of the input variables, while the object of uncertainties, in this game, is to maximize the problem regardless of constraint fulfillment, as indicated in [21]. Here it has also been demonstrated that if the system is linear $x(t+1) = \mathbf{A}(\omega(t))x(t) + \mathbf{B}(\omega(t))u(t) + Ev(t)$, with the uncertainty vector $\theta(t)^T = [\omega(t)^T v(t)^T]$, and the stage cost of the objective function defined as: $L(x(t+j), u(t+j)) \triangleq \|\mathbf{Q}x(t+j)\|_p + \|\mathbf{R}u(t+j)\|_p$ with the terminal cost defined as $J^*_{t+N}(x(t+N)) \triangleq \|\mathbf{P}x(t+N)\|_p$, the solution is a piecewise affine function of the state. This will be seen in Chapter 11.

Another way of solving the problem is to approximate functions $J^*_t(x(t+j))$ in a grid over the state as suggested in [117]. The idea is to impose a grid on the state space and then to compute $J^*_{t+N-1}(x(t+N-1))$ for all points in that grid. At the next stage function $J^*_{t+N-2}(x(t+N-2))$ is computed for

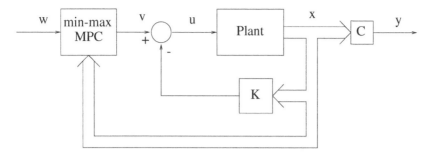

Fig. 8.4. Min-max with a linear feedback structure

the points in the grid using an interpolation of $J^*_{t+N-1}(x(t+N-1))$ when $x(t+N-1) = f(x(t+N-2), u(t+N-2), \theta(t+N-2))$ does not coincide with one of the points in the grid. The main drawback of this method is that only problems with a small dimension of the state space can be implemented as the number of one-stage min-max problems to be solved will be $N \times N_G^{\dim(x)}$

8.7.4 Linear Feedback

An approach used to reduce the uncertainty prediction bands is to consider that the future controller is going to perform some kind of controlling action based on the future process state. That is, that the controller in the future will have information about the uncertainties (or the effects on the process output or state) that have already occurred. A simple way of considering the reaction of the closed-loop system to the uncertainties in the prediction is to add a stabilizing regulator in a cascade fashion [17],[52]. Hence, the control actions of the MPC controller are the increments to the input provided by the stabilizing regulator. Let us consider, for the sake of simplicity, that the control objective is to take the state to zero; this can be accomplished by making

$$u(t+k) = -Kx(t+k \mid t) + v(t+k) \tag{8.43}$$

where K is a linear feedback gain that stabilizes the system and the auxiliary variable $v(t)$ is the reference signal for the inner loop controller, as can be seen in Figure 8.4. Consider a process described by:

$$x(t+1) = Ax(t) + Bu(t) + \vartheta(k) \tag{8.44}$$
$$y(t) = Cx(t)$$

If the linear feedback is introduced, the new equations are

$$x(t+1) = A_Kx(t) + Bv(t) + \vartheta(t) \tag{8.45}$$
$$y(t) = Cx(t)$$

with $A_K = A - BK$. That is, we have a new system with a new manipulated variable $v(t)$ instead of the previous manipulated variable $u(t)$ and a new matrix A_K. The prediction equations for both systems can be obtained by recurring expression (8.44) and (8.45):

$$x(t+k) = A^k x(t) + \sum_{j=0}^{k-1} A^{k-1-j} B u(t+j) + \sum_{j=0}^{k-1} A^{k-1-j} \vartheta(t+j) \qquad (8.46)$$

$$x(t+k) = A_K{}^k x(t) + \sum_{j=0}^{k-1} A_K{}^{k-1-j} B v(t+j) + \sum_{j=0}^{k-1} A_K{}^{k-1-j} \vartheta(t+j)$$

The first two terms of the right-hand side of Expressions (8.46) correspond to the nominal trajectory (i.e., when the uncertainties are zero). The error caused by uncertainties in the open-loop ($\tilde{x}_o(t+k)$) and closed-loop ($\tilde{x}_c(t+k)$) structures are given by:

$$\tilde{x}_o(t+k) = \sum_{j=0}^{k-1} A^{k-1-j} \vartheta(t+j)$$

$$\tilde{x}_c(t+k) = \sum_{j=0}^{k-1} A_K{}^{k-1-j} \vartheta(t+j)$$

Let us consider that the uncertainties are bounded by $\|\vartheta(t)\|_p \leq 1$. Therefore:

$$\|\tilde{x}_o(t+k)\|_p \leq \sum_{j=0}^{k-1} \|A^{k-1-j}\|_p$$

$$\|\tilde{x}_c(t+k)\|_p = \sum_{j=0}^{k-1} \|A_K{}^{k-1-j}\|_p$$

Notice that if the feedback gain is chosen such that $\|A_K\|_p < \|A\|_p$, the uncertainty bounds for predictions of the closed-loop systems will also be smaller than the corresponding bounds for the open loop.

Another interpretation of this is that by introducing a stabilizing regulator in a cascade fashion we have reduced the reaction of the closed-loop system to the uncertainties in the prediction. The effect of this controller can be seen as a reduction of the Lipschitz constant of the system. The Lipschitz constant is a gauge of the effect of the uncertainty on the prediction of the state at the next sample time. As shown in [125], the discrepancy between the nominal predicted trajectory and the uncertain evolution of the system is reduced if the Lipschitz constant is lower. Consequently, the predictions, and therefore the obtained MPC controller, are less conservative than the open-loop ones.

Notice that if there are some constraints on $u(t)$, these constraints have to be translated into the new manipulated variables $v(t)$. Let us consider that the original problem constraints were expressed by:

$$\mathbf{R}_u \mathbf{u} + \mathbf{R}_\vartheta \vartheta \leq \mathbf{r} + \mathbf{R}_x x(t) \tag{8.47}$$

The manipulated variable is computed as: $u(t + k) = -Kx(t + k \mid t) + v(t+k)$. The manipulated variable vector $\mathbf{u}$ for the complete control horizon can be expressed as

$$\mathbf{u} = \mathbf{M}_x x(t) + (\mathbf{I} + \mathbf{M}_v)\mathbf{v} + \mathbf{M}_\vartheta \vartheta \tag{8.48}$$

with

$$\mathbf{u} = \begin{bmatrix} u(t) \\ u(t+1) \\ \vdots \\ u(t+N-1) \end{bmatrix} \qquad \mathbf{M}_x = - \begin{bmatrix} K \\ KA^* \\ \vdots \\ KA^{*N-1} \end{bmatrix}$$

$$\mathbf{M}_v = - \begin{bmatrix} 0 & 0 & \cdots & 0 \\ KB & 0 & \cdots & 0 \\ \vdots & \vdots & \ddots & \vdots \\ KA^{*N-2}B & KA^{*N-3}B & \cdots & 0 \end{bmatrix} \qquad \mathbf{v} = \begin{bmatrix} v(t) \\ v(t+1) \\ \vdots \\ v(t+N-1) \end{bmatrix}$$

$$\mathbf{M}_\vartheta = - \begin{bmatrix} 0 & 0 & \cdots & 0 \\ K & 0 & \cdots & 0 \\ \vdots & \vdots & \ddots & \vdots \\ KA^{*N-2} & KA^{*N-3} & \cdots & 0 \end{bmatrix} \qquad \vartheta = \begin{bmatrix} \vartheta(t) \\ \vartheta(t+1) \\ \vdots \\ \vartheta(t+N-1) \end{bmatrix}$$

By introducing (8.48) into (8.47) we get the constraints expressed as a function of $\mathbf{v}$:

$$\mathbf{R}_u(\mathbf{I} + \mathbf{M}_v)\mathbf{v} + (\mathbf{R}_u \mathbf{M}_\vartheta + \mathbf{R}_\vartheta)\vartheta \leq \mathbf{r} + (\mathbf{R}_x - \mathbf{R}_u \mathbf{M}_x)x$$

8.7.5 An Illustrative Example

Consider the system described by $x(t + 1) = ax(t) + bu(t) + \vartheta(t)$ with $\underline{\vartheta} \leq \vartheta(t) \leq \overline{\vartheta}$, $\underline{u} \leq u(t) \leq \overline{u}$, and $\underline{x} \leq x(t) \leq \overline{x}$. Let us consider a control horizon $N = 3$ and an objective function $J = \sum_{j=1}^{N} x(t+j \mid t)^2 + \lambda u(t+j-1)^2$. The prediction equations are

$$\begin{bmatrix} x(t+1) \\ x(t+2) \\ x(t+3) \end{bmatrix} = \begin{bmatrix} a \\ a^2 \\ a^3 \end{bmatrix} x(t) + \begin{bmatrix} b & 0 & 0 \\ ab & b & 0 \\ a^2b & ab & b \end{bmatrix} \begin{bmatrix} u(t) \\ u(t+1) \\ u(t+2) \end{bmatrix} + \begin{bmatrix} 1 & 0 & 0 \\ a & 1 & 0 \\ a^2 & a & 1 \end{bmatrix} \begin{bmatrix} \vartheta(t) \\ \vartheta(t+1) \\ \vartheta(t+2) \end{bmatrix} \tag{8.49}$$

which can be expressed in a more compact way as:

$$\mathbf{x} = \mathbf{G}_x x(t) + \mathbf{G}_u \mathbf{u} + \mathbf{G}_\vartheta \vartheta \tag{8.50}$$

The constraints can be expressed as (8.47) with:

$$
\mathbf{R}_u = \begin{bmatrix} I \\ -I \\ \mathbf{G}_u \\ -\mathbf{G}_u \end{bmatrix} \quad \mathbf{R}_\vartheta = \begin{bmatrix} 0 \\ 0 \\ \mathbf{G}_\vartheta \\ -\mathbf{G}_\vartheta \end{bmatrix} \quad r = \begin{bmatrix} 1\overline{u} \\ -1\underline{u} \\ 1\overline{x} \\ -1\underline{x} \end{bmatrix} \quad \mathbf{R}_x = \begin{bmatrix} 0 \\ 0 \\ -\mathbf{G}_x \\ \mathbf{G}_x \end{bmatrix} \tag{8.51}
$$

The min-max MPC is reduced to solving the problem:

$$
\min_{\mathbf{u}} \max_{\vartheta} \; \mathbf{x}^T \mathbf{x} + \lambda \mathbf{u}^T \mathbf{u} \tag{8.52}
$$

$$
\text{s.t.} \quad \left. \begin{array}{l} \mathbf{x} = \mathbf{G}_x x(t) + \mathbf{G}_u \mathbf{u} + \mathbf{G}_\vartheta \vartheta \\ \mathbf{R}_u \mathbf{u} + \mathbf{R}_\vartheta \vartheta \leq r + \mathbf{R}_x x(t) \end{array} \right\} \; \forall \vartheta \in \Theta
$$

Notice that the constraints of Problem (8.52) have to be fulfilled $\forall \vartheta \in \Theta$ which would represent an infinite number of constraints (one for each point inside Θ). However, if the constraints are satisfied at the vertices of Θ they will also be satisfied in the interior of Θ and therefore it is sufficient to satisfy a finite number of constraints.

If the process model parameters are $a = 0.95$, $b = 0.1$, $-20 \leq u(t) \leq 20$, $-1.2 \leq x(t) \leq 1.2$, $r(t) = 0$, $\lambda = 2$, and $-0.5 \leq \vartheta(t) \leq 0.5$, the constraint matrices are defined as in (8.51) with:

$$
\mathbf{G}_x = \begin{bmatrix} 0.9500 \\ 0.9025 \\ 0.8574 \end{bmatrix} \quad \mathbf{G}_u = \begin{bmatrix} 0.1 & 0 & 0 \\ 0.095 & 0.1 & 0 \\ 0.0902 & 0.095 & 0.1 \end{bmatrix} \quad \mathbf{G}_\vartheta = \begin{bmatrix} 1 & 0 & 0 \\ 0.95 & 1 & 0 \\ 0.9025 & 0.95 & 1 \end{bmatrix} \tag{8.53}
$$

It can easily be seen that if $x(t) = 0$, for any control sequence, the error due to the uncertainty at $t+3$ is given by: $\tilde{x}(t+3) = 0.9025\vartheta(t) + 0.95\vartheta(t+1) + \vartheta(t+2)$. By making $\vartheta(t) = \vartheta(t+1) = \vartheta(t+2) = \overline{\vartheta}$, or $\vartheta(t) = \vartheta(t+1) = \vartheta(t+2) = \underline{\vartheta}$, we can see that by using these uncertainty values, the error band can be made as big as $2.8525\overline{\vartheta} = 1.4263$ and $2.8525\underline{\vartheta} = -1.4263$. That is, if the nominal trajectory makes $\hat{x}(t+3) \geq 0$, just by choosing the uncertainties to be $\overline{\vartheta}$, the state vector will be higher than the admissible value $(x(t) \leq 1.2)$. The same situation happens when $\hat{x}(t+3) \leq 0$, where choosing the uncertainties to be $\underline{\vartheta}$ will cause the state vector to have a lower value than allowed. That is, the problem is not feasible for any point in the state space.

Now suppose that the following linear feedback is considered: $u(t) = -8.5x(t) + v(t)$. The resulting system dynamics are now described by: $x(t+1) = 0.95x(t) + 0.1(-8.5x(t) + v(t)) + \vartheta(t) = 0.1x(t) + 0.1v(t) + \vartheta(t)$. The error due to uncertainties can be computed as:

$$
\begin{bmatrix} \tilde{x}(t+1) \\ \tilde{x}(t+2) \\ \tilde{x}(t+3) \end{bmatrix} = \begin{bmatrix} 1 & 0 & 0 \\ 0.1 & 1 & 0 \\ 0.01 & 0.1 & 1 \end{bmatrix} \begin{bmatrix} \vartheta(t) \\ \vartheta(t+1) \\ \vartheta(t+2) \end{bmatrix} \tag{8.54}
$$

The uncertainty band at $t+3$ is given by $\tilde{x}(t+3) = 0.01\vartheta(t) + 0.1\vartheta(t+1) + \vartheta(t+2)$. It can be seen that the errors are bounded by $-0.5 \leq \tilde{x}(t+1) \leq 0.5$, $-0.55 \leq \tilde{x}(t+2) \leq 0.55$, and $-0.555 \leq \tilde{x}(t+3) \leq 0.555$. Therefore, the problem is feasible for any initial state, such that a nominal trajectory can be computed separated from the bounds by $0.5, 0.55$, and 0.555. That is, $-0.7 \leq \hat{x}(t+1) \leq 0.7$, $-0.65 \leq \hat{x}(t+2) \leq 0.65$, and $-0.645 \leq \hat{x}(t+3) \leq 0.645$. In this case, a feasible solution exists for all $x(t) \in [-1.2, \ 1.2]$.

In conclusion, when open-loop MPC was used, no feasible solution could be found such that constraints would be fulfilled in spite of future uncertainties or perturbations. When we consider that information about the future process state will be taken into account (by a linear feedback in this example), the problem is feasible for any admissible value of $x(t)$.

8.8 Exercises

8.1. Consider the second-order system described by the following equation

$$y(t+1) = y(t) - 0.09y(t-1) + 0.09u(t) + \varepsilon(t)$$

with $-1 \leq u(t) \leq 1, -1 \leq y(t) \leq 1, 0.01 \leq \varepsilon(t) \leq 0.01$. The system is modelled by the following first-order model:

$$y(t+1) = ay(t) + bu(t) + \theta(t)$$

1. If the model parameters are chosen as $a = 0.9$, $b = 0.1$, determine a bound $\overline{\theta}$ for the uncertainty such that $-\overline{\theta} \leq \theta(t) \leq \overline{\theta}$.
2. Explain how you would find $\overline{\theta}$ experimentally if you did not know the equations of the *real* system but you could experiment on the *real* system itself.
3. Find a bound for the output prediction trajectory; i.e., $\overline{y}(t+j)$ such that $|y(t+j|t)| \leq \overline{y}(t+j)$.
4. Explain how you would calculate a and b so that the uncertainty bound $\overline{\theta}$ is minimized. Find the minimizing model parameters a and b and the minimal bound $\overline{\theta}$.
5. Find a bound for the output prediction trajectory with these new bounds and model. Compare the results with those obtained in number 3.
6. Formulate a min-max MPC using different types of objective functions (quadratic, 1-norm, ∞-norm).
7. Solve the min-max MPC problems of number 6 and simulate the responses with different control horizons.

8.2. Given the system $y(t+1) = ay(t) + bu(t) + \theta(t)$ with $a = 0.9$, $b = 0.1$, $-1 \leq u(t) \leq 1$, $-1 \leq y(t) \leq 1$, the uncertainty $\theta(t)$ bounded by: $-0.05 \leq \theta(t) \leq 0.05$ and a terminal region defined by the following box around the origin $-0.1 \leq y(t+N) \leq 0.1$.

1. Formulate a robust min-max MPC with $N = 3$ that takes the system to the terminal region for any realization of the uncertainties with different objective functions (quadratic, 1-norm, ∞-norm). Comment on the nature and difficulties of the optimization problems encountered.

2. Repeat the exercise of number 1 for a robust MPC but minimizing the objective function for the nominal system instead of the min-max MPC.
3. Solve the problems of numbers 1 and 2 for $N = 1$, $N = 2$ and $N = 3$. Discuss the feasibility of each.
4. Formulate the problems of numbers 1 and 2 for $N = 3$ but use a linear feedback as indicated in Section 8.7.4. Discuss the feasibility.

8.3. Given the system

$$x(t+1) = Ax(t) + Bu(t) + D\theta(t)$$

with

$$A = \begin{bmatrix} 1 & 1 \\ 0 & 1 \end{bmatrix}, \quad B = \begin{bmatrix} 0 \\ 1 \end{bmatrix}, \quad D = \begin{bmatrix} 1 & 0 \end{bmatrix}, \quad -0.1 \le \theta(t) \le 0.1, \quad -1 \le u(t) \le 1$$

The control objective consists of taking (and maintaining) the state vector as close to zero as possible by solving the following min-max problem.

$$\min_{u \in [-1,1]} \max_{\theta \in [-0.1,0.1]} \sum_{j=1}^{N} x(t+j)^T x(t+j) + 10\, u(t+j-1)^2$$

1. Formulate a robust min-max MPC with $N = 3$.
2. Repeat the exercise of number 1 for a robust MPC minimizing the objective function for the nominal system instead of the min-max MPC.
3. Repeat the exercises in numbers 1 and 2 considering the following terminal region: $\|x(t+N)\|_\infty \le 0.2$.
4. Solve and simulate the MPC of numbers 1, 2, and 3 for $N = 1$ and $N = 2$. Discuss the feasibility of each.

9

Nonlinear Model Predictive Control

In general, industrial processes are nonlinear, but, as has been shown in this book, most MPC applications are based on the use of linear models. There are two main reasons for this: on one hand, the identification of a linear model based on process data is relatively easy and, on the other hand, linear models provide good results when the plant is operating in the neighbourhood of the operating point. In the process industries, where linear MPC is widespread, the objective is to keep the process around the stationary state rather than perform frequent changes from one operation point to another and, therefore, a precise linear model is enough. Besides, the use of a linear model together with a quadratic objective function gives rise to a convex problem (Quadratic Programming) whose solution is well studied with many commercial products available. The existence of algorithms that can guarantee a convergent solution in a time shorter than the sampling time is crucial in processes where a great number of variables appear.

However, the dynamic response of the resulting linear controllers is unacceptable when applied to processes that are nonlinear to varying degrees of severity. Although in many situations the process will be operating in the neighbourhood of a steady state, and therefore a linear representation will be adequate, there are some very important situations where this does not occur. On one hand, there are processes for which the nonlinearities are so severe (even in the vicinity of steady states) and so crucial to the closed-loop stability that a linear model is not sufficient. On the other hand, there are some processes that experience continuous transitions (startups, shutdowns, etc.) and spend a great deal of time away from a steady-state operating region or even processes which are never in steady-state operation, as is the case of batch processes, where the whole operation is carried out in transient mode. For these processes a linear control law will not be very effective, so nonlinear controllers will be essential for improved performance or stable operation.

Although the number of applications of Nonlinear Model Predictive Control (NMPC) is still very limited (see [14], [171]), its potential is really great since nonlinear MPC has to make headway in those areas where process nonlinearities are strong and market demands require frequent changes in operation regimes.

This chapter is dedicated to NMPC, showing recent developments and new trends, in both theoretical and practical aspects. Notice that both aspects are important since there are open theoretical issues such as modelling or stability that are as important as practical ones such as identification or computational complexity.

9.1 Nonlinear MPC Versus Linear MPC

It is evident that the main advantage of NMPC with respect to MPC is the possibility of dealing with nonlinear dynamics. As new tools that facilitate attainment and representation of nonlinear models, either from first principles or from experimental data, appear on the market, interest in their use in NMPC is growing.

There is nothing in the basic concepts of MPC against the use of a nonlinear model. Therefore, the extension of MPC ideas to nonlinear processes is straightforward, at least conceptually. However, this is not a trivial matter, and there are many difficulties derived from the use of this kind of model such as:

- The availability of nonlinear models from experimental data is an open issue. There is a lack of identification techniques for nonlinear processes. The use of Neural Networks or Volterra series does not seem to solve the problem in a general form. On the other hand, model attainment from first principles (mass and energy balances) is not always feasible.
- The optimization problem is nonconvex, its resolution is much more difficult than the QP problem. Problems relative to local optimum appear, not only influencing control quality but also deriving in stability problems.
- The difficulty of the optimization problem translates into an important increase in computation time. This can constrain the use of this technique to *slow* processes.
- The study of crucial subjects such as stability and robustness is more complex in the case of nonlinear systems. It constitutes an open field of great interest for researchers.

Some of these problems are partially being solved, and NMPC is becoming a field of intense research and will become more common as users demand higher performance.

As an introductory example, consider the nonlinear system

$$y(t+1) = 0.9y(t) + u(t)^{\frac{1}{4}}$$

with $0 \leq u(t) \leq 1$. Figure 9.1 shows the response obtained when controlled with a linear MPC and a nonlinear MPC with $\lambda = 0$ and $N = 10$.

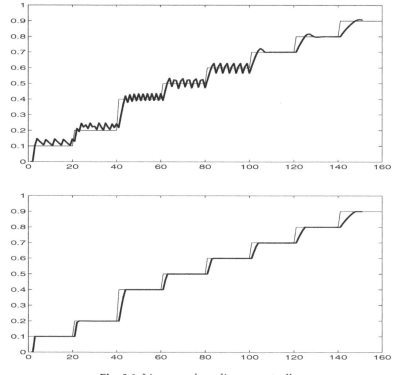

Fig. 9.1. Linear and nonlinear controller

For the linear case, the system is approximated by the linear model:

$$y(t+1) = 0.9y(t) + u(t)$$

In the nonlinear case it is easy to see that by making $v(t) = u(t)^{\frac{1}{4}}$, a linear system is obtained, and it can be solved using any linear MPC tool with this new manipulated variable. $v(t)^1$

As can be seen, the response obtained with the linear MPC oscillates for low values of the setpoints while the response obtained with the nonlinear MPC is very good for all setpoint changes.

9.2 Nonlinear Models

Developing adequate nonlinear models may be very difficult and there is no model form that is clearly suitable to represent general nonlinear processes. Part of the success of standard MPC was due to the relative ease with which step and impulse responses or low-order transfer functions could be

---

[1] Note that if $\lambda \neq 0$, the objective function would not be a quadratic function.

obtained. Nonlinear models are much more difficult to construct, either from input/output data correlation or by the use of first principles from well-known mass and energy conservation laws.

If the deviation from linearity is not too large, some approximations can be made which acknowledge that certain system characteristics change from operating point to operating point, but linearity is assumed in the neighbourhood of a specific operating point. There are some approximations to the problem, such as a scheduled linearized MPC (see Chapter 12), in which the model is linearized around several operating points and appropriately used within the linear MPC strategy as the process moves from one operating point to another. There is also the extended linear MPC in which a basic linear model is used in combination with an explicit nonlinear model which captures the nonlinearities.

However, when the nonlinearities are more severe, nonlinear models must be employed to describe process dynamics. The three main types of models that are used in this area are: empirical, fundamental, and grey box, which are discussed later.

9.2.1 Empirical Models

A major mathematical obstacle to a complete theory of nonlinear processes is the lack of a superposition principle for nonlinear systems. Because of this, the determination of models from process input/output data becomes a very difficult task. The number of plant tests required to identify a nonlinear plant is much greater than that needed for a linear plant. If the plant is linear, in an ideal situation, only a step test has to be performed in order to know the step response of the plant. Because of the superposition principle, the response to a different size step can be obtained by multiplying the response to the step test by the ratio of both step sizes. This is not the case for nonlinear processes where tests with many different-size steps must be performed to get the step response of the nonlinear plant. If the process is multivariable, the difference in the number of tests required is even greater. In general, if a linear system is tested with signals $u_1(t), u_2(t), \ldots u_n(t)$ and the corresponding responses are $y_1(t), y_2(t), \ldots y_n(t)$, the response to a signal which can be expressed as a linear combination of the tested input signals

$$u(t) = \alpha_1 u_1(t) + \alpha_2 u_2(t) + \cdots + \alpha_n u_n(t)$$

is

$$y(t) = \alpha_1 y_1(t) + \alpha_2 y_2(t) + \cdots + \alpha_n y_n(t)$$

That is, a linear system does not need to be tested for any input signal sequence that is a linear combination of previously tested input sequences, whilst this is not the case for a nonlinear system that must be analyzed for all possible input signals.

A fundamental difficulty associated with the empirical modelling approach is the selection of a suitable model form. The available nonlinear models used for NMPC are described later.

Input-output Models

The nonlinear discrete-time models used for control can be viewed as mappings between those variables that are available for predicting system behaviour up to the current time and those to be predicted at or after that instant.

This kind of model can be represented as a nonlinear autoregressive moving average model with exogenous input (NARMAX), which, for single-input single-output processes, is given by the general equation

$$y(t) = \Phi[y(t-1), \ldots, y(t-n_y), u(t-1), \ldots, u(t-n_u), e(t), \ldots, e(t-n_e+1)]$$
(9.1)

where Φ is a nonlinear mapping, y is the output, u is the input, and e is the noise input. The suitability of this model depends on the choice of the function Φ and the order parameters. Notice that this equation covers a wide range of descriptions, depending mainly on function Φ. Different choices of this function give rise to certain of the following models that can provide attractive formulations for predictive control. Volterra and related models, local model networks and neural networks are detailed in this subsection.

Volterra Models

If only Finite Impulse Response (FIR) models are considered (which is equivalent to $n_y = 0$ in equation (9.1)) and Φ is restricted to analytic functions, it follows that this function exhibits a Taylor series expansion which defines the class of discrete-time Volterra models.

These models are analogous to the continuous-time Volterra models, with the convolution integrals replaced by discrete convolution sums. The model response is given by

$$y(t) = y_0 + y_1(t) + y_2(t) + y_3(t) + \ldots y_n(t)$$

where the first term is an *offset* and the second is given by:

$$y_1(t) = \sum_{i=0}^{\infty} h_1(i)u(k-i)$$

which corresponds to the linear convolution model used in many linear MPC strategies (such as MAC or DMC) and the higher-order terms are given by:

$$y_2(t) = \sum_{i=0}^{\infty} \sum_{j=0}^{\infty} h_2(i,j)u(t-i)u(t-j)$$

$$y_3(t) = \sum_{i=0}^{\infty} \sum_{j=0}^{\infty} \sum_{l=0}^{\infty} h_3(i,j,l)u(t-i)u(t-j)u(t-l)$$

$$y_n(t) = \sum_{i_1=0}^{\infty} \cdots \sum_{i_n=0}^{\infty} h_n(i_1,\dots,i_n)u(t-i_1)\dots u(t-i_n)$$

In general, finite-dimensional discrete-time Volterra models can be written as:

$$y(t) = y_0 + \sum_{i_1=0}^{M} \cdots \sum_{i_n=0}^{M} a_n(i_1,\dots,i_n)u(t-i_1)\dots u(t-i_n) \qquad (9.2)$$

Although Volterra models have their limitations, they represent a simple logical extension of the convolution models that have been so successful in linear MPC. These models are generically well-behaved and their structure can be exploited in the design of the controller, especially for second-order models, as will be shown later. In this particular and useful case, and when the infinite terms are truncated finite to values, the process model is given by

$$y(t) = y_0 + \sum_{i=0}^{N} h_1(i)u(k-i) + \sum_{i=0}^{M}\sum_{j=0}^{M} h_2(i,j)u(t-i)u(t-j) \qquad (9.3)$$

which corresponds to the widely used linear convolution model with the nonlinearity appearing as an extra term, that is, the nonlinearity is additive.

The number of parameters needed to define a Volterra model is usually large. The dynamic order of the structure has to be chosen in relation to the settling time of the process. Since the settling time is typically 10 to 50 times the sampling time, a realistic choice of the order lies between 10 and 50, which gives rise to a great number of parameters. This order can be reduced by the use of a parametric Volterra model that also considers past output values:

$$y(t) = y_0 + \sum_{i=0}^{N} a_1(i)y(k-i) + \sum_{i=0}^{N} h_1(i)u(k-i) + \sum_{i=0}^{M}\sum_{j=0}^{M} h_2(i,j)u(t-i)u(t-j)$$

$$(9.4)$$

This additional linear output feedback helps reduce the dynamic order compared to the basic model.

Two special subclasses of the basic model are employed which reduce the complexity of the basic Volterra approach and have a reduced number of parameters. These are the Hammerstein and Wiener models.

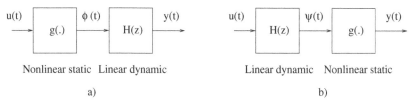

Fig. 9.2. Hammerstein and Wiener models

Hammerstein models belong to the family of block-oriented nonlinear models, built from the combination of linear dynamic models and static non-linearities. They consist of a single static nonlinearity $g(.)$ connected in cascade to a single linear dynamic model defined by a transfer function $H(z^{-1})$, as shown in Figure 9.2(a).

Notice that the intermediate input $\Phi(t)$ is given by

$$\Phi(t) = g(u(t)) = \sum_{j=0}^{N} \gamma_j u^j(t)$$

and the overall model output is expressed as:

$$y(t) = \sum_{i=0}^{M} h(i)\Phi(t-i)$$

Both expressions can be combined to give a single equation of the general finite-dimensional Volterra form (9.2), where the constant term is given by

$$y_0 = \gamma_0 \sum_{i=0}^{M} h(i)$$

and the coefficients α_n equal $\gamma_n h(i_1)$ for $i_1 = i_2 = \ldots = i_n$ and are otherwise zero.

Because of this, Hammerstein models can be considered *diagonal* Volterra models, since the off-diagonal coefficients are all zero. Notice that this means that the behaviour that can be represented by this type of model is restricted. This is the price to pay for a reduced number of parameters.

The Wiener model can be considered as the dual of the Hammerstein model, since it is composed of the same components connected in reverse order, as seen in Figure 9.2(b). The input sequence is first transformed by the linear part $H(z^{-1})$ to obtain $\Psi(t)$, which is transformed by the static non-linearity $g(.)$ to get the overall model output. Therefore the intermediate sequence is given by

$$\psi(t) = \sum_{i=0}^{M} h(i)u(t-i)$$

and the overall output by

$$y(t) = \sum_{j=0}^{N} \gamma_j \Psi^j(t)$$

These equations can be combined to give an expression of the form (9.2) considering $y_0 = \gamma_0$ and

$$\alpha_n(i_1, \ldots, i_n) = \gamma_n h(i_1) \cdots h(i_n)$$

The properties of Volterra and related models are extensively discussed in [64].

Closely related to Volterra models are bilinear models. The main difference between this kind of model and the Volterra approach is that crossed products between inputs and outputs appear in the model. A second-order finite-dimensional bilinear model is described as

$$y(t) = \sum_{i=0}^{N} h_1(i) u(k - i) + \sum_{i=0}^{M} \sum_{j=0}^{M} h_2(i, j) y(t - i) u(t - j)$$

which can also include linear output feedback as in Equation (9.4). Bilinear models have been successfully used to model and control heat exchangers, distillation columns, chemical reactors, waste treatment plants, and pH neutralization reactors [98]. It has been demonstrated that this type of model can be represented by a Volterra series [122].

Local Model Networks

Another way of using input-output models to represent nonlinear behaviour is to use a local model network representation. The idea is to use a set of local models to accommodate local operating regimes [99]. A global plant representation is formed using multiple models over the whole operating space of the nonlinear process. The plant model used for control provides an explicit, transparent plant representation which can be considered an advantage over *black-box* approaches such as neural networks (that will be described later).

The basics of this operating regime approach are to decompose the space into zones where linear models are adequate approximations to the dynamical behaviour within that regime, with a trade-off between the number of regimes and the complexity of the local model. The output of each submodel is passed through a local processing function that generates a window of validity of that particular submodel. The complete model output is then given by

$$y(t + 1) = F(\Psi(t), \Phi(t)) = \sum_{i=1}^{M} f_i(\Psi(t)) \rho_i(\Phi(t))$$

where the M local models $f_i(\Psi(t))$ are linear ARX functions of the measurement vector Ψ (inputs and outputs) and are multiplied by basis functions $\rho_i(\Phi(t))$ of the current operating vector. These basis functions are chosen to give a value close to 1 in regimes where f_i is a good approximation to the unknown F and a value close to 0 in other cases.

If the local models are of the affine ARX form:

$$f_i(\Psi) = a_{1i}y(t) + a_{2i}y(t-1) + \ldots a_{n_a i}y(t-n_a)$$
$$+ b_{0i}u(t) + b_{1i}u(t-1) + \ldots + b_{n_b i}u(t-n_b) + c_i$$

then the nonlinear network is given by

$$y(t+1) = A_1 y(t) + A_2 y(t-1) + \ldots + A_{na}y(t-n_a) +$$
$$+ B_0 u(t) + B_1 u(t-1) + \ldots + B_{nb}u(t-n_b) + C$$

which is an ARX model with its parameters defined at each operating regime:

$$A_i = \sum_{j=1}^{M} \rho_j(\Phi(t))a_{ij} \qquad B_i = \sum_{j=1}^{M} \rho_j(\Phi(t))b_{ij} \qquad C = \sum_{j=1}^{M} \rho_j(\Phi(t))c_j$$

Notice that this technique allows the use of a linear predictive controller, avoiding the problems associated to computation time and optimality of the nonlinear solution. However, identification of local operating regimes can be a difficult task.

Furthermore, a family of model-controller pairs can be used together with a scheduler in the same way as a gain scheduling strategy. The controllers are tuned about a model obtained from experiments around an operating point and the scheduler decides which controller, or combination of controllers, is applied to the plant. Notice that both the model and the controller are linear. These latter strategies have been successfully tested on a pH neutralization plant (see [200] for details).

Neural Networks

The nonlinear dynamics of the process can also be captured by an Artificial Neural Network (ANN). Neural networks are attractive tools to construct the model of nonlinear processes since they have an inherent ability to approximate any nonlinear function to an arbitrary degree of accuracy [92]. This, together with the availability of training techniques, has made them very successful in many predictive control applications and commercial products. More details of the use of ANN for control can be found in [152].

Neural networks are composed of many neuronlike processing elements called *nodes* which are interconnected to form a network. Input signals to the node are weighted and added together with a bias term (which is required to take care of the offset in the process model). The node output is obtained by passing this summed term through a nonlinear activation function $s(.)$ that is usually sigmoidal.

The input to the artificial neural network is the measurement vector $\Psi = [y(t-1), y(t-2), \ldots, y(t-n), u(t-1), u(t-2), \ldots u(t-m)]$ which is processed at every node of the input layer, that gives the following output

$$z_i = s(W_1\Psi(t-1) + b_i). \tag{9.5}$$

The activation function usually takes the form $s(x) = (1 - e^{-2x})/(1 + e^{-2x})$, W_1 is the vector of connection weights, and b_i is the bias term. The model output (considering a monovariable process) is given by the output layer which is usually a linear weighted combination of the vector Z of hidden nodes outputs z_i:

$$y(t) = W_2Z + b_o \tag{9.6}$$

The weights of the hidden layer W_1 and the output node W_2 as well as the hidden and output biases have to be estimated from experimental data. This is done by means of the *training* of the neural network using time series of inputs and outputs. In this case the process model is constituted by the weights and bias without any physical meaning, forming a *black-box* model.

Once the model has been obtained, the ANN can be used for prediction. Combining Equations (9.5) and (9.6), the following expression is obtained

$$y(t) = W_2[s(W_1\Psi(t-1)) + b_i] + b_o$$

which gives the output prediction along the horizon

$$\hat{y}(t+k) = W_2[s(W_1\Psi(t+k-1)) + b_i] + b_o$$

which will be used in the cost function to be minimized.

Neural Networks are usually combined with linear models in practical applications, since they are not able to extrapolate beyond the range of their training data set. Based on a model confidence index, the ANN is gradually turned off when its prediction looks unreliable, the predictions relying on the linear part.

Some commercial predictive controllers such as the MVC algorithm from Continental Controls and *Process Perfecter* from Pavillion Technologies use input-output models where a static nonlinear model is combined with a linear dynamic model [14]. The latter uses an ANN to describe the nonlinear part. The dynamic part is described by the deviation variables defined as

$$\delta u(t) = u(t) - u_s \qquad \delta y(t) = y(t) - y_s$$

where the steady-state values for input and output fulfill:

$$y_s = h_s(u_s)$$

and the deviation variables follow the linear dynamic relationship (usually with $n = 2$):

$$\delta y(t) = \sum_{i=1}^{n} a_i \delta y(t-i) + b_i \delta u(t-i) \tag{9.7}$$

Therefore, this is a model composed of a linear dynamic part and a nonlinear static one. Identification of the linear dynamic model is made based on plant tests, and the nonlinear static model is given by a neural network built from historical data. This procedure facilitates identification since historical data usually contain rich steady-state information and plant tests are only needed for dynamic submodel adjustment.

The basic idea is to update the dynamic model coefficients of Equation (9.7) using a static gain that is a linear interpolation of the initial and final steady-state gains:

$$K_s(u(t)) = K_s^i + \frac{K_s^f - K_s^i}{u_s^f - u_s^i} \delta u(t) \tag{9.8}$$

Notice that this gain depends on $u(t)$. The gains used for the interpolation are evaluated from the static nonlinear model as:

$$K_s^i = \frac{dy_s}{du_s}\Big|_{u_s^i} \qquad\qquad K_s^f = \frac{dy_s}{du_s}\Big|_{u_s^f}$$

It is assumed that the process dynamics remain linear over the entire range of operation (that is, coefficients a_i do not change). Therefore, b_i coefficients are adjusted to the new value of K_s given by (9.8), giving rise to new coefficients that also depend on $\delta u^2(t-i)$ since K_s depends on $\delta u(t)$. Now the dynamic equation turns to

$$\delta y(t) = \sum_{i=1}^{n} a_i \delta y(t-i) + \bar{b}_i \delta u(t-i) + g_i \delta u^2(t-i) \tag{9.9}$$

where

$$\bar{b}_i = \frac{b_i K_s^i (1 - \sum_{j=1}^{n} a_j)}{\sum_{j=1}^{n} b_j} \qquad g_i = \frac{b_i (1 - \sum_{j=1}^{n} a_j)}{\sum_{j=1}^{n} b_j \frac{K_s^f - K_s^i}{u_s^f - u_s^i}}$$

It can be checked that the static gain of (9.9) is $K_s(u(t))$, since

$$\frac{\sum_{i=1}^{n} \bar{b}_i + g_i}{\sum_{i=1}^{n} a_i} = K_s(u(t))$$

This dynamic model is the one used for controller calculation once a nonlinear optimization program has computed the best input and output values (u_s^f and y_s^f) using the static model.

It can be observed that the coefficients of the dynamic part in (9.9) change from one sample period to the next since they are rescaled to fit the new local gain of the static nonlinear model. This strategy can be considered a linear interpolation of the linearized gains done after a successive linearization at the initial and final state, in a similar formulation to gain scheduling, but with a different local model. Since it is assumed that the process dynamics remain linear over the operating range, asymmetric dynamics such as different time constants cannot be represented by this model.

State Space Models

The linear state space model can naturally be extended to include nonlinear dynamics. The following state space model can be used to describe a nonlinear plant

$$x(t+1) = f(x(t), u(t)) \tag{9.10}$$
$$y(t) = g(x(t))$$

where $x(t)$ is the state vector and f and g are generic nonlinear functions. Notice that the same equation can be used for monovariable and multivariable processes. Notice also that this model can easily be derived from the differential equations that describe the model (if they are known) by converting them into a set of first-order equations. Model attainment in this case is straightforward but the procedure is very difficult to obtain from experimental data when no differential equations are available.

This kind of model is the most widely extended for nonlinear plants since it has given rise to a lot of theoretical results: the majority of results about stability and robustness have been developed inside this framework. It is also used in commercial tools such as NOVA NLC or nonlinear PFC.

The use of this kind of model for predictive control needs the state to be accessible through measurements or the inclusion of a state observer. The choice of an appropriate observer may have influence on the closed-loop performance and stability. Although this is not considered in detail in this book, there exists an estimation approach that is dual to the NMPC problem. It is called Moving Horizon Estimation (MHE) and is formulated as an on-line optimization. It uses a moving horizon of old measurements to obtain an optimization-based estimate of the system state; see, for example, [3].

Although both functions f and g in Equation (9.10) are, in general, nonlinear, in the case that the first one is linear, a model composed of a combination of a linear state equation with a nonlinear output equation of the following form can be used. This is used in Aspen Target, and the plant model is given by

$$x(t+1) = Ax(t) + Bu(t)$$
$$y(t) = g(x(t))$$

Output nonlinearity can be modelled by the superposition of a linear relationship and a nonlinear neural network of the form:

$$g(x(t)) = Cx(t) + NN(x(t))$$

This model is generic and allows for the consideration of nonlinear effects other than measurement nonlinearity since the state vector is not limited to physical variables. Then the system is identified as a linear one and the output residual terms are adjusted to the states by means of the neural network. An extended Kalman filter (EKF) can be added to correct modelling errors and nonmeasurable disturbances, in this way substituting the constant feedback error that is usually used in MPC.

Nonlinear System Identification

A fundamental issue in NMPC is not only the choice of the type of model but also of an identification method that is both reliable and robust enough. Although one major characteristic of linear systems is that almost every nonlinear system is unique, many tools exist that have been developed to allow the use of the same approach for a broad variety of cases.

As stated by Henson [88], nonlinear system identification involves the following five tasks, which represent theoretical and practical challenges:

1. structure selection. This requires the choice of the kind of model to be used and the selection of the model parameters, such as coefficients n_y and n_u in equation (9.1) or the order of a Volterra model.
2. input sequence design. The determination of the input sequence to be injected into the plant to obtain the output sequence. Notice that this is an open point because of the lack of a superposition principle for nonlinear systems.
3. noise modelling. This is the determination of the dynamic model for the noise input.
4. parameter estimation. This is the estimation of the remaining model parameters from input/output data and the noise input.
5. model validation. The comparison of model predictions with plant data not used in model development.

This is currently a field of very active research with many open issues where optimization techniques play an important role. A thorough overview of the existing tools can be found in the book by Nelles [148].

9.2.2 Fundamental Models

Since it is sometimes difficult to develop liable empirical models based on empirical data, the possibility of using models that come directly from balance equations, usually called first principle models, exists. The equations are obtained from the knowledge of the underlying process. These models are derived by applying mass, energy, and momentum balances to the process. In this case, the prediction is made by a simulation of the nonlinear equations describing the process dynamics.

For complex industrial processes, this type of model is very difficult and expensive to obtain since it needs a lot of expertise. Nevertheless, in some cases a detailed model exists that has been developed for other purposes, as is the case of a process simulator for operator training. This simulator can also be used for control purposes, although on some occasions this dynamical model can be too complex to be useful for controller design. This encourages the use of reduction techniques such as singular perturbation to derive a simplified model which retains the basic dynamic behaviour of the complex model.

Fundamental models need less data than empirical models since model parameters have a physical meaning and can be estimated from laboratory experiments or operating data. Besides, they can extrapolate to operating regions which are not represented in the data used for model development. This is a particularly important advantage when the plant operates over a wide range of conditions.

9.2.3 Grey-box Models

These models are developed by combining the empirical and fundamental approaches, exploiting the advantages of each type of model. In this hybrid approach, basic first principle information is augmented by empirical data. This term must not be confused with hybrid processes; a hybrid or greybox model is a dynamical model that is obtained using both empirical and theoretical information, while a hybrid process is one that has time-driven and event-driven dynamics, as will be analyzed in Chapter 10.

There are two common ways of developing hybrid models [88]. The first is to use empirical models to estimate unknown functions in the fundamental model (such as reaction rates in a chemical reactor), in which case steady-state empirical models are sufficient. The second is to utilize a fundamental model to capture the basic process dynamics and then to describe the residual between the plant and the model using a nonlinear empirical model. Both techniques allow the integration of the physical knowledge of the plant without the need for a rigorous model of it. Although hybrid models offer promising potential, their use in NMPC has still to be exploited.

9.2.4 Modelling Example

This example shows the adjustment of several nonlinear models to capture the dynamics of a real process: a gypsum kiln. The behaviour of this kind of process is affected by nonlinear effects caused by the existence of disturbances and the coupling among several variables. The use of second-order

Volterra and Hammerstein models as appropriate solutions to describe process dynamics is analyzed. Choosing the type of model is a trade-off between model complexity and modelling error.

The state of the process is not depicted by a single variable but rather by several variables that are measured in the plant (for a complete process description, see [33]). This is a process in which a nonnegligible delay d exists due to mass transport along the rotary kiln. Outlet product temperature can be considered as the process output, although it is highly influenced by other temperatures, mainly by the calcination temperature. Plant operators know that a relationship between these variables exists, since the behaviour of the calcination temperature anticipates the outlet temperature evolution. Therefore, a good model of this part of the process has to be obtained in order to better control the plant.

In spite of the advantages associated with fundamental modelling, the difficulties associated with obtaining such a relationship for this case lead to the alternative of achieving a model from input/output data. Two models have been chosen for this study: quadratic Hammerstein and second-order Volterra models.

The quadratic Hammerstein model has been chosen as follows

$$y(k+d) = h_0 + \sum_{i=0}^{N} h_{1i} u(k-i) + \sum_{i=0}^{N} h_{2i} u^2(k-i) \qquad (9.11)$$

while the second-order Volterra model with truncation of order N is given by the expression

$$y(k+d) = h_0 + \sum_{i=0}^{N} h_{1i} u(k-i) + \sum_{i=0}^{N} \sum_{j=i}^{N} h_{2ij} u(k-i) u(k-j) \qquad (9.12)$$

The triangular form is used without any loss of generality, since the second-order parameters are symmetric for the Volterra model. In both cases the term h_0 is a bias, the h_{1i} terms are the impulse response coefficients, and the h_{2i} and h_{2ij} terms are the second-order Hammerstein and Volterra terms, respectively.

Once the model type has been decided there are some issues to be treated, such as the number of lagged data to be considered (truncation of the models) and determining what delay best describes the process when a particular model is being studied.

Real data from a gypsum kiln has been analyzed. To calculate the parameters for the models, *ad hoc* Matlab functions have been developed. They get as an input the sampled data and the model structure (Volterra, Hammerstein, and truncation order). Coefficients are calculated according to the least-squares method.

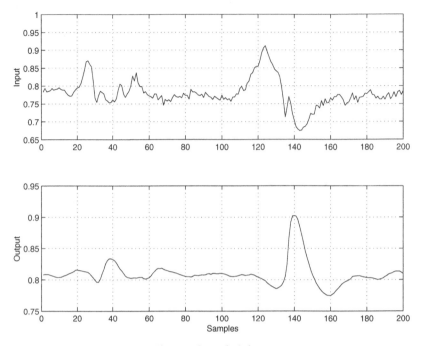

Fig. 9.3. Sampled data

In Figure 9.3, original work data are presented. Temperature values have already been scaled to working values for identification and estimation. The sample period for the data was 2 minutes.

One important issue not always well treated is the delay determination for processes. In this example, a search for the best combination delay/model order has been made for the three models studied: linear (to show the advances that can be obtained by using nonlinear models), Hammerstein, and Volterra. They are first presented and then compared using the root mean square error.

A Hammerstein model, as described in Equation (9.11) was adjusted to the data, modifying both the delay and the model truncation order. The optimal delay for Hammerstein models remains stable centered on one value (16 samples), presenting no variations even when the truncation order increased. Accuracy of the model is accomplished through higher-order models. However, this accuracy is not significantly improved after the fifth order model.

A Volterra model as shown in Equation (9.12) has proven to be the best choice to get an accurate description of the data. However, the most significant difference to the two cases already presented lies in the fact that optimal delay for identification shifts to lower values as the model order increases. This is caused by the fact that a crossed product between lagged inputs offers

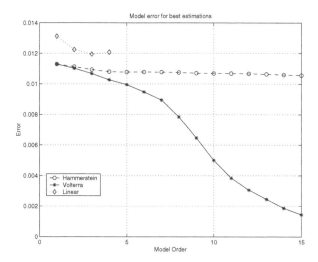

Fig. 9.4. Error comparison

richer information about the system than the other models. Optimal delay is placed for low-order systems close to 16 samples, but it comes to be about 3 or 4 samples when the Volterra model is truncated to higher values, such as the fifteenth term. Once more, accuracy is improved by increasing the model order, and in this case *information saturation* is not achieved as quickly as in Hammerstein models.

In Figure 9.4, plots for the minimum values of the model errors (for the best value of the delay) are presented. For the linear case, order is related to the AR polynomial (and thus the lagged outputs considered), while for the nonlinear models this means the truncation order N. For the first two cases, very high-order models do not necessarily mean significant improvement in model error, while Volterra models always achieve the best correlation and increasing its order makes performance improve more than in the other cases.

One of the Volterra model drawbacks is the need for a large number of parameters to be calculated. For the fitch order Hammerstein model, only five coefficients are needed to describe the quadratic part, while 55 are required for the Volterra model. When it comes to $N = 15$, Hammerstein still needs 15 and Volterra yields 120 coefficients.

Real data fitting for the models obtained through this search are presented in Figure 9.5. The best linear model is a third-order model in $A(q^{-1})$ polynomial. For the Hammerstein model, $N = 5$ has been chosen as the best trade-off between order and accuracy, and for Volterra $N=10$ was chosen.

Table 9.1 shows the estimation error for the models considered. Results are improved with the nonlinear models, especially with use of a Volterra structure which particularly fits the given data.

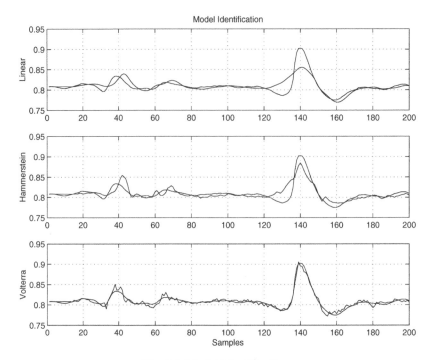

Fig. 9.5. Model identification

Table 9.1. Model error

Model	Order	Estimation Error
Linear	$N_y = 3$	0.0120
Hammerstein	$N = 5$	0.0108
Volterra	$N = 10$	0.0050

Taking into account the trade-off between complexity and accuracy, a tenth order Volterra model is proposed as the best choice that will allow the use of a nonlinear predictive controller in the plant. It is clear that this type of model can give better results since it has more degrees of freedom, although the number of parameters can be prohibitive if a high-order model is chosen.

9.3 Solution of the NMPC Problem

In spite of the difficulties associated with nonlinear modelling, the choice of appropriate model is not the only important issue. Using a nonlinear model changes the control problem from a convex quadratic program to a nonconvex nonlinear problem, which is much more difficult to solve. Furthermore, in this situation there is no guarantee that the global optimum can be found, especially in real-time control when the optimum has to be obtained in a prescribed time.

9.3.1 Problem Formulation

The problem to be solved at every sampling time is computation of the control sequence $\mathbf{u}$ that takes the process to the desired regime. This desired operating point $(\mathbf{y}_s, \mathbf{x}_s, \mathbf{u}_s)$ may be determined by a steady-state optimization which is usually based on economic objectives. The cost function to be minimized may take the general form

$$J = \sum_{j=1}^{N} \|\mathbf{y}(t+j) - \mathbf{y}_s\|_{\mathbf{R}}^q + \sum_{j=1}^{M-1} \|\triangle\mathbf{u}(t+j)\|_{\mathbf{P}}^q + \sum_{j=1}^{M-1} \|\mathbf{u}(t+j) - \mathbf{u}_s\|_{\mathbf{Q}}^q \quad (9.13)$$

where q can be 1 or 2, depending on the type of norm and $\mathbf{P}$, $\mathbf{Q}$, and $\mathbf{R}$ are weighting matrices. The minimization is subject to model constraint that, if a state space model (9.10) is used, is given by

$$\mathbf{x}(t+j) - f(\mathbf{x}(t+j-1), \mathbf{u}(t+j-1)) = 0 \qquad y(t+j) - g(\mathbf{x}(t+j)) = 0$$

and is subject to the rest of the inequality constraints that can be considered:

$$\underline{y} \leq y(t+j) \leq \overline{y} \quad \forall j = 1, N \tag{9.14}$$
$$\underline{u} \leq u(t+j) \leq \overline{u} \ \forall j = 1, M-1 \tag{9.15}$$
$$\triangle\underline{u} \leq \triangle u(t+j) \leq \triangle\overline{u} \ \forall j = 1, M-1 \tag{9.16}$$

Soft constraints in the output variable can easily be considered in the formulation by adding the term $\|s\|_T^q$ to the cost function, where T is a matrix that penalizes the violation of the output limits by a small amount s. The inequality constraint now results in:

$$\underline{y} - s \leq y(t+j) \leq \overline{y} + s \quad \forall j = 1, N$$

Notice that the former expressions are derived for a monovariable process; in the case of a MIMO plant, variables y and u must be substituted by vectors $\mathbf{y}$ and $\mathbf{u}$.

9.3.2 Solution

The solution of this problem requires the consideration (and at least a partial solution) of a nonconvex, nonlinear problem (NLP) which gives rise to a lot of computational difficulties related to the expense and reliability of solving the NLP online.

The problem is often solved using Sequential Quadratic Programming (SQP) techniques. These are extensions of Newton-type methods for converging to the solution of the Karush-Kuhn-Tucker (KKT) conditions of the optimization problem. The method must guarantee fast convergence and must be able to deal with ill conditioning and extreme nonlinearities.

SQP is an iterative technique in which the solution at each step is obtained by solving an approximation to the nonlinear problem in which the objective is replaced by a quadratic approximation and the nonlinear constraints are replaced by linear approximations.

The NMPC problem can be written as a general nonlinear programming problem with $\mathbf{w}^T = [\mathbf{u}^T \mathbf{x}^T \mathbf{y}^T]$, that is,

$$\min_{\mathbf{w}} J(\mathbf{w})$$

$$\text{subject to: } \mathbf{c}(\mathbf{w}) = 0, \quad \mathbf{h}(\mathbf{w}) \leq 0$$

where the equality constraint vector $\mathbf{c}$ corresponds to the model constraints

$$f(\mathbf{x}, \mathbf{u}) = 0 \qquad \mathbf{y} - g(\mathbf{x}) = 0$$

and has n_c components, and the inequality constraint vector $\mathbf{h}$ corresponds to (9.14) and has n_h components. The following Lagrangian function can be defined for this problem

$$\mathcal{L}(\mathbf{w}, \lambda_1, \lambda_2) = J(\mathbf{w}) + \lambda_1^T \mathbf{c}(\mathbf{w}) + \lambda_2^T \mathbf{h}(\mathbf{w})$$

where the optimum first-order KKT conditions state that Lagrange multipliers $\lambda_1 \in R^{n_c}$ and $\lambda_2 \in R^{n_h}$ must exist such that

$$\nabla_{\mathbf{w}} \mathcal{L}(\mathbf{w}, \lambda_1, \lambda_2) = 0 \qquad (9.17)$$

$$\mathbf{c}(\mathbf{w}) = 0 \qquad (9.18)$$

$$\mathbf{h}(\mathbf{w}) + \mathbf{s} \leq 0 \qquad (9.19)$$

$$\mathbf{s} \geq 0, \quad \lambda_2 \geq 0, \quad \mathbf{s}^T \lambda = 0 \qquad (9.20)$$

At each iteration k, a quadratic programming (QP) subproblem is created and solved, giving rise to a new value of the unknown vector such that the new direction $\mathbf{d}^T = [\mathbf{d}_u^T \; \mathbf{d}_x^T \; \mathbf{d}_y^T]$ is used as a search direction to converge to the original problem solution. It starts from $\mathbf{w}_k$ and supplies the new value $\mathbf{w}_{k+1} = \mathbf{w}_k + \mathbf{d}_k$, solving the following quadratic problem

$$\min_{\mathbf{d}} \nabla J(\mathbf{w}_k)^T \mathbf{d} + \frac{1}{2} \mathbf{d}^T B_k \mathbf{d}$$

where B_k is an approximation to the Hessian of the Lagrangian

$$B_k = \nabla^2_{\mathbf{ww}} \mathcal{L}(\mathbf{w}, \lambda_1, \lambda_2) = \nabla^2 J(\mathbf{w}) + \sum_{i=1}^{n_c} \lambda_{1i} \nabla^2 c_i(\mathbf{w}) + \sum_{i=1}^{n_h} \lambda_{2i} \nabla^2 h_i(\mathbf{w})$$

subject to the linearized constraints

$$\mathbf{c}(\mathbf{w}_k) + \nabla\mathbf{c}(\mathbf{w}_k)^T\mathbf{d} = 0, \qquad \mathbf{h}(\mathbf{w}_k) + \nabla\mathbf{h}(\mathbf{w}_k)^T\mathbf{d} \leq 0$$

Inequality constraints are in general linear and do not need to be linearized, while the equality constraints that correspond to the model can be written as:

$$f(\mathbf{x}_k, \mathbf{u}_k) + \nabla_{\mathbf{x}}f(\mathbf{x}_k, \mathbf{u}_k)^T\mathbf{d}_x + \nabla_{\mathbf{u}}f(\mathbf{x}_k, \mathbf{u}_k)^T\mathbf{d}_u = 0$$
$$\mathbf{y}_k + \mathbf{d}_y - g(\mathbf{x}_k) - \nabla_{\mathbf{x}}g(\mathbf{x}_k)^T\mathbf{d}_x = 0$$

Many problems may appear when applying the method, such as the availability of the second derivatives or the feasibility of the intermediate solution. This last condition is very important in real-time optimization since, if time is insufficient, the last iteration $\mathbf{u}_k$, which satisfies the local linear approximation to the constraints, is sent to the plant, although it may violate the original constraints. Several variations of the original method exist that try to overcome the main problems. Convergence properties and variants of the method that enhance efficiency are thoroughly discussed by Biegler [26].

It should be noticed that an iterative algorithm consisting of solving a QP problem (which is itself iterative) is used at each sampling instant. Therefore the computational cost is very high, and this justifies the development of special formulations to solve particular problems or approximate solutions in spite of losing optimality.

There has been a rapid development of efficient nonlinear programming algorithms capable of handling large numbers of variables and constraints. In spite of the significant advances which have been made in this field and in the computing power of modern control equipment, most of the issues related to NMPC are as yet unresolved and there is still much work to be done from the point of view of both theoretical analysis and practical implementation.

9.4 Techniques for Nonlinear Predictive Control

As has been shown in the previous section, the exact solution of the optimization problem at every sampling instant is a difficult task. Therefore, a set of efficient formulations that try to avoid the problems associated to nonconvex optimization has appeared in recent years. They are briefly depicted here.

9.4.1 Extended Linear MPC

This is one of the simplest ways of dealing with process nonlinearities; it was originally devised for the DMC controller [90]. The idea is to add a new term to the output prediction that tries to take nonlinearities into account. This term is added to the prediction equation (see Chapter 3) giving:

$$\hat{\mathbf{y}}_{el} = \mathbf{Gu} + \mathbf{f} + \mathbf{d}_{nl}$$

The elements of vector $\mathbf{d}_{nl}$ are computed by minimizing the difference between the prediction obtained from the extended linear model, $\hat{\mathbf{y}}_{el}$, and the prediction obtained from a full-scale nonlinear model of the plant. This way the process nonlinearities as captured by the nonlinear model are incorporated directly into the linear MPC formulation, preserving the original QP framework.

9.4.2 Local Models

A simple way to deal with the nonlinear model equation is to perform successive linearization about a nominal operating point. This yields linear MPC and allows an online solution, since the problem to be solved at every sampling instant is a QP. Usually the current operating point is used for linearization although the accuracy can be improved by linearizing several times over the sampling period. The linearized model is used for output predictions while the original nonlinear model can be used to compute the effect of past input moves.

This idea is extended in [111], which uses linearization about the predicted trajectories that would be obtained if the extension to the current time of the previously computed optimal control sequence were used. This extension is referred to as the *tail* of the previous time optimal input trajectory. The linearized model is time-varying but it can be discretized and used with a linear MPC with endpoint terminal constraint as SGPC (see Section (4.11)). The effect of linearization errors can be reduced by the appropriate use of linear interpolation, compensating in this way the loss of optimality. The resulting algorithm guarantees stability on account of the implicit use of endpoint constraints on the linearized models and allows an online solution.

This concept is closely related to the use of a local model network (as described in Section 9.2). This network of linear models is used to capture the process dynamics at different operating points. Two alternative methods of exploiting this idea are presented in [200] using a Generalized Predictive Controller. The first consists of a network of GPCs, each designed around one of the local models. The control action is formed by a combination of the output of the linear controllers in a similar way to gain scheduling. A scheduler uses interpolation and smooths the transition between local controllers. The second method uses a single GPC, with a linear model obtained as a combination of the local linear models. An interpolation function generates activation weights for each model. High weights are given to models that give a good approximation in the regime where the model is operating while weights approaching zeros are given to the others. This network of local models produces an overall nonlinear ARX model of the plant, which may be assumed to be a locally valid representation of the plant. It is clear that in both cases the problem is reduced to a QP.

9.4.3 Suboptimal NPMC

This approach avoids the need to find the minimum of a nonconvex cost function by considering the satisfaction of constraints to be the primary objective. If an optimization strategy that delivers feasible solutions at every sub-iteration (inside a sampling period) is used and a decrease in the cost function is achieved, optimization can be stopped when the time is over and stability can still be guaranteed. It can be demonstrated that it is sufficient to achieve a continuous decrease in the cost function to guarantee stability.

The main technique that uses this concept was proposed by Scokaert *et al.* [190], and consists of a dual-mode strategy (see next section for details) which steers the state towards a terminal set Ω and, once the state has entered the set, a local controller drives the state to the origin. Now, the first controller does not try to minimize the cost function J, but to find a predicted control trajectory which gives a sufficient reduction of the cost.

If the cost function is given by

$$J(t) = \sum_{k=0}^{N-1} L(x(t+k), u(t+k))$$

then any control action that satisfies constraints and fulfills the following condition (with $0 < \mu < 1$) is acceptable.

$$J(t) \leq J(t-1) - \mu L(x(t-1), u(t-1))$$

Once a solution is found, the search for solutions that give a bigger reduction can continue if there is available time. Notice that the choice of μ has a great influence over the difficulty of finding the solution. Small values of this parameter make it easy to find a solution, but at the cost of poor performance since the reduction of the cost is small. Such small values are useful when it is difficult to find feasible trajectories, as in the case of large disturbances or modelling errors.

9.4.4 Use of Short Horizons

It is clear that short horizons are desirable from a computational point of view, since the number of decision variables of the optimization problem is reduced. However, long horizons are required to achieve the desired closed-loop performance and stability (as will be shown in the next section). Some approaches have been proposed that try to overcome this problem.

In [214] an algorithm which combines the best features of exact optimization and a low computational demand is presented. The key idea is to calculate exactly the first control move which is actually implemented, and to approximate the rest of the control sequence which is not implemented. Therefore the number of decision variables is one (or the number of inputs in the MIMO case), regardless of the control horizon. The idea is that if there is not enough time to calculate the control sequence $u(t \mid t), u(t + 1 \mid t), \ldots u(t - M - 1 \mid t)$, then compute only the first one and approximate the rest as well as possible. An approximation to the next value $u(t + 1 \mid t)$ is obtained by linearizing the nonlinear system at $\{x(t + 1 \mid t), u(t \mid t)\}$ which corresponds to the control signal that would be implemented if the system were linear. This is done along the rest of the control horizon to approximate the rest of the sequence. The values thus obtained appear as constraints in the optimization problem which has to calculate only one decision variable. Clearly it results in significant savings in online computational time as the online computational time grows exponentially with the number of decision variables in the worst case.

An algorithm that uses only a single degree of freedom is proposed in [110] for nonlinear, control affine plants. A univariate online optimization is derived by interpolating between a control law which is optimal in the absence of constraints (although it may violate constraints and may not be stabilizing) and a sub-optimal control law with a large associated stabilizable set. The interpolation law inherits the desirable optimality and feasibility from these control laws. The stabilizing control law uses an optimization based on only a single degree of freedom and can be performed by including a suitable penalty in the cost or an artificial convergence constraint.

9.4.5 Decomposition of the Control Sequence

One of the key ideas in linear MPC is the use of free and forced response concepts. Although this is no longer valid for nonlinear processes, since the superposition principle does not hold in this case, variation of the idea can be used to obtain implementable formulations of NMPC.

In [42], the prediction of process output is made by adding the *free* response (future response obtained if the system input is maintained at a constant value during the control and prediction horizons) obtained from a nonlinear model of the plant and the *forced* response (the response obtained due to future control moves) obtained from an incremental linear model of the plant. The predictions obtained this way are only an approximation because the superposition principle, which permits the mentioned division in *free* and *forced* responses, only applies to linear systems. However, the approximation obtained in this way is shown to be better than those obtained using a linearized process model to compute both responses. If a quadratic cost function is used, the objective function is a quadratic function in the decision variables (future control moves) and the future control sequence can be computed in the unconstrained case, as the solution of a set of linear equations, leading to a simple control law. The only difference from standard linear MPC is that the *free* response is computed by a nonlinear model of the process. As the superposition principle does not hold for the nonlinear models, the approximation is only valid when the sequence of future control moves is small. Notice that this occurs when the process is operating in steady state with small perturbations. When the process is being changed from operating conditions or the external perturbations are high, the future control moves are usually high and the approximation is not very good.

A way to overcome this problem has been suggested in [106] for EPSAC. The key idea is that the manipulated variable sequence can be considered to be the addition of a base control sequence $(u_b(t+j))$ plus a sequence of increments of the manipulated variables $(u_i(t+j))$, that is,

$$u(t+j) = u_b(t+j) + u_i(t+j)$$

The process output j step ahead prediction is computed as the sum of the response of the process $(y_b(t+j))$ due to the base input sequence plus the response of the process $(y_i(t+j))$ due to the future control increments on the base input sequence $u_i(t+j)$:

$$y(t+j) = y_b(t+j) + y_i(t+j)$$

As a nonlinear model is used to compute $y_b(t+j)$ while $y_i(t+j))$ is computed from a linear model of the plant, the cost function is quadratic in the decision variables $(u_i(t+j))$ and it can be solved by a QP algorithm as in linear MPC. The superposition principle does not hold for nonlinear processes and the process output generated this way and the process output generated by the nonlinear controller will only coincide in the case when the sequence of future control moves is zero.

If this is not the case, the base is made equal to the last base control sequence plus the optimal control increments found by the QP algorithm. The procedure is repeated until the sequence of future controls is driven close enough to zero.

The initial conditions for the base control sequence can first be made equal to the last control signal applied to the process. Notice that this is the case when computing the *free* response in linear MPC. A better initial guess can be made by making the base sequence equal to the optimal control sequence determined for the last sampling instant with the corresponding time shift.

The convergence conditions of the algorithm are very difficult to obtain as they depend on the severity of the nonlinear characteristics of the process, on past inputs and outputs, on the future reference sequence and on perturbations.

9.4.6 Feedback Linearization

In some cases, the nonlinear model can be transformed into a linear model by appropriate transformations. Consider, for example, the process described by the following state space model:

$$x(t + 1) = f(x(t), u(t))$$
$$y(t) = g(x(t))$$

The method consists of finding state and input transformation functions $z(t) = h(x(t))$ and $u(t) = p(x(t), v(t))$ such that:

$$z(t + 1) = Az(t) + Bv(t)$$
$$y(t) = Cz(t)$$

The method has two important drawbacks:

- The transformation functions $z(t) = h(x(t))$ and $u(t) = p(x(t), v(t))$ can be obtained for few cases.
- The constraints, which are usually linear, are transformed into a nonlinear set of constraints.

That is, even in the cases where the model can be linearized by suitable transformations, the constrained problem is easy to solve due to the nonlinear constraints. The objective function is usually transformed into a nonlinear function, since it was quadratic in $u(t)$ but not necessarily in $v(t)$. If the nonlinear constraints are approximated and the objective functions remain quadratic, only a quadratic program needs to be solved at each sampling instant. However, linear approximation is only valid when both state and input do not deviate too far from the operating regime. This implies that the control actions have to be close to their linearized values to preserve stability, and performance may be sacrificed for computational simplicity.

9.4.7 MPC Based on Volterra Models

In some cases, the NLP shows a special structure that can be exploited to achieve an online feasible solution to the general optimization problem. If the process is described by a Volterra model (9.2) efficient solutions can be found, especially for second-order models.

The general nonlinear, nonconvex optimization problem has some peculiarities in this case that permit an easier solution. Consider the most used case of a second-order model (9.3). If this quadratic model is combined with the quadratic cost function, it gives rise to a fourth-order programming problem regardless of the values of the horizons (also for Auto-Regressive models). This kind of problem (a fourth-order objective function) is easier to solve than a general NLP, where the cost function is usually of higher order (for instance twentieth-order) in the decision variables. Notice that in this case the nonlinearities are always polynomial, while in other type of models (fundamental models, for instance) they can be exponential. Since the typical NLP solvers approximate the nonlinear cost function by a quadratic one (as has been shown for SQP), it is clear that the problem is easier to solve if the cost function is of low order.

Apart from this complexity reduction in the direct solution, a control strategy can be devised that can solve the nonlinear problem by iteration of the linear solution, based on the particular structure of Volterra models. This iterative procedure proposed by Doyle *et al.* [64] gives rise to an analytical solution in the unconstrained case or a QP solution if constraints exist and allows an easy solution to the nonlinear problem. If a second-order model is used, the prediction can be written as an extension of the linear process $y = Gu + f + c$, where f includes the terms that depends on past and known values and the new term c takes into account new terms that depend on crossed products between past and future control actions. If p is the prediction horizon and m the control horizon, the predictions can be written as:

$$
\begin{bmatrix} \hat{y}(t+1 \mid t) \\ \hat{y}(t+2 \mid t) \\ \vdots \\ \hat{y}(t+p \mid t) \end{bmatrix} = \begin{bmatrix} h_1(1) & 0 & \cdots & 0 \\ h_1(2) & h_1(1) & \cdots & 0 \\ \vdots & \vdots & \cdots & h_1(1) \\ \vdots & \vdots & \cdots & h_1(1)+h_1(2) \\ \vdots & \vdots & \cdots & \vdots \\ h_1(p) & h_1(p-1) & \cdots & \sum_{i=1}^{p-m+1} h_1(i) \end{bmatrix} \begin{bmatrix} u(t) \\ u(t+1) \\ \vdots \\ u(t+m-1) \end{bmatrix}
$$
$$
+ \begin{bmatrix} f(t+1) \\ f(t+2) \\ \vdots \\ f(t+p) \end{bmatrix} + \begin{bmatrix} c(t+1) \\ c(t+2) \\ \vdots \\ c(t+p) \end{bmatrix}
$$

The terms $h_1(t)$ are the elements of the linear part of the Volterra model (similar to the impulse response in a linear system), and vector f (similar to the free response) is given by (where N is the truncation order of the linear part of the model):

$$
\begin{bmatrix} f(t+1) \\ f(t+2) \\ \vdots \\ f(t+p) \end{bmatrix} = \begin{bmatrix} h_1(2) & h_1(3) \cdots & \cdots & h_1(p) & 0 \\ h_1(3) & h_1(4) \cdots & h_{(p)} & 0 & 0 \\ \vdots & \vdots & \vdots & \vdots & \vdots & \vdots \\ h_1(p-1) & h_1(p) & \vdots & \vdots & \vdots & \vdots \\ h_1(p) & 0 & \vdots & \vdots & \vdots & \vdots \\ 0 & 0 & \vdots & 0 & 0 & 0 \end{bmatrix} \begin{bmatrix} u(t-1) \\ u(t-2) \\ \vdots \\ u(t-N) \end{bmatrix} +
$$

$$
+ \begin{bmatrix} d(t+1) \\ d(t+2) \\ \vdots \\ d(t+p) \end{bmatrix} + \begin{bmatrix} g(t+1) \\ g(t+2) \\ \vdots \\ g(t+p) \end{bmatrix}
$$

Vector $\mathbf{d}$ is the feedback term, that is, the difference between the measured output value and the predicted output at time t, and usually all its elements are equal to $d(t+1)$ (the error is considered to be constant along the horizon and equal to the current value). Vector $\mathbf{g}$ is formed by the crossed products of past outputs (and is therefore known). Vector $\mathbf{c}$ is composed of the products of future control actions with past and future control actions (quadratic terms) and is therefore unknown. Then the prediction

$$
\mathbf{y} = \mathbf{G}\mathbf{u} + \underbrace{\mathbf{H}\mathbf{u}_{\text{past}} + \mathbf{d} + \mathbf{g}}_{\mathbf{f}} + \mathbf{c}(\mathbf{u})
$$

depends on the unknowns ($\mathbf{u}$) both in a linear form ($\mathbf{G}\,\mathbf{u}$) and a quadratic form ($\mathbf{c}(\mathbf{u})$) and cannot be solved analytically as in the linear unconstrained case. However, the iterative procedure proposed in [64] starts with an initial value of $\mathbf{c}$ and solves the problem, obtaining the solution:

$$
\mathbf{u} = (\mathbf{G}^T\mathbf{G} + \lambda\mathbf{I})^{-1}\mathbf{G}^T(\mathbf{w} - \mathbf{f} - \mathbf{c}) \tag{9.21}
$$

The new solution is used to recalculate $\mathbf{c}$ and the problem is solved again until the iterated solution is close enough to the previous one. In the constrained case, $\mathbf{u}$ is computed solving a QP instead of Equation (9.21). Due to the feasibility of its being implemented in real time, this method has been successfully applied to real plants, such as polymerization processes [133] or biochemical reactors [64].

A suboptimal approach for Volterra models using a GPC with input signal parametrization (control increments equal to zero after the first control action, that is, control horizon equal to one, or all control increments with the same value) is proposed in [86]. In the absence of constraints, the authors provide an analytical solution (without iterations) by solving a third-order polynomial.

Hammerstein and Wiener Models

If the process can be modelled by a Hammerstein model (see Figure 9.2), the problem can be easily transformed into a linear one by inverting the nonlinear static part, $g(.)$. This way, the solution can be obtained by minimizing a cost function that depends on the intermediate variable $\Phi(t) = g(u(t))$:

$$J_H = \sum_{i=1}^{p} \|e(t+i)\|_Q + \sum_{i=1}^{m} \|\Phi(t+i-1)\|_R$$

Notice that this cost function may lose part of its physical meaning, since the real control action $u(t)$ is not considered. The minimization is performed with respect to Φ, which is not the physical control action. If the transformed input constraints are convex, the minimization is reduced to a QP problem that supplies the best values for the intermediate variable Φ. The real control signal to be implemented is obtained by the inversion of the static part: $u(t) = g^{-1}(\Phi(t))$.

Although the calculated $\Phi(t)$ is optimal with respect to J_H, the resulting $u(t)$ does not have to be optimal with respect to the original problem (formulated as a function of y and u). Besides, on many occasions it is not possible to find the inverse, or this is not unique. In the latter case, the solution giving the lowest cost is the one that is actually implemented. Instead of using the inverse, the static input nonlinearity can be transformed into a polytopic description, as in [28], where the use of a robust linear MPC algorithm is proposed, consisting of a convex optimization problem with nominal closed-loop stability.

The same idea can be applied to Wiener models, where the static nonlinearity goes after the linear dynamics. In [158] a pH neutralization process is controlled in this way.

9.4.8 Neural Networks

Artificial Neural Networks, apart from providing a modelling tool that enables accurate nonlinear model attainment from input-output data, can also be used for control. Since ANNs are universal approximators, they can *learn* the behaviour of a nonlinear controller and calculate the control signal online with few calculations, since the time-consuming part of the ANN (training) is done beforehand. This has been applied to several processes in the process industry [9], [24] and to systems with short sampling intervals (in the range of milliseconds) such as internal combustion engines [150]. These issues will be addressed again in Chapter 11, and an application of an NN controller to a mobile robot is detailed in Chapter 12.

9.4.9 Commercial Products

In order to exploit the success of linear MPC in industry, many products have recently appeared on the market that try to solve the nonlinear problem.

They are mainly aimed at the chemical and petrochemical sectors and provide integrated solutions that usually include modelling and commissioning facilities.

The most extended products are:

- PFC by Adersa,
- *Aspen Target* by Aspen Technology,
- MVC by Continental Controls,
- NOVA NLC by DOT Products,
- *Process Perfecter* by Pavillion Technologies, and
- INCA by IPCOS.

These products try to provide a solution in real time using some of the techniques previously shown, looking for a feasible solution although this implies a loss of optimality. They share some characteristics, which are described here.

Different types of models are used. The most extended is state space, but all of the models described in Section 9.2 appear in one product or another. All of them also provide a steady-state optimization, which computes optimal targets to be used in the dynamic optimization. The dynamic objective function is quadratic except in NOVA NLC which can also use 1-norm. The optimization of this function includes constraints that are usually hard for the inputs but can be softened for the outputs, since output hard constraints can easily lead to unfeasibility. *Process Perfecter* offers the possibility of applying soft constraints using a *frustum* method, which gives more freedom to the output at the beginning of the horizon than at the end but no error is allowed outside the *frustum*.

Usually the prediction horizon is finite but very large, in order to capture the output dynamics up to the permanent regime. This can be interpreted as an approximation to the infinite horizon method used to guarantee closed-loop stability (see next section) and can explain why terminal constraints are not included in any of the products.

In order to reduce the complexity of the problem, some products use a control horizon of one while PFCs use the concept of basis functions, described in Section 3.3 for the linear case. This product reduces the solution complexity by solving the unconstrained problem with a nonlinear least squares method and clipping the inputs to their limits if constraints are violated. Logically, the solution is not optimal but the speed of execution is enhanced, allowing the use of this method in fast processes such as missile control. The remaining products use several algorithms for the solution, ranging from multistep Newton-type methods to generalized reduced gradient methods and proprietary nonlinear programming techniques.

A more complete review of the existing products and applications can be found in [14].

9.5 Stability and Nonlinear MPC

The efficient solution of the optimal control problem is important for any application of NMPC to real processes, but stability of the closed loop is also of crucial importance. Some significant results related to closed-loop stability have recently appeared. Even in the case that the optimization algorithm finds a solution, this fact does not guarantee closed-loop stability (even with perfect model match). The use of terminal penalties and/or constraints, Lyapunov functions or invariant sets has given rise to a wide family of techniques that guarantee the stability of the controlled system.

This problem has been tackled from different points of view, and several contributions have appeared in recent years, always analyzing the regulator problem (drive the state to zero) in a state space framework. The main proposals are the following:

- infinite horizon. This solution was proposed by Keerthi and Gilbert [103] and consists of increasing the control and prediction horizons to infinity, $P, M \rightarrow \infty$. In this case, the objective function can be considered a Lyapunov function, providing nominal stability. This is an important concept, but it cannot be directly implemented since an infinite set of decision variables should be computed at each sampling time.
- terminal constraint. The same authors proposed another solution considering a finite horizon and ensuring stability by adding a state terminal constraint of the form:

$$x(k + P) = x_s$$

  With this constraint, the state is zero at the end of the finite horizon and therefore the control action is also zero; consequently (if there are no disturbances) the system stays at the origin. Notice that this adds extra computational cost and gives rise to a restrictive operating region, which makes it very difficult to implement in practice.
- dual control. This last difficulty made Michalska and Mayne [138] look for a less restrictive constraint. The idea was to define a region around the final state inside which the system could be driven to the final state by means of a linear state feedback controller. Now the constraint is:

$$x(t + P) \in \Omega$$

  The nonlinear MPC algorithm is used outside the region in such a way that the prediction horizon is considered as a decision variable and is decreased at each sampling time. Once the state enters Ω, the controller switches to a previously computed linear strategy.
- quasi-infinite horizon. Chen and Allgöwer [49] extended this concept, using the idea of terminal region and stabilizing control, but only for the computation of the terminal cost. The control action is determined by solving a finite horizon problem without switching to the linear controller

even inside the terminal region. The method adds the term $\|x(t + T_p)\|_P^2$ to the cost function. This term is an upper bound of the cost needed to drive the nonlinear system to the origin starting from a state in the terminal region and therefore this finite horizon cost function approximates the infinite- horizon one.

These formulations and others with guaranteed stability were summarized in the survey paper by Mayne *et al.* [137]. In this reference, the authors present general sufficient conditions to design a stabilizing constrained MPC and demonstrate that all the aforementioned formulations are particular cases of them.

The key ingredients of the stabilizing MPC are a terminal set and a terminal cost. The terminal state denotes the state of the system predicted at the end of the prediction horizon. This terminal state is forced to reach a terminal set that contains the steady state. This state has an associated cost denoted as terminal cost, which is added to the cost function.

It is assumed that the system is locally stabilizable by a control law $u = h(x)$. This control law must satisfy the following conditions:

- There is a region Ω such that for all $x(t) \in \Omega$, then $h(x(t)) \in U$ (set of admissible control actions) and the state of the closed loop system at the next sample time $x(t + 1) \in \Omega$.
- For all $x(t) \in \Omega$, there exists a Lyapunov function $V(x)$ such that

$$V(x(t)) - V(x(t + 1)) \geq x(t)^T R x(t) + h(x(t))^T S h(x(t))$$

If these conditions are verified, then considering Ω as terminal set and $V(x)$ as terminal cost, the MPC controller (with equal values of prediction and control horizons) asymptotically stabilizes all initial states which are feasible. Therefore, if the initial state is such that the optimization problem has a solution, then the system is steered to the steady state asymptotically and satisfies the constraints along its evolution.

The condition imposed on Ω ensures constraint fulfillment. Effectively, consider that $x(t)$ is a feasible state and $\mathbf{u}^*(t)$ the optimal solution; then a feasible solution can be obtained for $x(t + 1)$. This is the composition of the remaining tail of $\mathbf{u}^*(t)$ finished with the control action derived from the local control law $h(x)$. Therefore, since no uncertainty is assumed, $x(t + j|t + 1) = x(t+j|t)$ for all $j \geq 1$. Then the predicted evolution satisfies the constraints and $x(t+P|t+1) \in \Omega$, being P the prediction horizon. Thus, applying $h(x(t+P|t + 1))$, the system remains in the terminal set Ω. Consequently, if $x(t)$ is feasible, then $x(t+1)$ is feasible too. Since all feasible states are in X, then the system fulfills the constraints.

The second condition ensures that the optimal cost is a Lyapunov function. Hence, it is necessary for the asymptotic convergence of the system to the steady state. Furthermore, the terminal cost is an upper bound of the optimal cost of the terminal state, in a similar way to the quasi-infinite horizon formulation of MPC.

These are mild conditions and the terminal cost and terminal set are not difficult to compute. If the linearized system in the steady state is stabilizable then they can be computed based on it [132],[163].

Since the optimization problem to be solved at each sample time may not be convex, the optimal solution may not be unique and may be very difficult to obtain. Thus, different approaches have been proposed to relax this fact. The main contribution in this topic is the proof of the asymptotic stability of the controller in the case of suboptimal solutions [190]: it suffices to consider any feasible solution with an associated cost strictly lower than the one of the previous sample time. In effect, any feasible solution ensures feasibility, and hence constraint satisfaction, and the strictly decreasing cost guarantees asymptotic stability. It is worth remarking that suboptimality is not desirable, since it implies a loss of performance.

Another technique for reducing the computational burden of the optimization problem is the removal of the terminal constraint. It is especially interesting when the system is unconstrained on the state. In this case, the computational burden of the optimization problem does not have to be increased by introducing terminal state constraints due to stabilizing reasons. This topic has been analyzed in [93],[96], and [126]. In [96] it is proven that a region exists around the terminal set where the terminal constraint may be removed without effecting the asymptotic stability. In [93] it is shown that MPC without a terminal constraint asymptotically stabilizes any initial state so that the optimal solution steers the system to the terminal set. In a recent paper [126] it is demonstrated that by simply weighting the terminal cost, an MPC without a terminal constraint stabilizes any initial state that can be stabilized by the controller with the terminal constraint. Furthermore, a procedure to compute the terminal weighting for a given initial state is presented.

The prediction horizon is the design parameter of the MPC with an important effect on the computational cost of the controller. If a long prediction horizon is considered, then the domain of attraction of the controller is bigger and the performance is improved. However, the number of decision variables increases and hence the complexity of the optimization problem to be solved increases. The necessary condition that must be considered for choosing the prediction horizon is the feasibility of the initial state. Thus, this horizon can be reduced by enlarging the terminal region. In [132] a controller is proposed with a prediction horizon larger than the control horizon. This allows the region of feasible initial states with the same control horizon to be enlarged. In [124] an MPC with a contractive terminal constraint based on invariant sets is presented. This provides a larger domain of attraction for a given prediction horizon.

The conditions previously presented are based on a state space representation of the system and full state information available at each sample time. However, most of the time the only available information is the measurement of the system output. In this case the controller can be reformulated using the outputs and under certain observability and controllability conditions [103], closed-loop stability can be proved. However, the most common way of applying MPC in the input-output formulation is by estimating the state by means of an observer. It is well known that even when the state space MPC and the observer are both stable, there is no guarantee that the cascaded closed-loop system is stable. Thus, additional stabilizing conditions must be considered [67].

If stability analysis in NMPC is a complex task, robustness analysis (that is, stability when modelling errors appear) is logically worse. The previously shown stability results are valid only in the case of a perfect model, which is not the case in practice. This can be considered as an open field with only preliminary results. Formulations in the form of a min-max problem or an H_∞-NMPC have been proposed, although the computational requirements are prohibitive.

9.6 Case Study: pH Neutralization Process

This section deals with the application of NMPC to the control of a typical nonlinear plant: a pH neutralization process. The control of pH is common in the chemical and biotechnological industries. Examples of this kind of plant can be found in wastewater treatment plants, the production of pharmaceuticals, and fermentation processes. Controlling the pH value of these processes is difficult due to the highly nonlinear response of the pH to the addition of acid or base. These processes can exhibit severe static nonlinear behaviour because the gain can vary several orders of magnitude for a slight range of pH values. Many control strategies have been applied to the pH process [89], ranging from linear or nonlinear PIDs to adaptive nonlinear control strategies. Predictive and fuzzy controllers have also been used.

This example corresponds to a laboratory fermentation process taken from [13]. The pH value ranks as one of the most important factors that influence a fermentation process. A pH value out of its optimum often inhibits the growth of the essential micro-organisms, alters the bacterial population and inhibits the desirable enzymatic activities. The result is a delay in the fermentation process or even the death of the micro-organisms. The controller must achieve the prescribed accuracy (around 0.05 pH unit) despite the severe process nonlinearities.

The process is shown in Figure 9.6 and consists of a tank where three streams are mixed:

- an acid ($H\,Cl$) stream (q_1),
- a buffer ($K\,H_2\,P\,O_4$ and $Na_2\,H\,P\,O_4$) stream (q_2), and
- a base ($Na\,O\,H$) stream (q_3).

The process output is the pH value of the solution which can be controlled acting on the valves.

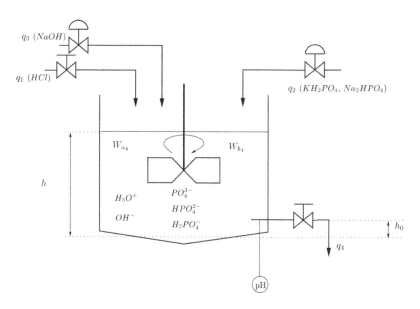

Fig. 9.6. pH fermentation process

A continuous model can be obtained by first principles, assuming perfect mixing, constant density, and complete solubility of the ions. This model can be described by the following equations (see [13] for details and numerical values):

$$Ah\frac{dW_{a_4}}{dt} = q_1\left(W_{a_1} - W_{a_4}\right) + q_2\left(W_{a_2} - W_{a_4}\right) + q_3\left(W_{a_3} - W_{a_4}\right) \quad (9.22)$$

$$Ah\frac{dW_{b_4}}{dt} = q_1\left(W_{b_1} - W_{b_4}\right) + q_2\left(W_{b_2} - W_{b_4}\right) + q_3\left(W_{b_3} - W_{b_4}\right) \quad (9.23)$$

$$A\frac{dh}{dt} = q_1 + q_2 + q_3 - q_4 \quad (9.24)$$

$$W_{b_4} \frac{K_{a_4}/[H_3O^+] + 2K_{a_4}K_{a_5}/[H_3O^+]^2 + 3K_{a_4}K_{a_5}K_{a_6}/[H_3O^+]^3}{1 + K_{a_4}/[H_3O^+] + K_{a_4}K_{a_5}/[H_3O^+]^2 + K_{a_4}K_{a_5}K_{a_6}/[H_3O^+]^3}$$
$$+ W_{a_4} + \frac{K_w}{[H_3O^+]} - [H_3O^+] = 0 \quad (9.25)$$

$$pH = -\log\left[H_3O^+\right] \quad (9.26)$$

where W_{a4} and W_{b4} are the chemical reaction invariants of the output stream q_4, and the equilibrium constants are denoted by K_x. The static Equations (9.25) and (9.26) make the process highly nonlinear and can be used to

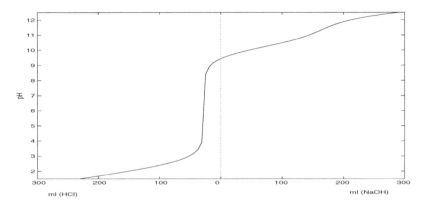

Fig. 9.7. Titration curve

obtain the theoretical pH curve of a solution of KH_2PO_4 and Na_2HPO_4 to a changing flow in q_3. The parameters of the model are given in Table 9.2.

Table 9.2. Model parameters

Variable	Value	Units
A	0.7854	dm^2
h	1.5915	dm
K_{a_4}	$7.5857 10^{-3}$	
K_{a_5}	$6.1659 10^{-11}$	
K_{a_6}	$2.1379 10^{-13}$	
K_w	$1 10^{-14}$	
W_{a_2}	$-177.3 10^{-3}$	mol/l
W_{a_3}	-1	mol/l
W_{b_2}	$154.9 10^{-3}$	mol/l
W_{b_3}	0	mol/l

This curve is shown in Figure 9.7 and is obtained by integrating Equations (9.22)-(9.24) with zero initial conditions. The figure shows the static nonlinear pH characteristic of a 10-ml phosphoric acid buffering solution to the addition of base and acid. It is clearly shown how the slope of the pH curve presents high variations along the curve.

9.6.1 Process Model

In order to devise an NMPC, a nonlinear model of the process must be obtained. A Hammerstein model seems attractive since this type of representation is adequate to model the effect of the nonlinear gain. The process output is the pH value, which will be controlled by adjusting q_3, while the other streams are kept at the values of $q_1 = 0.0059$ l/s and $q_2 = 0.3333$ l/s. A dynamic model formed by a static nonlinear part

$$\Phi(t) = g(u(t)) = \sum_{j=0}^{8} \gamma_j u^j(t) \tag{9.27}$$

and a first-order dynamic linear part

$$A(z^{-1})y(t) = B(z^{-1})\Phi(t-1)$$

is used, where the parameters are given in Table 9.3.

Table 9.3. Hammerstein model parameters

Variable	Value
γ_0	8.71
γ_1	148.36
γ_2	-750.54
γ_3	23860.37
γ_4	-3.9055×10^6
γ_5	3.52051×10^7
γ_6	-1.782601×10^8
γ_7	4.767372×10^8
γ_8	-5.249928×10^8
$A(z^{-1})$	$1 - 0.7165z^{-1}$
$B(z^{-1})$	$0.2835z^{-1}$

Figure 9.8 shows the polynomial adjustment to the nonlinear response. This model can capture the process dynamics at different operating points, although it fits better for values of pH around 9.5, as shown in Figure 9.9.

9.6.2 Results

Using the Hammerstein model, a suboptimal strategy can be used, as described in Section 9.4.7, which gives rise to a simple solution by decomposing the problem into a linear MPC and the inversion of (9.27). The computational cost is therefore negligible. The results obtained for this control strategy are shown in Figure 9.10, which shows the controller behaviour when a setpoint change from 10 to 8.5 is produced. The response for several values of λ is shown, as well as disturbance rejection to a step change of 10% in q_1. The fastest response is obtained for $\lambda = 200$, and the other is obtained for $\lambda = 500$. Notice that the high values of the control weighting factor are due to the fact that the variables are not normalized; the output is around a value of 10 and the input around 0.01.

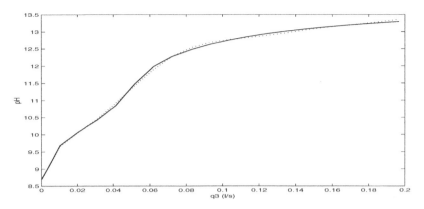

Fig. 9.8. Hammerstein model adjustment

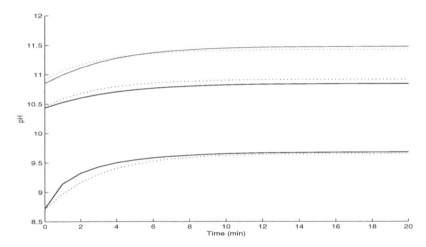

Fig. 9.9. Response to step changes in the manipulated variable

The test of the controller is done using the simulated nonlinear model of the plant given by Equations (9.22)-(9.26). It is clear that, as the nonlinear model is not perfect, the controller is not able to eliminate all the nonlinear effects. This can be seen in Figure 9.11, which shows the same setpoint change but in the opposite direction.

Another example of an NMPC strategy is also tested on the simulated plant. An EPSAC is used to control the plant, as shown in Figure 9.12. The system response to setpoint changes is similar to that obtained with the Hammerstein controller. This controller uses an iterative algorithm with a slightly greater computational cost (around one-third greater in this simulated exam-

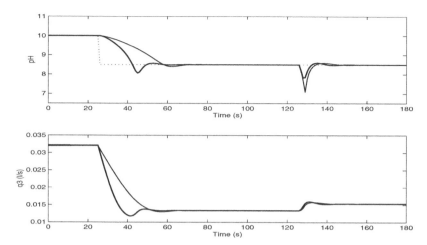

Fig. 9.10. pH control with Hammerstein model: $\lambda = 200$ (thick line) and $\lambda = 500$ (thin line)

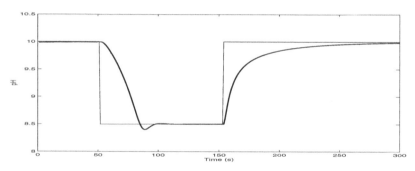

Fig. 9.11. Nonlinear effect

ple). However, EPSAC formulation is valid for any type of nonlinear model, while the Hammerstein formulation works well in the specific case where the process can be modelled by a static nonlinear gain and a linear dynamic part.

9.7 Exercises

9.1. Write the Hammerstein model of the neutralization process as a general Volterra model in the form of Equation (9.3) and in the form of equation (9.4).

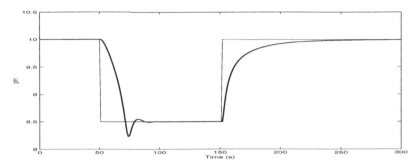

Fig. 9.12. pH control by nonlinear EPSAC

9.2. Adjust a third-order Hammerstein model to the previous process. This can be done generating the curve shown in Figure 9.8 from the process equations and obtaining the best coefficients by a least-squares fit.

9.3. Apply a linear GPC to control the neutralization process given by its eight-order Hammerstein model.

9.4. Change this linear GPC in such a way that the free response is calculated using the complete Hammerstein model. Compare the result with the previous case.

9.5. Develop step by step a GPC with $N_1 = 1$, $N_2 = N_u = 3$ that calculates $\Psi(t)$ and then obtain $u(t)$ by inverting Equation (9.27).

9.6. Use an SQP routine to solve the general nonlinear control of the neutralization process.

9.7. Train a Neural Network to reproduce the behaviour of the previous controller. Compare the results in both performance and computation time.

10

Model Predictive Control and Hybrid Systems

In most processes there are not only continuous variables but also variables that have a discrete nature. For a long time, the control of processes with discrete variables and the control of processes with continuous variables were considered to be two completely different things. On the one hand, the theories of finite state machines were used to control processes with discrete variables, and on the other hand, linear and nonlinear control theory was used for the control of continuous variables. The techniques for modelling and analysis of these types of systems are different. In the case of continuous systems, differential equations, transfer functions, etc., are used as modelling tools, while in the discrete counterpart, state transition graphs, Petri Nets, etc., are employed (see Figure 10.1). From the beginning of the 1990s there has been great interest in processes that have both discrete and continuous parts. Hybrid systems are dynamic systems with both continuous-state and discrete-state and event variables. That is, the plant has time-driven and event-driven dynamics, the controller affects both time-driven and event-driven components, and it may deal with continuous and/or discrete signals.

This chapter is devoted to introducing how model predictive control is able to cope with hybrid systems.

10.1 Hybrid System Modelling

Hybrid modelling techniques used for control purposes have to be descriptive enough to capture the behavior of the various parts of the system; i.e., continuous dynamics (physical laws) and logic components (switches, automata, software code), and to take into account interconnections between logic and continuous dynamics. At the same time, the model has to be simple enough to solve analysis and synthesis problems.

The discrete (usually logical) parts of the systems can be described by discrete automata. An automata is defined by the following 5-tuple $[X, U, Y, \phi, \gamma]$,

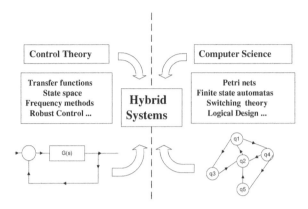

Fig. 10.1. Hybrid systems

where X is the finite set of possible states and U and Y are the finite sets of possible inputs and outputs. The transition function $\delta : X \times U \to X$ defines the next state of the system $(x(t + \delta t) = \phi(x(t), u(t)))$. The output function $\gamma : X \times U \to Y$ describes the output of the system as a function of the state and input (when the Mealy model is used) or just the state (when the Moore model is used).

Functions ϕ and γ are logical functions and can be defined by logical expressions or truth tables. The time interval δt appearing in the state transition equation may be dictated by a clock in the case of synchronous systems or by the occurrence of a particular event in the case of asynchronous systems (see [91], a classical book on the subject). Events are usually produced when one of the input signals changes state. These abstract descriptions of an automata resemble the state space description of continuous[1] systems. However, other descriptions and models, such as Petri nets, can be more appropriate in many cases as they allow parallelism, synchronization, shared resources, etc. to be modelled.

These types of models have been widely used in the computer science community but rarely used in the control community. The approach taken by the computer science community to model hybrid systems has been the hybrid automata. That is an extension of the automata models to cope with continuous signals and equations. The main idea in this context is that the continuous parts evolve within each of the states.

Hybrid automata are finite state machines where a continuous dynamics is associated to each discrete state. Hybrid automata are composed of a set of discrete states $\mathcal{X}_d$ and a set of continuous states $\mathcal{X}_c$. A switch is a change of the discrete state of hybrid automata. When a switch takes place the continuous state is reset. Events are associated to the instant of time when a switch

---

[1] The word *continuous* is used here to refer to systems with variables that take real values.

takes place. Discrete state changes can be produced by changes of the discrete inputs, or when the continuous state (augmented by the input signals) enters or exits determined regions (usually defined by polyhedra).

People coming from the continuous control field have used a more algebraic approach. As will be seen later, the logical relationships coming from the discrete part of the hybrid system are described by real algebraic equations and constraints. That is, mainly algebraic equations and constraints extended to cope with logical variables and predicates. The main ideas are similar to those used in the Constraint Logic Programming (CLP) paradigm where some resemblance to traditional Operation Research approaches are sought to solve logical problems. Constraint programming is the study of computational systems based on constraints. The idea of constraint programming is to solve problems by stating constraints (conditions, properties) which must be satisfied by the solution. Then an algorithm is used to find a feasible solution to the problem.

Let us consider a hybrid dynamical system composed of continuous and discrete input, output, and state variables. The dynamics can be described as follows

$$\dot{x}_c(t) = f_c(x_c(t), u_c(t), x_d(T_k), u_d(t), t, \delta t) \quad T_k \le t < T_k + \delta t$$
$$x_d(T_k + \delta t) = f_d(x_c(t), u_c(t), x_d(T_k), u_d(t), t, \delta t)$$
$$y_c(t) = h_(x_c(t), u_c(t), x_d(T_k), u_d(t), t)$$
$$y_d(t) = h_d(x_c(t), u_c(t), x_d(T_k), u_d(t), t)$$
$$\delta t = \zeta(x_c(t), u_c(t), x_d(T_k), u_d(t), t)$$

where x_c, u_c, and y_c are the continuous state, inputs, and outputs; functions f_c and h_c are the state transition and output functions of the continuous parts. The discrete state, input, and output variables are x_d, u_d, and y_d. Functions f_d and h_d are the state transition and output function of the discrete parts. The function ζ determines the time interval δt until the next event. Notice that function f_c, which determines the next state, can usually be obtained by integrating the continuous equations. Determining δt (time to next event) is one of the most difficult issues in hybrid control, especially in applications which are time-critical. When this is not the case, a good approximation can be obtained by making the times between events δ constant and equal to the sampling time if the sampling time is sufficiently small. This is the approach normally used when applying MPC to hybrid systems an approach that will be used in the book.

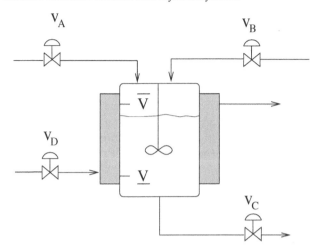

Fig. 10.2. Jacket cooled batch reactor

10.2 Example: A Jacket Cooled Batch Reactor

Consider the batch reactor shown in Figure 10.2. The manipulated variables are the two product input valves v_A and v_B, the output valve v_C, and the valve controlling the coolant inflow to the refrigeration jacket v_D.

The state variables are the concentration C, the temperature of the batch T, and the volume V. That is, $x(t)^T = [C(t) \quad T(t) \quad V(t)]^T \in \mathbb{R}^3$. The sensors installed in the reactor are two on-off volume sensors $S_{\underline{V}}$ and $S_{\overline{V}}$ which take the "ON" value if $V(t) \leq \underline{V}$ and $V(t) \geq \overline{V}$ respectively and the "OFF" value otherwise. $\underline{V}$ and $\overline{V}$ are fixed values corresponding to the empty and full values of the product volume in the reactor. There is also a temperature sensor that gives an analogue reading of the product temperature (it is assumed that the temperature is uniform because the product is well stirred); that is, the output vector can be expressed as $y(t) = [\underline{V}(t) \quad \overline{V}(t) \quad T(t)]$. If we use the typical assignment of the ON value to 1 and OFF to 0, the output vector $y(t) \in \{0,1\}^2 \times \mathbb{R}$.

The manipulated variables are $u(t) = [v_A(t) \quad v_B(t) \quad v_C(t) \quad v_D(t)]^T$. The product inlet and outlet valves can be either totally closed or open while the valve controlling the coolant inflow to the refrigeration jacket can be continuously manipulated to the intermediate position; that is, $u(t) \in \{0,1\}^3 \times \mathbb{R}$.

The dynamic equations describing the process variables can be approximated as

$$\dot{C}(t) = K_R(1 - C(t))(T(t) - T_0)$$
$$\dot{T}(t) = K_{T1}(1 - C(t))(T(t) - T_0) - K_{T2}v_D(t)/V(t)$$
$$\dot{V}(t) = K_A v_A(t) + K_B v_B(t) - K_C v_C(t)$$

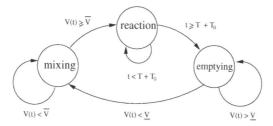

Fig. 10.3. Jacket cooled batch reactor state diagram

where the reaction speed $\dot{C}(t)$ is considered to be proportional to the product of the temperature by the amount of product yet to react. The heat produced by the reaction is considered to be proportional to the reaction speed. K_R, T_0, K_{T1}, K_{T2}, K_A, K_B, and K_C are the corresponding proportional constants.

The reactor is first loaded with the two products, then a temperature-controlled reaction takes place, and after a fixed reaction time, the product is emptied and a new cycle starts. In order to accomplish this, the controller of the process works as indicated in Figure 10.3. First the product input valves are fully open, while the product outlet valve and the cooling valve are fully closed ($u_1(t) = v_A(t) = 1, u_2(t) = v_B(t) = 1, u_3(t) = v_C(t) = 0, u_4(t) = v_D(t) = 0$). Once the volume in the reactor reaches $\overline{V}$ ($y_2(t) = S_{\overline{V}}(t) = 1$), the inlet product valves are closed and the coolant inlet valve opening is determined by a PID temperature controller; that is, $u_1(t) = v_A(t) = 0$, $u_2(t) = v_B(t) = 0, u_3(t) = v_C(t) = 0, u_4(t) = v_D(t) = u_{PID}(t)$. The process remains in this state for the prescribed reaction time. After this, the product outlet valves opens until the reactor is empty ($y_1(t) = S_{\underline{V}}(t) = 1$) and a new cycle starts.

As can be seen, the process is characterized by continuous and discrete process and controller variables. The process controller is composed of a continuous part (PID) and a discrete part (automata) controlling the transitions between states. The process is also event-driven as changes of states are dictated by the occurrence of certain events such as when $V(t) \geq \overline{V}$ and the system goes from the filling-up state to the reaction state.

10.2.1 Mixed Logical Dynamical Systems

This type of model was proposed by Bemporad and Morari [22]. Mixed Logical Dynamical (MLD) systems are described by

$$x(t+1) = Ax(t) + B_1U(t) + B_2\delta(t) + B_3z(t) \tag{10.1}$$
$$y(t) = Cx(t) + D_1u(t) + D_2\delta(t) + D_3z(t) \tag{10.2}$$
$$E_1x(t) + E_2u(t) + E_3\delta(t) + E_4z(t) \leq g \tag{10.3}$$

where $x(t) = [x_r^T(t) \quad x_b^T(t)]$ with $x_r(t) \in \mathbb{R}^n$ is the continuous part of the state vector and $x_b(t) \in \{0,1\}^{n_b}$ is the part of the state vector corresponding to the discrete part. Notice that if the state is discrete, not necessarily Boolean, but finite, it can be coded into a set of Boolean variables. The output signals also have a similar structure $y(t) = [y_r^T(t) \quad y_b^T(t)]$ with $y_r(t) \in \mathbb{R}^m$ is the continuous part of the output and $y_b(t) \in \{0,1\}^{m_b}$ is the discrete part. The input vector $u(t) = [u_r^T(t) \quad u_b^T(t)]$ is composed of a continuous part $u_r(t) \in \mathbb{R}^l$ and a discrete part $u_b(t) \in \{0,1\}^{l_b}$. Some auxiliary continuous $z(t) \in \mathbb{R}^r$ and discrete variables $\delta(t) \in \{0,1\}^{r_b}$ are usually needed.

The key idea of the method is that logical expressions can be transformed into algebraic constraints. Consider the logical predicates L_1 and L_2. A logical predicate (or literal) is a sentence that can be either true or false. For example, L_1 can be something like: *the flow is higher than 0.2 liters/second*, which can be true or false. Boolean algebra allows simple predicates to be combined using connectives and/or modifiers such as *and* ($\wedge$), *or* ($\vee$), *not* ($\sim$), *implies* ($\rightarrow$) or *if and only if* ($\leftrightarrow$). Predicates take logical values *true* or *false*. Any logical predicate can be written using a subset of connectives. The most common are: $\{\wedge, \vee, \sim\}$. There are simple transformations that can be used to express any logical predicate in terms of the basic connectives. For example:[2]

$$L_1 \rightarrow L_2 \Leftrightarrow \sim L_1 \vee L_2 \tag{10.4}$$

$$L_1 \leftrightarrow L_2 \Leftrightarrow L_1 = L_2 \tag{10.5}$$

Expressions (10.4) and (10.5) can be proved by a truth table (computing both side of the equation for all possibilities) as can be seen in the following truth table comparing the corresponding columns, where *true* and *false* are denoted by T and F, respectively.

$L1$	$L2$	$\sim L1$	$L1 \vee L2$	$L1 \wedge L2$	$L1 \rightarrow L2$	$\sim L1 \vee L2$	$L1 \leftrightarrow L3$	$L1 = L2$
F	F	T	F	F	T	T	T	T
F	T	T	T	F	T	T	F	F
T	F	F	T	F	F	F	F	F
T	T	F	T	T	T	T	T	T

Although it is normal to associate values 1 and 0 to the *true* and *false* values, they are not numbers. Let us associate an integer variable δ to these predicates such that $\delta = 1$ if $L = T$ (true) and $\delta = 0$ if $L = F$ (false). It is very easy to see that the basic logical operators have the following corresponding algebraic inequalities:

---

[2] The symbol $\Leftrightarrow$ will be used in the text as a meta symbol to express that the expressions on both sides of the symbol $\Leftrightarrow$ are equivalent.

$$\sim L_1 \Leftrightarrow \delta_1 = 0 \tag{10.6}$$

$$L_1 \vee L_2 \Leftrightarrow \delta_1 + \delta_2 \geq 1 \tag{10.7}$$

$$L_1 \wedge L_2 \Leftrightarrow \delta_1 = 1, \delta_2 = 1 \tag{10.8}$$

$$L_1 \rightarrow L_2 \Leftrightarrow \delta_1 - \delta_2 \leq 0 \tag{10.9}$$

$$L_1 \leftrightarrow L_2 \Leftrightarrow \delta_1 - \delta_2 = 0 \tag{10.10}$$

In order to connect the logical and the continuous parts, consider a function $f : \mathbb{R}^n \rightarrow \mathbb{R}$ and a logical variable $l \in \{0, 1\}$. Consider that f is bounded in its domain by $\underline{f} \leq f(x) \leq \overline{f}$. Then, consider a logical predicate about the continuous function such as $f(x) \leq 0$. It can be seen that the following logical statements and constraints are equivalent

$$(f(x) \leq 0) \wedge (\delta = 1) \Leftrightarrow f(x) - \delta \leq -1 + \underline{f}(1 - \delta) \tag{10.11}$$

$$(f(x) \leq 0) \vee (\delta = 1) \Leftrightarrow f(x) \leq \overline{f}\delta \tag{10.12}$$

$$\sim (f(x) \leq 0) \Leftrightarrow f(x) > 0 \text{ or } f(x) \geq \epsilon \tag{10.13}$$

where ϵ is the smallest number in the computer. From expression (10.4), it follows that:

$$(f(x) \leq 0) \rightarrow (\delta = 1) \Leftrightarrow \sim (f(x) \leq 0) \vee (\delta = 1) \tag{10.14}$$

Taking into account (10.13);

$$(f(x) \leq 0) \rightarrow (\delta = 1) \Leftrightarrow (f(x) \geq \epsilon) \vee (\delta = 1)$$
$$\Leftrightarrow (g(x) \leq 0) \vee (\delta = 1) \text{ with } g(x) = \epsilon - f(x)$$
$$\Leftrightarrow g(x) \leq \overline{g}\delta \text{ with } \overline{g} = \max_{dom(g)} g(x) = \epsilon - \underline{f}$$
$$\Leftrightarrow \epsilon - f(x) \leq (\epsilon - \underline{f})\delta$$
$$\Leftrightarrow f(x) \geq \epsilon + \delta(\underline{f} - \epsilon)$$

that is,

$$(f(x) \leq 0) \rightarrow (\delta = 1) \Leftrightarrow f(x) \geq \epsilon + \delta(\underline{f} - \epsilon) \tag{10.15}$$

Similarly,

$$(\delta = 1) \rightarrow (f(x) \leq 0) \Leftrightarrow \sim (\delta = 1) \vee (f(x) \leq 0)$$
$$\Leftrightarrow (\delta' = 1) \vee (f(x) \leq 0) \text{ with } \delta' = 1 - \delta$$
$$\Leftrightarrow f(x) \leq \overline{f}\delta' = \overline{f}(1 - \delta)$$

that is,

$$(\delta = 1) \rightarrow (f(x) \leq 0) \Leftrightarrow f(x) \leq \overline{f}\delta' = \overline{f}(1 - \delta) \tag{10.16}$$

The *if and only if* condition ($\leftrightarrow$) necessary to establish the equivalence between the logical predicate and the set of constraints can be obtained by combining (10.15) and (10.16):

$$(\delta = 1) \leftrightarrow (f(x) \leq 0) \Leftrightarrow \begin{cases} f(x) \leq \overline{f}\delta' = \overline{f}(1 - \delta) \\ f(x) \geq \epsilon + \delta(\underline{f} - \epsilon) \end{cases} \tag{10.17}$$

The combinations of logical predicates and real functions can also be handled. As an example, consider a term in the form $\delta f(x)$. We can introduce an auxiliary real variable $z \triangleq \delta f(x)$ satisfying $z = 0$ when $\delta = 0$ and $z = f(x)$ when $\delta = 1$. By using the previous results, these predicates $((\delta = 0) \rightarrow (z = 0)$ and $(\delta = 1) \rightarrow (z = f(x)))$ can be translated into the following linear inequalities:

$$z \leq \overline{f}\delta$$
$$z \geq \underline{f}\delta$$
$$z \leq f(x) - \underline{f}(1 - \delta)$$
$$z \geq f(x) - \overline{f}(1 - \delta)$$

The previous inequalities can easily be obtained considering that $z = 0 \Leftrightarrow (z \leq 0) \wedge (z \geq 0)$, $z = f(x) \Leftrightarrow (z - f(x) \leq 0) \wedge (f(x) - z) \leq 0)$, $\delta' = 1 - \delta$ and then using (10.15) and (10.16).

The product of variables (δ_1, δ_2) associated to logical predicates can also be translated as a set of linear constraints by introducing an auxiliary $\delta_3 \triangleq \delta_1\delta_2$. This is equivalent to $(\delta_3 = 1) \leftrightarrow (\delta_1 = 1) \wedge (\delta_2 = 1)$. By using the previous results:

$$(\delta_3 = 1) \leftrightarrow (\delta_1 = 1) \wedge (\delta_2 = 1) \Leftrightarrow \begin{cases} -\delta_1 + \delta_3 \leq 0 \\ -\delta_2 + \delta_3 \leq 0 \\ \delta_1 + \delta_2 - \delta_3 \leq 1 \end{cases} \tag{10.18}$$

There are different ways of translating logic predicates into linear inequalities. The number of auxiliary variables and constraints will depend on which one is chosen.

10.2.2 Example

To illustrate how a hybrid system can be modelled by an MLD, consider the mixing process described in Figure 10.4.

The mixer is controlled by an automata and works as follows: the tank is filled (valves V_A and V_B open, valve V_C closed) and stirred at the same time until the level in the tank reaches the maximum height H. At this point the outlet valve opens, letting the product out until the level in the tank is below the *empty* level h. There are two Boolean states, the filling-mixing one and the emptying one. We can associate a Boolean state variable $x_b = 0$ when the process is in the filling-mixing state and $x_b = 1$ when the process is in

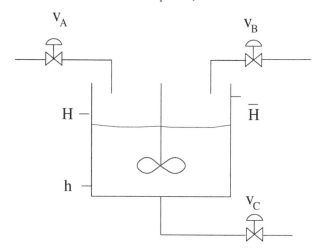

Fig. 10.4. Mixing process

the emptying state. The continuous state $x_c(t)$ is associated to the height in the mixing tank. It can easily be seen that the Boolean state variable will take the value one when the height in the mixing deposit is above H or when the system is already in state one and the level is above h. That is:

$$((x_b(t) = 1) \wedge (x_c(t) > h)) \vee (x_c(t) > H) \leftrightarrow x_b(t + 1) = 1$$

Let us introduce some logical auxiliary variables $\delta_1, \delta_2, \delta_3$, and δ_4 with the following meanings:

$$(\delta_1 = 1) \leftrightarrow (x_c(t) \leq H) \tag{10.19}$$

$$(\delta_2 = 1) \leftrightarrow (x_c(t) > h) \tag{10.20}$$

$$(\delta_3 = 1) \leftrightarrow ((x_b(t) = 1) \wedge \delta_2) \tag{10.21}$$

$$(\delta_4 = 1) \leftrightarrow (\delta_3 \vee \sim \delta_1) \tag{10.22}$$

By using the previous transformation, the following linear inequalities are obtained

$$(\delta_1 = 1) \leftrightarrow (x_c(t) \leq H) \Leftrightarrow \begin{cases} x_c(t) - H \leq (\overline{H} - H)(1 - \delta_1) \\ x_c(t) - H \geq \epsilon - (H + \epsilon)\delta_1 \end{cases} \tag{10.23}$$

$$(\delta_2 = 1) \leftrightarrow (x_c(t) > h) \Leftrightarrow \begin{cases} h - x_c(t) \leq h(1 - \delta_2) - \epsilon\delta_2 \\ h - x_c(t) \geq (h - \overline{H})\delta_2 \end{cases} \tag{10.24}$$

$$(\delta_3 = 1) \leftrightarrow ((x_b(t) = 1) \wedge \delta_2) \Leftrightarrow \begin{cases} -x_b(t) + \delta_3 \leq 0 \\ -\delta_2 + \delta_3 \leq 0 \\ x_b(t) + \delta_2 - \delta_3 \leq 1 \end{cases} \tag{10.25}$$

$$(\delta_4 = 1) \leftrightarrow (\delta_3 \vee \sim \delta_1) \Leftrightarrow \begin{cases} \delta_1 - \delta_3 + \delta_4 \leq 1 \\ \delta_1 - \delta_3 + 2\delta_4 \geq 1 \end{cases} \tag{10.26}$$

where $\overline{H}$ is the maximum height of the tank and ϵ is a very small number. The continuous part of the process can be described by

$$x_c(t+1) = x_c(t) + (1 - x_b(t))(Q_A + Q_B)/S - x_b(t)Q_C/S \qquad (10.27)$$

where S is the section of the tank and $Q_A, Q_B,$ and Q_C are the flows of products when valves $V_A, V_B,$ or V_C are open. Notice that when the process is at state $x_b(t) = 0$, corresponding to the filling state, the inlet flows correspond to the filling valves open; when the Boolean state is $x_b(t) = 1$, the net flow corresponds to the flow through the outlet valve. The Boolean state in the next sampling period is given by:

$$x_b(t+1) = \delta_4(t) \qquad (10.28)$$

Other operating constraints can be added to the process. For example, constraints about the minimum and maximum heights of the liquid in the deposit $x_c(t) \le \overline{H}$ and $x_c(t) \ge 0$. The model is described by Equations (10.27) and (10.28) and linear inequalities (10.23) and (10.26) plus other operational constraints. Notice that these expressions have the shape of Expressions (10.1)-(10.2). That is:

$$x_c(t+1) = x_c(t) + (1 - x_b(t))(Q_A + Q_B)/S - x_b(t)Q_C/S$$
$$x_b(t+1) = \delta_4(t)$$
$$\text{s. t.:} \quad x_c(t) - H \le (\overline{H} - H)(1 - \delta_1)$$
$$x_c(t) - H \ge \epsilon - (H + \epsilon)\delta_1$$
$$h - x_c(t) \le h(1 - \delta_2) - \epsilon\delta_2$$
$$h - x_c(t) \ge \epsilon(h - \overline{H})\delta_2$$
$$-x_b(t) + \delta_3 \le 0$$
$$-\delta_2 + \delta_3 \le 0$$
$$x_b(t) + \delta_2 - \delta_3 \le 1$$
$$\delta_1 - \delta_3 + \delta_4 \le 1$$
$$\delta_1 - \delta_3 + 2\delta_4 \ge 1$$
$$x_c(t) \le \overline{H}$$
$$x_c(t) \ge 0$$

10.3 Model Predictive Control of MLD Systems

The ideas of MPC (optimising an objective function over a finite and rolling control horizon) can also be applied to MLD systems. However the problem is far more difficult than in the case of processes with real variables. The problem can be formulated as:

$$\mathbf{u}^* = \arg\min_{\mathbf{u}} \|\mathbf{x} - \mathbf{r}_x\|_{\mathbf{Q}_x}^p + \|\mathbf{u} - \mathbf{r}_u\|_{\mathbf{Q}_u}^p + \|\delta - \mathbf{r}_\delta\|_{\mathbf{Q}_\delta}^p + \|\mathbf{z} - \mathbf{r}_z\|_{\mathbf{Q}_z}^p \qquad (10.29)$$

s. t.:

$$x(t + j) = Ax(t + j - 1) + B_1u(t + j - 1) + B_2\delta(t + j - 1) + B_3z(t + j - 1)$$
$$y(t + j) = Cx(t + j) + D_1u(t + j) + D_2\delta(t + j) + D_3z(t + j)$$
$$E_1x(t + j) + E_2u(t + j) + E_3\delta(t + j) + E_4z(t + j) \leq g$$

where $\|x\|_Q^p$ denotes x^TQx when $p = 2$ and $Q\|x\|_p$ for $p = 1$ or $p = \infty$ and $\mathbf{Q}_x$, $\mathbf{Q}_u$, $\mathbf{Q}_\delta$, and $\mathbf{Q}_z$ are weighting matrices or vectors of appropriate dimensions and all the signals in the future are predicted with the information available at time t, as is usual in MPC. The vectors $\mathbf{x}$, $\mathbf{u}$, δ, $\mathbf{z}$, $\mathbf{r}_x$, $\mathbf{r}_u$, $\mathbf{r}_\delta$ and $\mathbf{r}_z$ are the vectors of future predicted states, control moves, auxiliary Boolean variables, auxiliary real variables, and their corresponding future references. They are defined as

$$\mathbf{u} \triangleq \begin{bmatrix} u(t) \\ u(t+1) \\ \vdots \\ u(t+N-1) \end{bmatrix} \quad \mathbf{x} \triangleq \begin{bmatrix} x(t+1) \\ x(t+2) \\ \vdots \\ x(t+N) \end{bmatrix} \quad \delta \triangleq \begin{bmatrix} \delta(t+1) \\ \delta(t+2) \\ \vdots \\ \delta(t+N) \end{bmatrix} \quad \mathbf{z} \triangleq \begin{bmatrix} z(t+1) \\ z(t+2) \\ \vdots \\ z(t+N) \end{bmatrix}$$

with $u(t) \triangleq \begin{bmatrix} u_c(t) \\ u_b(t) \end{bmatrix}$, $x(t) \triangleq \begin{bmatrix} x_c(t) \\ x_b(t) \end{bmatrix}$

The MLD MPC (10.29) results in an optimization problem with a set of linear constraints and with real and integer (Boolean in this case) decision variables. These types of optimization problems are known, in general, as Mixed Integer Programming (MIP) problems. If the objective function is a linear function these problems are known as Mixed Integer Linear Programming (MILP) problems, or Mixed Integer Quadratic Programming (MIQP) problems when the objective function is quadratic. It has to be said that this is a much more difficult problem to solve than an LP or a QP problem (see, for example, [68] for an excellent introduction).

Notice that for each of the possible (feasible) combinations of the discrete decision variables, a QP problem (with the remaining continuous decision variables) can be solved. A brute force approach would be to solve all of these QPs; the solution will be the minimum of the solutions of all the QP problems. If all the discrete decision variables are Boolean, the number of possible QP problems is 2^{n_b}. Fortunately there are more efficient ways of solving this type of problem. They are usually based on branch and bound methods and solve only a portion of all QP problems.

10.3.1 Branch and Bound Mixed Integer Programming

Consider the optimization problem $P0$ with $\mathbf{u} \in \mathbb{R}^n$ and $\delta \in \{0,1\}^m$:

$$\min_{\mathbf{u},\delta} J(\mathbf{u}, \delta) \quad \text{subject to} \quad R \begin{bmatrix} \mathbf{u} \\ \delta \end{bmatrix} \leq \mathbf{r} \tag{10.30}$$

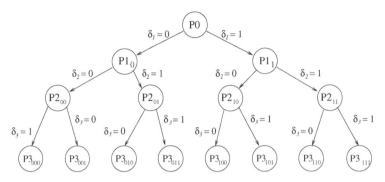

Fig. 10.5. Binary tree representation of a MIQP

A fundamental concept of the branch and bound algorithms is the enumeration of the optimization problems to be solved. These problems can be represented. If the integer variables are Boolean, the resulting tree is a binary tree (two branches per node), as the tree shown in Figure 10.5. The root node (the node on the top) represents the original MIP problem and is located at level 0. At the next level (depth = 1), there are two nodes (children). The first one $(P1_0)$ represents a subset of the *parent* problem, characterized by having the first component of the Boolean decision variables (δ_1) equal to zero. The other node at level 1 $(P1_1)$ represents the subset of the original problem characterized by $\delta_1 = 1$. Notice that the solution of the original problem $(P0)$ is the minimum of the solution of problems $P1_0$ and $P1_1$. That is, $\min(P0) = \min(\min(P1_0), \min(P1_1))$. At level 2, there are four nodes, nodes $P2_{\{00\}}$ and $P2_{\{01\}}$, descendants (children) from node $P2_0$ and nodes $P2_{\{10\}}$, and $P2_{\{11\}}$, descendants from node $P2_1$. The nodes at level m in the tree (leaves of the tree) correspond to each of the possible 2^m QP problems that would have to be solved if brute force were used. Figure 10.5 corresponds to a tree with three logical variables and the number of leaves and possible QPs is 8.

The key idea of branch and bound is to establish a partition of the feasible set into smaller subsets and then calculate certain bounds on the costs within some of the subsets to eliminate from further consideration other subsets and, therefore, to reduce the number of optimization problems to be solved.

Suppose that we have to solve $\min_u f(u)$ with $u \in \mathbb{U}$. Given two subsets $\mathbb{U}_1 \subseteq \mathbb{U}$ and $\mathbb{U}_2 \subseteq \mathbb{U}$, suppose that we have bounds $\underline{f_1} \leq \min_{u \in \mathbb{U}_1} f(u)$, $\overline{f_2} \geq \min_{u \in \mathbb{U}_2} f(u)$. Then, if $\overline{f_2} \leq \underline{f_1}$, the points in $\mathbb{U}_1$ may be disregarded since their cost cannot be smaller than the cost of any solution in $\mathbb{U}_2$. An easy way of computing bounds for the optimization problem is by relaxation of the original problem. Consider problem $P0$ defined in (10.30). If we relax the conditions $\delta \in \{0,1\}^m$ and substitute them by $0 \leq \delta \leq 1$, we have a relaxed problem $(RP0)$ with larger feasibility region and the following characteristics:

1. If $P0$ is feasible, so is $RP0$; if $RP0$ is infeasible so is $P0$.
2. The minimum of the relaxed problem is a lower bound of the minimum of the original problem (i.e., $\min RP0 \leq \min P0$).
3. If the optimal solution of $RP0$ is feasible for $P0$, it is also the optimal solution for $P0$.

More formally, we can define the problem Pk_j, with k equal to the depth in the tree and j a binary combination of k zeros and ones corresponding to a particular realization of the Boolean variables for one of the nodes being explored ($j \in \{0,1\}^k$). For example, $P2_{10}$ represents the problem corresponding to fixing the first two Boolean decision variables to 1 and 0, respectively ($\delta_1 = 1, \delta_2 = 0$) while leaving the remaining Boolean variables free $\delta_{k+i} \in \{0,1\}$ for $i = 1,\ldots,m-k$. The relaxed problem RPk_j is the problem obtained when the constraints $\delta_{k+i} \in \{0,1\}$ for $i = 1,\ldots,m-k$ are relaxed to $0 \leq \delta_{k+i} \leq 1$ for $i = 1,\ldots,m-k$.

Let us define $\mathbf{u}_r^* = \arg\min RPk_j$ (the value that minimizes the relaxed problem keeping the first k Boolean variables at the values corresponding to index j), $\mathbf{u}^* = \arg\min Pk_j$, $\mathcal{F}(Pk_j)$, and $\mathcal{F}(RPk_j)$ the feasible regions of problems Pk_j and RPk_j, respectively. Then it can be seen that:

1. $\mathcal{F}(Pk_j) \subseteq \mathcal{F}(RPk_j)$.

2. $\mathcal{F}(Pk_j) \neq \emptyset \Rightarrow \mathcal{F}(RPk_j) \neq \emptyset$.

3. $\mathcal{F}(RPk_j) = \emptyset \Rightarrow \mathcal{F}(Pk_j) = \emptyset$.

4. $\min RPk_j \leq \min Pk_j$.

5. If $\mathbf{u}_r^* \in \mathcal{F}(Pk_j) \Rightarrow \mathbf{u}^* = \mathbf{u}_r^*$.

Notice that problem RPk_j is an optimization problem with $n+m-k$ optimization variables. A basic branch and bound-based MIP algorithm works as follow:

1. Calculate an upper bound $\overline{J}(P0)$ and a lower bound $\underline{J}(P0)$ of $\min_\mathbf{u}(P0)$;[3] $\emptyset \rightarrow$ SOL. Put $P0$ and its associated lower bound $(\underline{J}(P0))$ into OPEN (List of Candidate Solutions).
2. If OPEN is empty, the optimum solution is in SOL and its value is $\overline{J}$, STOP.
3. Otherwise, get a problem P_i from OPEN. If the associated lower bound of P_i is bigger than $\overline{J}$, go to step 2 (disregard node as its lower bound is above the upper bound of the solution found so far).
4. Form the relaxed problem RP_i. Solve RP_i. Let SOL1 be the solution (it may be empty if the problem is unfeasible) and J_i be the minimum value of the objective function.

---

[3] These bounds can be always set to a very high number and a very low number.

5. If SOL1 is empty (problem RPi infeasible), go to step 2.
6. If $J_i \geq \overline{J}$ go to step 2 (disregard node as its lower bound is above the upper bound of the solution found so far).
7. If SOL1 is also feasible for P_i set $J_i^* = J_i$, set $\overline{J} = J_i$ and the solution of SOL=SOL1; go to step 2.
8. Generate problems P_{i_0} and P_{i_1} (children of problem P_i) and put them in OPEN with associated lower bounds $\underline{J}(P_{i_0}) = \underline{J}(P_{i_0}) = J_i$, go to step 2.

Some remarks can be made on this basic branch and bound algorithm:

- The initial upper and lower bounds on optimal cost can be fixed at very high and low values if no information is available. However, the number of visited nodes can be reduced if these bounds are tightly estimated.
- One of the most important aspects, regarding efficiency, of the algorithm is how the next node to be expanded is selected from the set of candidate solutions (OPEN). The two classical alternatives for exploring a tree are the depth-first search, and breadth-first search. The depth-first search consists in selecting the node with greatest depth. The second strategy consists of expanding the nodes with the smallest depth. This strategy looks more systematic but requires more nodes in memory to be maintained. More efficient strategies use information about the objective function and expand more promising nodes by selecting nodes with lower bounds of the objective function, for example.
- The order in which the Boolean variables are selected for branching can be determinative. In Figure 10.5, the first variable selected was δ_1, then δ_2 and δ_3. There is nothing to keep us from choosing another order, for example, $(\delta_3, \delta_1, \delta_2)$. A logical way is to order these variables according to the influence they will have on the objective function.

10.3.2 An Illustrative Example

Consider the following MIQP

$$\min_{u,\delta} \frac{1}{2} \ (2u^2 + \delta_1^2 + 5\delta_2^2 + 4\delta_3^2 + u\delta_1) + u - \delta_1 - 3\delta_2 - 2\delta_3 \quad (10.31)$$
$$\text{subject to } -1 \leq u \leq 1$$
$$1 \leq \delta_1 + \delta_2 + \delta_3 \leq 2$$
$$\delta \in \{0,1\}^3$$

with $\delta^T = [\delta_1 \ \delta_2 \ \delta_3]$. The first step of the algorithm is to assign a high value to $\overline{J}(P0)$ (7.5 is more than enough in this case) and a low value to $\underline{J}(P0)$ (it can easily be computed as -7 in this case by adding all the negative terms of J). Then, form the relaxed problem obtained when relaxing the integer Boolean conditions for the P0 problem; that is, form the $RP0$ problem replacing the last constraints of Problem (10.31) by $0 \leq \delta \leq 1$.

The solution of $RP0$ is obtained for $u = -0.6792$, $\delta_1 = 0.717$, $\delta_2 = 0.5943$, $\delta_3 = 0.6887$, and a value of the objective function of -2.9009. Notice that this value is a lower bound of the objective function. The relaxed problems $RP1_0$ and $RP1_1$ are generated and put into OPEN with the lower bound (equal for both) of the parent node of -2.9009.

The problem $RP1_1$ is chosen from OPEN for expansion because the solution of $RP0$ is for $\delta_1 = 0.717$, which is nearer to 1 than to 0 and one should expect that the integer solution should approximate the real solution. Problem $RP1_1$ is solved, resulting in $u = -0.75$, $\delta_1 = 1$, $\delta_2 = 0.5$, $\delta_3 = 0.5$, and a value of the objective function of -2.8125. Problems $RP2_{10}$ and $RP2_{11}$ are generated and put into OPEN with a lower bound of the parent node $RP1_1$ of -2.8125.

The next problem from OPEN to be expanded is $PRP1_0$ as it is the node with the smallest lower bound. Problem $PRP1_0$ is solved for $u = -0.5$, $\delta_1 = 0$, $\delta_2 = 0.75$, $\delta_3 = 1$, and a value of the objective function of -2.375. The problems $RP2_{00}$ and $RP2_{01}$ are formed and included in OPEN with a lower bound of -2.375.

Node $RP2_{10}$ is selected from OPEN as it has the lowest lower bound. $RP2_{10}$ is solved, resulting in $u = -0.75$, $\delta_1 = 1$, $\delta_2 = 0$, $\delta_3 = 1$, and a value of the objective function of -2.0625. As the solution is also feasible for $P2_{10}$ no more nodes are generated. Furthermore, as $-2.0625 < \overline{J}$ then $-2.0625 \rightarrow \overline{J}$ and SOL is labelled by $RP2_{10}$ (this is the best solution found so far).

The next problem to be expanded from OPEN is $RP2_{11}$ which is solved, resulting in $u = -0.75$, $\delta_1 = 1$, $\delta_2 = 1$, $\delta_3 = 0$, and a value of the objective function of -2.0625. As the solution is also feasible for $P2_{11}$, no more nodes are generated.

$RP2_{00}$ is selected from OPEN and its optimal solution is $u = -0.5$, $\delta_1 = 0$, $\delta_2 = 0$, $\delta_3 = 1$, and a value of the objective function of -1.25. As the solution is also feasible for $P2_{00}$, no more nodes are generated. Since the value of the objective function does not improve the solution found so far no more actions are taken by the algorithm.

The next node to be expanded from OPEN is $RP2_{01}$, which has the solution $u = -0.5$, $\delta_1 = 0$, $\delta_2 = 1$, $\delta_3 = 1$, and a value of the objective function of -2.25. As the solution is also feasible for $P2_{01}$, no more nodes are generated. Furthermore, as $-2.25 < \overline{J} (-2.0625)$ then $-2.25 \rightarrow \overline{J}$ and SOL is labelled by $RP2_{01}$ (this is the best solution found so far). No more problems can be expanded so the final solution is the solution of node $P2_{01}$ given by $u = -0.5$, $\delta_1 = 0$, $\delta_2 = 1$, $\delta_3 = 1$, and a value of the objective function of -2.25.

10.4 Piecewise Affine Systems

Another way to model hybrid systems is by piecewise affine systems (PWA). It has been shown that MLD systems, and other types of Hybrid System descriptions, are equivalent to a PWA description [87]. PWA systems have other advantages, such as being able to approximate nonlinear dynamics arbitrarily well, and are suitable for stability analysis and reachability analysis [195]. A PWA system is defined as

$$x(t+1) = A^i x(t) + B^i u(t) + f^i \quad \text{for} \quad \begin{bmatrix} x(t) \\ u(t) \end{bmatrix} \in \mathcal{X}_i \qquad (10.32)$$
$$y(t) = C^i x(t) + g^i$$

where $\{\mathcal{X}_i\}_{i=1}^s$ is a polyhedral partition of the states and input space. Each $\mathcal{X}_i$ is given by

$$\mathcal{X}_i \triangleq \left\{ \begin{bmatrix} x(t) \\ u(t) \end{bmatrix} | \mathbf{R}^i \begin{bmatrix} x(t) \\ u(t) \end{bmatrix} \leq \mathbf{r}^i \right\}$$

where $x(t)$, $u(t)$, and $y(t)$ denote the state, input, and output vectors, respectively. Each subsystem $\mathbb{S}^i$ defined by the 7-tuple $(A^i, B^i, C^i, f^i, g^i, \mathbf{R}^i, \mathbf{r}^i)$, $i \in \{1, 2, \ldots, s\}$, is termed a component of the PWA system (10.32). $A^i \in \mathbb{R}^{n \times n}$, $B^i \in \mathbb{R}^{n \times m}$, and (A^i, B^i) is a controllable pair. $C^i \in \mathbb{R}^{r \times n}$ and $\mathbf{R}^i \in \mathbb{R}^{p_i \times (n+m)}$ and $f^i, g^i, \mathbf{r}^i$ are suitable constant vectors. Note that n is the number of states, m is the number of inputs, r is the number of outputs, and p_i is the number of hyperplanes that define the i polyhedral.

Assume that a full measurement of the state is available at the current time k. The formulation of MPC for a PWA system can be expressed as:

$$\mathbf{u} = \arg(\min_{\mathbf{u}} J) \qquad (10.33)$$

$$\text{s.t.} \; : \; J = \sum_{i=1}^{N} q_{ii}(y(t+i \mid t) - w(t+i))^2 + \sum_{i=0}^{N-1} r_{ii} u(t+i)^2 \qquad (10.34)$$

$$u_{\min} \leq u(t+i) \leq u_{\max} \quad i = 0, \ldots, N-1 \qquad (10.35)$$

Let us consider the prediction problems associated to the MPC in the case of a PWA system. The subsystem describing the process is known if $x(t)$ is known, but the following subsystems depend on the applied control sequence. It can be considered that a change (transition) of model is produced between one sampling instant and the next. In general, a sequence of subsystems $I = \{I(t) \; I(t+1) \; \ldots \; I(t+N)\}$ is activated. Only the initial value $I(t) = I(t)(x(t))$ of this sequence is known. If no constraints are considered, the number of possible sequences for a prediction horizon N is s^{N-1}. In order to solve the MPC problem (10.34) the optimization sequence is added to the decision variables. The resulting optimization problem can be stated as

$$\mathbf{u}^* = \arg(\min_{\mathbf{u}, I} J) \qquad (10.36)$$

where constraints relating the dependencies of the possible sequences $\mathbf{u}$ and I have to be added, i.e.:

$$\mathbf{R}^{I(t+j)} x(t+j) \leq \mathbf{r}^{I(t+j)}, \; j = \{1, \ldots, N\} \qquad (10.37)$$

Due to the integer nature of sequence I, the problem of finding the optimum can be solved by finding the optimum of the solutions for all possible sequences of I, i.e.

$$\mathbf{u}^* = \arg\left(\min_I\left(\min_{\mathbf{u}}\left(\frac{J}{\mathbf{R}^{IU}\mathbf{u} \leq \mathbf{q}^{IU}}\right)\right)\right) \tag{10.38}$$

where $\mathbf{R}^{IU}\mathbf{u} \leq \mathbf{r}^{IU}$ indicate the constraints due to dependencies between I and U.

Equation (10.34) can be written as

$$J = (\mathbf{y} - \mathbf{w})^T\mathbf{Q}(\mathbf{y} - \mathbf{w}) + \mathbf{u}^T\mathbf{Q}_u\mathbf{u} \tag{10.39}$$

where $(\mathbf{Q}_u = \mathbf{Q}_u^T \succ 0)$ and $(\mathbf{Q} = \mathbf{Q}^T \succ 0)$ are weight matrices penalizing the control effort and the tracking errors and $\mathbf{y} = \left[y(t+1)^T \cdots y(t+N)^T\right]^T$, $\mathbf{w} = \left[w(t+1)^T \cdots w(t+N)^T\right]^T$, $\mathbf{u} = \left[u(t)^T \cdots u(t+N-1)^T\right]^T$. The predicted output vector can be written as

$$\mathbf{y} = \mathbf{F}_y x(t) + \mathbf{H}_y\mathbf{u} + \mathbf{f}_{o_y} \tag{10.40}$$
$$\mathbf{F}_y = \mathbf{C}_y\mathbf{F}_x, \ \mathbf{H}_y = \mathbf{C}_y\mathbf{H}_x, \ \mathbf{f}_{o_y} = \mathbf{C}_y\mathbf{f}_{o_x} + \mathbf{g}_o \tag{10.41}$$

where

$$\mathbf{C}_y = diag(C^{I(t+1)}, C^{I(t+2)}, \cdots, C^{I(t+N)})$$
$$\mathbf{H}_x = \left[h_1 \ h_2 \ \cdots \ h_N\right]$$

$$\mathbf{F}_x = \begin{bmatrix} A^{I(t)} \\ A^{I(t+1)}A^{I(t)} \\ \vdots \\ A^{I(t+N-1)}A^{I(t+1)}\cdots A^{I(t)} \end{bmatrix}$$

$$h_1 = \begin{bmatrix} B^{I(t)} \\ A^{I(t+1)}B^{I(t)} \\ \vdots \\ A^{I(t+N-1)}A^{I(t+N-2)}\cdots A^{I(t+1)}B^{I(t)} \end{bmatrix}$$

$$h_2 = \begin{bmatrix} 0 \\ B^{I(t+1)} \\ A^{I(t+2)}B^{I(t+1)} \\ \vdots \\ A^{I(t+N-1)}A^{I(t+N-2)}\cdots A^{I(t+1)}B^{I(t+1)} \end{bmatrix}$$

$$h_N = \left[0 \ 0 \ 0 \cdots \ (B^{I(t+N-1)})^T\right]^T$$

$$\mathbf{f}_{o_x}{}^T = \begin{bmatrix} f_1\ f_2 \cdots f_N \end{bmatrix}^T \quad \mathbf{g}_o{}^T = \begin{bmatrix} g^{I(t)T}\ g^{I(t+1)T} \cdots g^{I(t+N-1)T} \end{bmatrix}^T$$

$$f_1 = \begin{bmatrix} I \\ A^{I(t+1)} \\ \vdots \\ A^{I(t+N-1)} A^{I(t+N-2)} \cdots A^{I(t+1)} \end{bmatrix}$$

$$f_2 = \begin{bmatrix} 0 \\ I \\ A^{I(t+2)} \\ \vdots \\ A^{I(t+N-1)} A^{I(t+N-2)} \cdots A^{I(t+2)} \end{bmatrix}$$

$$f_N = \begin{bmatrix} 0\ 0\ 0 \cdots I \end{bmatrix}^T$$

Note that the following equalities are fulfilled:

$$\mathbf{x} = \mathbf{F}_x x(t) + \mathbf{H}_x \mathbf{u} + \mathbf{f}_{o_x} \tag{10.42}$$
$$\mathbf{y} = \mathbf{C}_y \mathbf{x} + \mathbf{g}_o$$
$$\mathbf{x} = \begin{bmatrix} x(t+1)^T\ x(t+2)^T \cdots x(t+N)^T \end{bmatrix}^T$$

Replacing (10.40) in (10.39),

$$J(I, \mathbf{u}) = \mathbf{u}\mathbf{H}_{QP}\mathbf{u} + \mathbf{f}_{QP}^T \mathbf{u} + g_{QP} \tag{10.43}$$

where

$$\mathbf{H}_{QP} = \begin{bmatrix} \mathbf{H}_y^T \mathbf{Q}\mathbf{H}_y + \mathbf{Q}_u \end{bmatrix}$$
$$\mathbf{f}_{QP}^T = \begin{bmatrix} 2x(t)\mathbf{F}_y^T \mathbf{Q}\mathbf{H}_y + 2\mathbf{f}_{o_y}^T \mathbf{Q}\mathbf{H}_y - 2\mathbf{w}^T \mathbf{Q}\mathbf{H}_y \end{bmatrix}$$
$$g_{QP} = x(t)^T \mathbf{F}_y^T \mathbf{Q}\mathbf{F}_y x(t) + 2\mathbf{f}_{o_y}^T \mathbf{Q}\mathbf{F}_y x(t) - 2\mathbf{f}_{o_y}^T \mathbf{Q}\mathbf{w}$$
$$+ \mathbf{f}_{o_y}^T \mathbf{Q}\mathbf{f}_{o_y} - 2\mathbf{w}^T \mathbf{Q}\mathbf{F}_y x(t) + \mathbf{w}^T \mathbf{Q}\mathbf{w}$$

The constraints over the control (10.35) can be written as $\mathbf{R}^u \mathbf{u} \leq \mathbf{r}^u$ (normally $\mathbf{R}^u = [-I_{N \times N}\ I_{N \times N}]^T$, $\mathbf{r}^u = \begin{bmatrix} -(u_{\min})^T\ (u_{\max})^T \end{bmatrix}^T$), and the constraints due to I and $\mathbf{u}$ dependency (10.37) can be written as

$$\mathbf{R}^{I_x} \mathbf{x}_I \leq \mathbf{r}^{I_x} \tag{10.44}$$

where
$\mathbf{x}_I = \begin{bmatrix} x(t+1)^T, \ldots, x(t+N-1)^T \end{bmatrix}^T$, $\mathbf{R}^{I_x} = diag(\mathbf{R}^{I(t+1)}, \ldots, \mathbf{R}^{I(t+N-1)})$,
$\mathbf{r}^{I_x} = \begin{bmatrix} (\mathbf{r}^{I(t+1)})^T \cdots, (\mathbf{r}^{I(t+N-1)})^T \end{bmatrix}^T$.

Note that $x(t)$ and $x(t+N)$ are not taken into account in these constraints because the original region is known and the final region where $x(N)$ will be

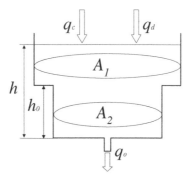

Fig. 10.6. Tank with discontinuous section

located is either fixed by stability or operational constraints or is a part of the MPC problem for the next sampling instant. The vector $\mathbf{x}_I$ can be written by

$$\mathbf{x}_I = C_x \mathbf{x} \tag{10.45}$$

where $C_x = \begin{bmatrix} I_{(n*(N-1))\times(n*(N-1))} & 0 \end{bmatrix}$. Replacing (10.45) and (10.42) in (10.44), the constraints due to the dependency between U and I result in $\mathbf{R}^{IU}\mathbf{u} \leq \mathbf{r}^{IU}$, $\mathbf{R}^{IU} = \mathbf{R}^{I_x}C_x\mathbf{H}_x$, $\mathbf{r}^{IU} = \mathbf{r}^{I_x} - \mathbf{R}^{I_x}C_xF_x x(t) - \mathbf{R}^{I_x}C_x f_{o_x}$. If constraints on the control actions are also considered then

$$\mathbf{R}_{QP}\mathbf{u} \leq \mathbf{r}_{QP} \tag{10.46}$$

$\mathbf{R}_{QP} = \left[(\mathbf{R}^u)^T \ (\mathbf{R}^{IU})^T \right]^T$, $\mathbf{r}_{QP} = \left[(\mathbf{r}^u)^T \ (\mathbf{r}^{IU})^T \right]^T$.

Therefore, once the sequence I is fixed, the problem can be solved by minimizing (10.43) subject to the constraints in (10.46).

10.4.1 Example: Tank with Different Area Sections

Consider the tank process shown in Figure 10.6. As can be seen, the area section of the tank changes abruptly at height h_0. The tank has two manipulated input variables. One is the flow $q_c \in [0, Q_c]$ of a continuous regulated pump and the other is the flow $q_d \in \{0, Q_d\}$ of an ON-OFF pump. The area below and above h_0 are denoted by A_1 and A_2, respectively. The dynamics can be described by the following equations:

$$\dot{h} = \frac{q_c + q_d - q_o}{A_1} \quad \text{for } h_0 < h \leq h_{\max} \tag{10.47}$$

$$\dot{h} = \frac{q_c + q_d - q_o}{A_2} \quad \text{for } h_{\min} \leq h \leq h_0 \tag{10.48}$$

This is a model described by two different physical models (one for $h_0 < h$ and the other one for $h \geq h_0$), each described by a continuous differential equation. The model can be approximated by the following discrete time model:

$$h(t+1) = h(t) + \frac{T_0(q_c + q_d - q_o)}{A_1} \quad \text{for } h_0 < h(t) \le h_{max} \qquad (10.49)$$

$$h(t+1) = h(t) + \frac{T_0(q_c + q_d - q_o)}{A_2} \quad \text{for } h_{min} \le h(t) \le h_0 \qquad (10.50)$$

where T_0 is the sampling time. Notice that the discrete-time approximated equations are exact except when the height crosses h_0 between two sampling times. Consider that $x(t)$ and $u_c(t)$ correspond to the continuous state and manipulated variables sampled with T_0, that is, $x(t) = h(t)$, $u_c(t) = q_c(t)$, $u_d(t) = q_d(t)$, the model can be described by the following PWA systems with the following four regions:

1. $(x(t) \le h_0) \wedge (u_d(t) = 0) \Rightarrow x(t+1) = x(t) + (T_0(u_c(t) - q_o))/A_2$
2. $(x(t) \le h_0) \wedge (u_d(t) = Q_d) \Rightarrow x(t+1) = x(t) + (T_0(u_c(t) + Q_d - q_o))/A_2$
3. $(x(t) > h_0) \wedge (u_d(t) = 0) \Rightarrow x(t+1) = x(t) + (T_0(u_c(t) - q_o))/A_1$
4. $(x(t) > h_0) \wedge (u_d(t) = Q_d) \Rightarrow x(t+1) = x(t) + (T_0(u_c(t) + Q_d - q_o))/A_1$

The sampling time is set to $T_0 = 0.2$ second and the rest of the model parameters are set to: $A_1 = 1$, $A_2 = 0.5$, $h_0 = 1$, $h_{max} = 2$, $Q_c = 0.2$, $Q_d = 0.2$, and $q_o = 0, 3$. The water level in the tank was originally set to $h(0) = 0.1$.

Figure 10.7 shows the results obtained by applying an MPC with the following objective function

$$\sum_{j=1}^{4} r_x(r(j+t) - x(t+j))^2 + r_{ud}u_d(t+j-1)^2 + r_{uc}u_c(t+j-1)^2$$

and the weighting factors set to: $r_x = 3$, $r_{ud} = 0.1$, and $r_{uc} = 2$. The different sectors reached can be observed in the bottom part of Figure 10.7.

10.4.2 Reach Set, Controllable Set, and STG Algorithm

Although the MPC problem for the PWA system (i.e., minimising (10.43)) subject to constraints (10.46) can be solved by an MIQP algorithm, the number of possible combinations of the integer variables, and therefore the maximum number of QP problems, may be very high. The maximum number of QP problems is S^N (with S the number of polyhedral regions and N the control horizon). Consider, for example, a PWA system with a polyhedral partition of 100 regions and a horizon of 10. The maximum number of QP problems to be solved at each sampling instant would be 10^{20}, too many for real-time applications.

In [164] an algorithm was proposed to reduce the maximum number of QP problems to be solved. The method reduces the number of possible sequences of the integer variables using knowledge about the process dynamics. The key idea of the algorithm is to determine the set of possible regions that can be reached from the actual region at the next few sampling times.

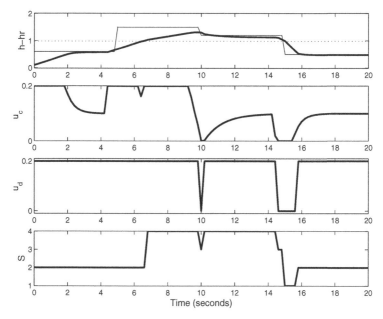

Fig. 10.7. Tank with discontinuous section: simulation results

The reach set concept [104] is used for this purpose. The set of possible regions to be reached from a particular one can be organized as a state transition graph. A search directed by this graph can then be implemented; that is, every sequence that cannot be obtained following the transition graph is not considered.

10.5 Exercises

10.1. Consider the tank of Figure 10.6; find a hybrid model and formulate an MPC in the following cases:

1. a new ON-OFF feeding pump is added.
2. a new area section A_3 is added. The tank area section is A_3 when the height of the water in the tank is below h_1 ($0 \le h \le h_1 < h_0 < h_{max}$), A_2 when h is above h_1 and below h_0, and A_1 when h is above h_0.

10.2. Solve the MPC for the three level-tank of the previous exercise when the control horizon is $N = 3$. What is the number of QP problems to be solved? Is there any way of reducing the number of QP problems to be solved?

10.3. Consider the system described by the following equations:

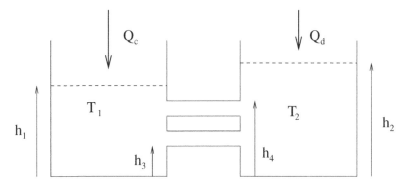

Fig. 10.8. Interconnected-tanks

$$x_c(t+1) = 0.5x_c(t) + u_c(t) + x_d(t)$$
$$x_d(t+1) = (x_d(t) \wedge (x_c(t) > x_0)) \vee (\backsim x_d(t) \wedge (x_c(t) > x_1))$$

with:
$$x_c(t) \in [-5, 5] \in \mathbb{R}, \ u_c(t) \in [-1, 1] \in \mathbb{R}, \ x_d(t) \in \{0, 1\}$$

1. Describe the system as a hybrid automaton; i.e., find the state transition diagram and switching conditions.
2. Find equivalent MLD and PWA descriptions.
3. Formulate an MPC problem.
4. Solve the previous MPC problem with different control horizons.

10.4. Consider two tanks T_1 and T_2 interconnected by two pipes at heights h_3 and h_3 as shown in Figure 10.8. The manipulated variables are the inflow Q_c to tank T_1 controlled by an analogue valve and the inflow Q_d to tank T_2 controlled by an ON-OFF valve.

1. Describe the system as a hybrid automaton.
2. Describe the system as an MLD system (linearize when necessary).
3. Formulate an MPC to control the level in both tanks (h_1 and h_2). Comment on the complexity of the problem to be solved.

Fast Methods for Implementing Model Predictive Control

One of the disadvantages of MPC is that the computation time required in some cases considerably limits the bandwidth of processes to which it can be applied. This is the case of MPC in the presence of constraints, adaptive MPC, robust MPC and MPC of nonlinear processes. This chapter is devoted to explaining some of the procedures used to reduce the amount of computation needed for the implementation of MPC. All of these procedures are based on doing most of the required computation off-line, leaving only part of the computation for the online part of the implementation.

11.1 Piecewise Affinity of MPC

As was shown in Chapter 7, MPC in the presence of constraints results in a QP problem when the objective function is quadratic and in an LP problem if the objective function is a 1-norm or ∞-norm type of function and the constraints are linear. Although very efficient algorithms exist for solving these types of problems, the computation time required is too high when MPC is applied to fast processes. It has been shown by Bemporad *et al.* that MPC for these cases can be considered as a multiparametric quadratic [23] or linear programming problem [20] and that MPC solution turns out to be a relatively easy-to-implement piecewise affine controller.

The idea is simple and was first pointed out in [212]: the optimum of a QP problem is reached for a set of active constraints (the set may be empty) and for all points in the state space with the same set of active constraints, the solution is an affine function of the state. Consider the optimization problem originated by a constrained MPC with a quadratic objective function

$$J = (\mathbf{w} - \mathbf{y})^T \mathbf{Q}(\mathbf{w} - \mathbf{y}) + \mathbf{u}^T \mathbf{Q}_u \mathbf{u} \qquad (11.1)$$

where $\mathbf{Q}$ and $\mathbf{Q}_u$ are weighting matrices penalizing the tracking errors and control effort, respectively. If the prediction equation $\mathbf{y} = \mathbf{Gu} + \mathbf{F}x(t)$ is introduced in (11.1), the constrained MPC can be expressed as

$$\min_{\mathbf{u}} \frac{1}{2}\mathbf{u}^T\mathbf{Hu} + \mathbf{b}^T\mathbf{u} + f_0 \qquad (11.2)$$

$$\text{s.\,t.}\ \ \mathbf{Ru} \le \mathbf{r} = \mathbf{V}x(t) + \mathbf{r}^b \qquad (11.3)$$

where $\mathbf{R}$ is an $m \times n$ matrix, $\mathbf{r}$ is an m vector, and:

$$\mathbf{H} = 2(\mathbf{G}^T\mathbf{QG} + \mathbf{Q}_u)$$
$$\mathbf{b} = 2(\mathbf{G}^T\mathbf{Q}(\mathbf{F}x(t) - \mathbf{w}))$$
$$f_0 = x(t)^T\mathbf{F}^T\mathbf{Q}\mathbf{F}x(t) + \mathbf{w}^T\mathbf{Qw} - 2\mathbf{w}^T\mathbf{QF}x(t)$$

Notice that the term f_0 is a quadratic function of the process state $x(t)$ and that vectors $\mathbf{b}$ and $\mathbf{r}$ depend in an affine way on the free response and therefore on the process state $x(t)$.

Let us consider all points in the state space where the optimization problem is feasible and let us denote its solution by $\mathbf{u}^*$. There are two possible situations for these points: the solution is either inside the polytope defined by the constraints or at its boundary. Let us define set Ω_0 as the points in the state space where the solution of the MPC optimization problem lies inside the polytope defining the constraints. For all points contained in Ω_0 the solution of the MPC optimization problem is equivalent to the unconstrained minimization of Function (11.2) of which the solution is:

$$\mathbf{u}^* = -\mathbf{H}^{-1}\mathbf{b} = 2\mathbf{H}^{-1}\mathbf{G}^T\mathbf{Q}(\mathbf{w} - \mathbf{F}x(t))$$

that is, $\mathbf{u}^*$ is an affine function of the state $\mathbf{x}(t)$ for all points in Ω_0. Let us now suppose that the solution lies at the boundary of the polytope. Define by Ω_p the region in the state space such that the solution of the MPC optimization problem lies in a set of constraints denoted by p (one of the multiple combinations of constraints that can be active).

The rows of $\mathbf{R}$ for a particular set of active constraints can be reordered in such a way that the constraint matrix $\mathbf{R}$ can be partitioned as:

$$\mathbf{R} = \begin{bmatrix} \mathbf{R_1} \\ \mathbf{R_2} \end{bmatrix} \quad \mathbf{r} = \begin{bmatrix} \mathbf{r_1} \\ \mathbf{r_2} \end{bmatrix} \ \text{with} \ \begin{matrix} \mathbf{R_1u} = \mathbf{r_1} \\ \mathbf{R_2u} < \mathbf{r_2} \end{matrix}$$

where $\mathbf{R}_1$ is an $m_1 \times n$ matrix and $\mathbf{r}_1$ is an m_1 vector. It is assumed that $m_1 < n$ and that $rank(\mathbf{R}_1) = m_1$. If this is not the case, keep the maximal set of linearly independent active constraints to form $\mathbf{R}_1$.

The MPC optimization problem for points in Ω_p is equivalent to:

$$\min_{\mathbf{u}} \frac{1}{2}\mathbf{u}^T\mathbf{Hu} + \mathbf{b}^T\mathbf{u} + f_0$$
$$\text{s.\,t.}\ \ \mathbf{R_1u} = \mathbf{r_1}$$

A direct way of solving this problem is to use constraints $\mathbf{R}_1\mathbf{u} = \mathbf{r}_1$ to express m_1 of the $\mathbf{u}$ variables as a function of the remaining $n - m_1$ variables

(u_b) and then to substitute them in the objective function. The problem is reduced to minimizing a quadratic function of $n - m_1$ variables without constraints. As was pointed out in Chapter 7, a generalized elimination method is normally used instead of a direct elimination procedure. The idea is to express u as a function of a reduced set of $n - m_1$ variables, $u = Yr_1 + Zv$, where Y and Z are $n \times m_1$ and $n \times (n - m_1)$ matrices such that $R_1Y = I$, $R_1Z = 0$ and the matrix $[Y \ Z]$ has full rank. If this substitution is made, the equality constraints hold and the objective function

$$J(v) = \frac{1}{2}[Yr_1 + Zv]^T H[Yr_1 + Zv] + b^T[Yr_1 + Zv] + f_0$$
$$= \frac{1}{2}v^T H_v v + b_v^T v + f_{v0}$$

with $H_v = Z^T HZ$, $b_v = Z^T(b + HYr_1)$ and $f_{v0} = [\frac{1}{2}r_1^T Y^T H + b^T]Yr_1 + f_0$; that is, an unconstrained QP problem of $n - m_1$ variables. As matrix $Z^T HZ$ is positive definite, there is only one global optimum point that can be found solving the linear set of equations $Z^T HZv = -Z^T(b+HYr_1)$ whose solution is:

$$v = -(Z^T HZ)^{-1}Z^T(b + HYr_1)$$

This expression shows that v is an affine function of b and r_1 which are also affine functions of $x(t)$. The solution of the MPC optimization problem (u^*), which is also an affine function of v, is, therefore, an affine function of the state for all points in region Ω_p. That is, an MPC with a quadratic objective function and linear constraints results in a controller which is a piecewise affine (PWA) function of the process state[1]. The optimal control moves for any $x(t) \in \Omega_p$ can now be computed as:

$$u = Yr_1 + Zv$$
$$= Yr_1 - Z[(Z^T HZ)^{-1}Z^T(b + HYr_1)]$$
$$= Y(V_1 x(t) + r_1^b) - Z[(Z^T HZ)^{-1}Z^T((2G^T Q^T(Fx(t) - w))$$
$$+ HY(V_1 x(t) + r_1^b))]$$

where V_1 and r_1^b are the rows of matrix V and vector r^b of the MPC constraints in (11.3). The previous expression shows that for all points in region Ω_p (i.e., all points where the optimum lies in the active sets of constraints defined by $R_1 u = r_1$) the controller is an affine function of the state $x(t)$ and future reference w

$$u = K_w^p w + K_x^p x(t) + K^p \tag{11.4}$$

---

[1] It is also an affine function of the future references and of the measurable disturbances, as will be seen later in this chapter.

where $\mathbf{K}_w^p = \mathbf{Z}(\mathbf{Z}^T\mathbf{HZ})^{-1}\mathbf{Z}^T$, $\mathbf{K}_x^p = -\mathbf{K}_w^p(2\mathbf{G}^T\mathbf{QF} + \mathbf{HYV}_1) + \mathbf{YV}_1$, and $\mathbf{K}^p = \mathbf{Yr}_1^b - \mathbf{K}_w^p\mathbf{HYr}_1^b$.

Notice that the controller constant matrices can be computed easily if the set of active constraints is known. Notice that if the number of constraints is L, the maximum number of possible combinations, and therefore of possible values of p and of possible regions, can be very high for normal values of the prediction horizon. Fortunately, the number of constraint combinations that generate a nonempty region is usually only a small fraction of the maximum number of possible combinations

The problem is how to characterize regions Ω_p for all possible constraint active sets. The solution of this problem can be obtained using multiparametric programming concepts, as will be seen in the next section.

Once regions Ω_p have been determined in an off-line manner, the controller will consist of reading (or estimating) the process state $x(t)$, then determining in which region Ω_p the current process state $x(t)$ lies and finally applying the corresponding affine controller (11.4).

11.2 MPC and Multiparametric Programming

The same results as those reached in the previous section can be obtained by showing that the MPC with linear constraints problem can be formulated as a multiparametric programming problem. This type of problem is intimately related to the sensitivity analysis of the solution of optimization problems. Sensitivity analysis consists in determining how the optimal solution varies when some uncertain coefficients of the problem change. There are basically two questions addressed by sensitivity analysis. One is how the optimal solution changes with the coefficients of the objective function. The other question is how the optimal solution changes with the right-hand side of the constraints. If the solution of the problem is expressed in terms of these changing parameters we are solving what is known as a multiparametric optimization problem. Multiparametric programming is a technique for obtaining the solution of an optimization problem as a function of the uncertain parameters. The advantage of the technique is that if the parameters change, the optimization problem need not be solved again since the solution has been obtained as a function of the uncertain parameters.

In the MPC context, the state is considered to be the vector of uncertainty parameters and the solution is made a function of the state. It will be seen that the solution of MPC problems with linear constraints and quadratic, 1-norm or ∞-norm types of objective functions turns out to be a piecewise affine function of the state.

Let us suppose that the future references are zero ($\mathbf{w} = 0$). Then the MPC optimization problem (11.3) results in

$$\min_{\mathbf{u}} \frac{1}{2}\mathbf{u}^T\mathbf{Hu} + 2x(t)^T\mathbf{F}^T\mathbf{QGu} + x(t)^T\mathbf{F}^T\mathbf{QF}x(t)$$

$$\text{s. t. } \mathbf{Ru} \leq \mathbf{V}x(t) + \mathbf{r}^b$$

which can be formulated as the multiparametric QP problem:

$$\mu(x(t)) \triangleq \min_{\mathbf{z}} \frac{1}{2}\mathbf{z}^T\mathbf{Hz}$$

$$\text{s.t. } \mathbf{Rz} \leq \mathbf{S}x(t) + \mathbf{r}^b$$

where $\mathbf{z} = \mathbf{u} + 2\mathbf{H}^{-1}\mathbf{G}^T\mathbf{QF}x(t)$ and $\mathbf{S} = \mathbf{V} + 2\mathbf{RH}^{-1}\mathbf{G}^T\mathbf{QF}$.

The first-order Karush-Kuhn-Tucker (KKT) optimality conditions can be expressed as:

$$\mathbf{Hz} + \mathbf{R}^T\lambda = 0 \tag{11.5}$$

$$\lambda_i(\mathbf{R}^i\mathbf{z} - \mathbf{r}_i^b - \mathbf{S}^ix(t)) = 0 \tag{11.6}$$

$$\lambda \geq 0 \tag{11.7}$$

$$\mathbf{Rz} \leq \mathbf{S}x(t) + \mathbf{r}^b \tag{11.8}$$

The superscript or subindex i indicates the ith row or components of the corresponding matrix or vector. Solving (11.5) for $\mathbf{z}$ results in

$$\mathbf{z} = -\mathbf{H}^{-1}\mathbf{R}^T\lambda \tag{11.9}$$

which, substituted into (11.6), leads to

$$\lambda_i(\mathbf{R}^i(-\mathbf{H}^{-1}\mathbf{R}^T\lambda) - \mathbf{r}_i^b - \mathbf{S}^ix(t)) = 0$$

Let λ_1 and λ_2 denote the sets of Lagrange multipliers corresponding to the active and inactive constraints, respectively. For the inactive constraints $\lambda_2 = 0$ while for active constraints $\lambda_1 > 0$ which implies that $\mathbf{R}^i(-\mathbf{H}^{-1}\mathbf{R}^T\lambda) - \mathbf{r}_i^b - \mathbf{S}^ix(t) = 0$ for all constraints belonging to the active set. These conditions can be expressed in condensed form as

$$\mathbf{R}_p(-\mathbf{H}^{-1}\mathbf{R}_p^T\lambda_1) - \mathbf{r}_p^b - \mathbf{S}_px(t) = 0$$

where $\mathbf{R}_p$, $\mathbf{r}_p^b$ and $\mathbf{S}_p$ are the matrices formed by the rows corresponding to active constraints of matrices $\mathbf{R}$, $\mathbf{r}^b$ and $\mathbf{S}$. Solving for λ_1,

$$\lambda_1 = -(\mathbf{R}_p\mathbf{H}^{-1}\mathbf{R}_p^T)^{-1}(\mathbf{r}_p^b + \mathbf{S}_px(t))$$

substituting λ_1 in (11.9),

$$\mathbf{z} = \mathbf{H}^{-1}\mathbf{R}_p^T(\mathbf{R}_p\mathbf{H}^{-1}\mathbf{R}_p^T)^{-1}(\mathbf{r}_p^b + \mathbf{S}_px(t)) \tag{11.10}$$

If (11.10) is inserted in $\mathbf{z} = \mathbf{u} + 2\mathbf{H}^{-1}\mathbf{G}^T\mathbf{QF}x(t)$, solving for $\mathbf{u}$ leads to

$$\mathbf{u} = \mathbf{K}_{xp}x(t) + \mathbf{k}_p \tag{11.11}$$

with

$$\mathbf{K}_{xp} = \mathbf{H}^{-1}[\mathbf{R}_p^T(\mathbf{R}_p\mathbf{H}^{-1}\mathbf{R}_p^T)^{-1}\mathbf{S}_p - 2\mathbf{G}^T\mathbf{Q}\mathbf{F}], \quad \mathbf{k}_p = \mathbf{H}^{-1}\mathbf{R}_p^T(\mathbf{R}_p\mathbf{H}^{-1}\mathbf{R}_p^T)^{-1}\mathbf{r}_p^b$$

which shows that $\mathbf{u}$ is an affine function of $x(t)$.

11.3 Piecewise Implementation of MPC

As has been seen in previous sections, the resulting controller is an affine function of the process state with gains depending on the set of active constraints at the optimum. In order to implement the controller, it is necessary to determine the regions in which a determined set of constraints is active. This, as shown in [23], can be done by imposing the remaining KKT conditions, $\lambda \geq 0$ and $\mathbf{R}\,\mathbf{z} \leq \mathbf{S}x(t) + \mathbf{r}^b$:

$$-(\mathbf{R}_p\mathbf{H}^{-1}\mathbf{R}_p^T)^{-1}(\mathbf{r}_p^b + \mathbf{S}_px(t)) \geq 0 \tag{11.12}$$

$$\mathbf{R}\mathbf{H}^{-1}\mathbf{R}_p^T(\mathbf{R}_p\mathbf{H}^{-1}\mathbf{R}_p^T)^{-1}(\mathbf{r}_p^b + \mathbf{S}_px(t)) \leq \mathbf{S}x(t) + \mathbf{r}^b \tag{11.13}$$

Inequalities (11.12) and (11.13) define a (possibly empty) region Ω_p in the space. For all points in this region, the solution of the optimization problem lies in the intersection of the set of active constraints p and the controller is given by Equation (11.11).

A brute force approach to determining all the regions would consist of using an enumerative algorithm to generate all possible sets of active constraints, then computing the corresponding $\mathbf{R}_p$, $\mathbf{S}_p$ and $\mathbf{r}_p^b$ and using Expressions (11.12) and (11.13) to define the regions. A more efficient method has been proposed in [23] to generate the regions. The procedure is based on the following theorem [23]

Theorem 11.1. *Let* $\mathbf{Y} \in \mathbb{R}^n$ *be a nonempty polyhedron, and* $\Omega_0 \triangleq \{x \in \mathbf{Y} : \mathbf{R}x \leq \mathbf{r}\}$ *a polyhedral subset of* $\mathbf{Y}$. *And let*

$$\Omega_i = \left\{ \begin{matrix} x \in \mathbf{Y} \\ \mathbf{R}_i x > r_i \\ \mathbf{R}_j x \leq r_j, \ \forall j < i \end{matrix} \right\}, \ i = 1, \cdots, m$$

where $m = dim(\mathbf{r})$ *and let* $\Omega^{rest} \triangleq \bigcup_{i=1}^m \Omega_i$. *Then: (i)* $\Omega^{rest} \bigcup \Omega_0 = \mathbf{Y}$, *(ii)* $\Omega^0 \bigcap \Omega_i = \emptyset$, *(iii)* $\Omega_i \bigcap \Omega_j = \emptyset, \ \forall i \neq j$, *i.e.* $\{ \Omega_0, \Omega_1, \cdots, \Omega_m \}$ *is a partition of* $\mathbf{Y}$.

The procedure starts by choosing a feasible point $x_0 \in \mathbf{X}$ in the state space and solving the associated QP problem[2]. The active constraints are obtained from the solution and, after eliminating the superfluous constraints

---

[2] The Chebyshev centre of $\mathbf{X}$ is proposed in [23] as x_0. The Chebyshev centre of a polytope can be found by solving an LP problem.

from (11.12) and (11.13), region Ω_0 is determined. Then all the inequalities defining Ω_0 are considered one by one as indicated in Theorem 11.1, making $\mathbf{Y} = \mathbf{X}$. That is, for $i = 1$ only the first row of the inequality matrix defining Ω_0 is chosen. This will establish a partition in $\mathbf{Y}$ formed by the regions Ω_0, $\Omega_1 \triangleq \{x \in \mathbf{Y}, \mathbf{R}_1 x > r_1\}$, $\Omega_2 \triangleq \{x \in \mathbf{Y}, \mathbf{R}_1 x \leq r_1, \mathbf{R}_2 x > r_2\}, \cdots$, $\Omega_m \triangleq \{x \in \mathbf{Y}, \mathbf{R}_1 x \leq r_1, \mathbf{R}_2 x \leq r_2, \cdots, \mathbf{R}_m x > r_m\}$. Where $\mathbf{R}_i$ and r_i are the ith row and ith component of matrix $\mathbf{R}$ and vector $\mathbf{r}$, respectively.

The procedure is applied recursively to all regions $\{\Omega_1, \cdots, \Omega_m\}$. That is, a feasible point is found in Ω_1 and the corresponding QP problem solved. The critical region Ω_{1_0} is determined (Ω_{1_0} is the region corresponding to the feasible point found in Ω_1). Then Theorem 11.1 is applied again with the constraints defining Ω_{1_0} and making $Y = \Omega_1$. Some precautions have to be taken to make sure that matrix $\mathbf{R}_p \mathbf{H}^{-1} \mathbf{R}_p^T$ is nonsingular [23].

11.3.1 Illustrative Example: The Double Integrator

Let us consider a double integrator process described by the following continuous transfer function:

$$Y(s) = \frac{1}{s^2} U(s)$$

If the process is sampled with a sampling time of one unit and assuming a sample and hold at both integrator inputs, the discrete state space representation is given by:

$$A = \begin{bmatrix} 1 & 1 \\ 0 & 1 \end{bmatrix} \quad B = \begin{bmatrix} 0 \\ 1 \end{bmatrix} \quad C = \begin{bmatrix} 1 & 0 \end{bmatrix}$$

Let us consider an objective function with an infinity control horizon given by:

$$J = \sum_{j=1}^{\infty} (y(t+j)^2 + \lambda u(t+j-1)^2)$$

Notice that using an infinite horizon guarantees stability for the nominal case when no constraints are present. When constraints have to be considered this problem cannot be solved because the number of control moves and, therefore, the number of decision variables is infinite. However, if we suppose that the constraints are always going to be fulfilled after a finite number of steps N, the objective function can be decomposed into the following terms

$$J = \lambda u(t)^2 + \sum_{j=1}^{N-1} (x(t+j)^T Q x(t+j) + u(t+j)^T R u(t+j))$$

$$+ \sum_{j=N}^{\infty} (x(t+j)^T Q x(t+j) + u(t+j)^T R u(t+j)) \tag{11.14}$$

with $R = \lambda$ and $Q = \begin{bmatrix} 1 & 0 \\ 0 & 0 \end{bmatrix}$ for the example. If the last part of cost function (11.14) is not affected by constraints, the optimal value for that part can be found solving a Riccati equation (see Appendix B). For $N = 2$ and $\lambda = 0.1$, the resulting control law (obtained by solving the Riccati equation) is $u(t) = Kx(t)$ with $K = [-0.8166 \ \ -1.7499]$. The value of the cost function when the optimal control law is applied (for the nominal model and no noise) is a function of the state and is given by $J(x) = x^T P x$, where P has been obtained solving the previous Riccati equation. In this case matrix P turns out to be:

$$P = \begin{bmatrix} 2.1429 & 1.2246 \\ 1.2246 & 1.3996 \end{bmatrix}$$

If the optimal control law is applied for $k > N$, the cost function (11.14) can be expressed as

$$J = \sum_{j=1}^{N} (x(t+j)^T Q_j x(t+j) + u(t+j-1)^T R u(t+j-1)) \qquad (11.15)$$

with $Q_j = Q$ for $j < N$ and $Q_N = P$. When the control moves are constrained by $-1 \leq u(t) \leq 1$ and $-1 \leq u(t+1) \leq 1$, there are nine resulting regions. These regions and corresponding controllers are given by the following,

Region 1:

If $\begin{bmatrix} -0.8166 & -1.7499 \\ 0.6124 & 0.4957 \\ 0.8166 & 1.7499 \\ -0.6124 & -0.4957 \end{bmatrix} x(t) \leq \begin{bmatrix} 1 \\ 1 \\ 1 \\ 1 \end{bmatrix}$ then $u(t) = [-0.8166 \ \ -1.7499]x(t)$

Region 2:

If $\begin{bmatrix} 2.4491 & 5.2482 \\ -0.8166 & -2.5665 \\ 0.8166 & 2.5665 \end{bmatrix} x(t) \leq \begin{bmatrix} -2.9991 \\ 2.7499 \\ -0.7499 \end{bmatrix}$ then $u(t) = 1$

Region 3:

If $\begin{bmatrix} -0.4521 & -0.3660 \\ -0.5528 & -1.5364 \\ 0.5528 & 1.5364 \end{bmatrix} x(t) \leq \begin{bmatrix} -0.7383 \\ 1.4308 \\ 0.5692 \end{bmatrix}$ then $u(t) = [-0.5528 \ \ -1.5364]x(t) - 0.4308$

Region 4:

If $\begin{bmatrix} -2.4491 & -5.2482 \\ -0.8166 & -2.5665 \\ 0.8166 & 2.5665 \end{bmatrix} x(t) \leq \begin{bmatrix} -2.9991 \\ -0.7499 \\ 2.7499 \end{bmatrix}$ then $u(t) = -1$

Region 5:

$$\text{If } \begin{bmatrix} 0.4521 & 0.3660 \\ -0.5528 & -1.5364 \\ 0.5528 & 1.5364 \end{bmatrix} x(t) \le \begin{bmatrix} -0.7383 \\ 0.5692 \\ 1.4308 \end{bmatrix} \begin{array}{l} \text{then} \\ u(t) = [-0.5528 - 1.5364]x(t) + 0.4308 \end{array}$$

Region 6:

$$\text{If } \begin{bmatrix} 6.7349 & 18.7181 \\ 2.4491 & 7.6974 \end{bmatrix} x(t) \le \begin{bmatrix} -17.4314 \\ -8.2474 \end{bmatrix} \text{ then } u(t) = 1$$

Region 7:

$$\text{If } \begin{bmatrix} 6.7349 & 18.7181 \\ -2.4491 & -7.6974 \end{bmatrix} x(t) \le \begin{bmatrix} -6.9349 \\ 2.2491 \end{bmatrix} \text{ then } u(t) = 1$$

Region 8:

$$\text{If } \begin{bmatrix} 2.4491 & 7.6974 \\ -6.7349 & -18.7181 \end{bmatrix} x(t) \le \begin{bmatrix} 2.2491 \\ -6.9349 \end{bmatrix} \text{ then } u(t) = -1$$

Region 9:

$$\text{If } \begin{bmatrix} -6.7349 & -18.7181 \\ -2.4491 & -7.6974 \end{bmatrix} x(t) \le \begin{bmatrix} -17.4314 \\ -8.2474 \end{bmatrix} \text{ then } u(t) = -1$$

The resulting control signal is unconstrained for all points in region 1 and coincides with the solution of the Riccati equation, as expected. The regions can be seen in Figure 11.1, which also shows the evolution of the system state when the optimal control law (the piecewise affine function described earlier) is applied and the initial state is given by $x(t)^T = [-5 \; 5]$.

The controller minimizes Expression (11.15), that is, a quadratic function which also penalizes the state at the end of the prediction horizon ($N = 2$ in the example). Minimizing this objective function will minimize the infinite horizon objective function when the control law is unconstrained for horizons bigger than 2 and this does not happen in all cases (see Exercise 11.1).

Notice that the online part of the controller would consist of reading or estimating the process state, then deciding what region the state is in and applying the corresponding control law. In this example, the number of operations needed to determine the region where the process state is will take, in the worst case, 48 multiplications, 24 additions and 24 comparisons. Computing the controller would require two multiplications and one addition.

Notice that the most time-consuming operation is determining the appropriate region and that the computational burden depends on the number of regions and number of constraints defining each region. If the number of regions is very high, the time required to determine the affine control law to be applied would also be very high and this would be a limiting factor. Some ideas have been proposed to overcome this difficulty. These ideas are

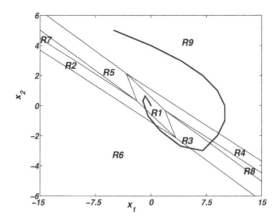

Fig. 11.1. Affine controller regions for the double integrator

based on the fact that each of the inequalities $(\mathbf{R}_i x \leq r_i)$ defining the critical regions divides the state space into two parts: $A_i = \{x \in X : \mathbf{R}_i x \leq r_i\}$ and $B_i = \{x \in X : \mathbf{R}_i x > r_i\}$). Some regions will be included in A_i, some will be included in B_i and some will intersect both A_i and B_i. Suppose that by checking $\mathbf{R}_i x \leq r_i$ we determine that x is in A_i; then all regions included in B_i can be discarded from further tests. An algorithm has been proposed in [199] to organize the inequalities in such a way that the number of constraint checks is minimized. The idea is to first choose those inequalities with more discriminating power. The ideal situation would be for half of the regions to be included in A_i and the rest in B_i.

11.3.2 Nonconstant References and Measurable Disturbances

In the case of a nonzero (nonconstant in general) reference trajectory and measurable disturbances, consider an augmented vector

$$\mathbf{v} = \begin{bmatrix} x(t) \\ \mathbf{w} \\ \mathbf{p} \end{bmatrix}$$

where $\mathbf{w}$ is the vector of future references and $\mathbf{p}$ is the vector of predicted future measurable disturbances. The objective Function (11.1) can then be expressed as [3]

$$\mathbf{J} = (\mathbf{w} - (\mathbf{Gu} + \mathbf{F}x(t) + \mathbf{F}_p\mathbf{p}))^T \mathbf{Q}(\mathbf{w} - (\mathbf{Gu} + \mathbf{F}x(t) + \mathbf{F}_p\mathbf{p})) + \mathbf{u}^T\mathbf{Ru}$$
$$= \mathbf{u}^T(\mathbf{G}^T\mathbf{QG} + \mathbf{R})\mathbf{u} + 2(\mathbf{F}_v\mathbf{v})^T\mathbf{QGu} + (\mathbf{F}_v\mathbf{v})^T\mathbf{Q}(\mathbf{F}_v\mathbf{v})$$

---

[3] Recall from Section 4.8 that the predicted output vector can be computed as $\mathbf{y} = \mathbf{Gu} + \mathbf{F}x(t) + \mathbf{F}_p\mathbf{p}$.

with $\mathbf{F}_v = [\mathbf{F} \quad -\mathbf{I} \quad \mathbf{F}_p]$. If the constraints on the problem are expressed as $\mathbf{Ru} \leq \mathbf{Vv} + \mathbf{r}^b$, the MPC can be expressed as the following multiparametric quadratic problem:

$$\mu(\mathbf{v}) \triangleq \min_{\mathbf{z}} \frac{1}{2}\mathbf{z}^T\mathbf{Hz} \tag{11.16}$$

$$\text{s. t. } \mathbf{Rz} \leq \mathbf{Sv} + \mathbf{r}^b$$

where $\mathbf{H} = 2(\mathbf{G}^T\mathbf{QG} + \mathbf{R})$, $\mathbf{z} = \mathbf{u} + 2\mathbf{H}^{-1}\mathbf{G}^T\mathbf{QF}_v\mathbf{v}$ and $\mathbf{S} = \mathbf{V} + 2\mathbf{RH}^{-1}\mathbf{G}^T\mathbf{QF}_v$.

The solution to the multiparametric problem is again a piecewise affine function. However, the dimension of the space where the control regions are defined is augmented by the dimension of the future reference trajectory w and the dimension of the predicted disturbances, that is, by the cost horizon multiplied by the dimension of the output and/or the cost horizon multiplied by the dimension of the measurable disturbances. If only setpoint changes are considered, the reference trajectory can be expressed as the desired setpoint multiplied by a vector whose entries are ones. In this case, the dimension of the augmented space where the regions are defined is only increased by the number of outputs.

11.3.3 Example

Consider the system described by

$$x(t+1) = 0.9x(t) + u(t)$$

with the manipulated variable constrained by $-1 \leq u(t) \leq 1$ and an objective function

$$J = \sum_{j=1}^{5} \left((x(t+j) - w(t+j))^2 + \lambda u(t+j-1)^2 \right)$$

Let us consider that the future reference is unknown and that it is made equal to the present reference, i.e., $w(t+j) = w(t)$ for $j = 1,\ldots,5$, and that the control weight λ is equal to one. The prediction of the state vector for the prediction horizon can be computed as

$$\mathbf{x} = \mathbf{Gu} + \mathbf{F}x(t)$$

with

$$\mathbf{G} = \begin{bmatrix} 1 & 0 & 0 & 0 & 0 \\ 0.9 & 1 & 0 & 0 & 0 \\ 0.81 & 0.9 & 1 & 0 & 0 \\ 0.729 & 0.81 & 0.9 & 1 & 0 \\ 0.6561 & 0.729 & 0.81 & 0.9 & 1 \end{bmatrix} \quad \mathbf{F} = \begin{bmatrix} 0.9 \\ 0.81 \\ 0.729 \\ 0.6561 \\ 0.5905 \end{bmatrix}$$

The objective function can be expressed as

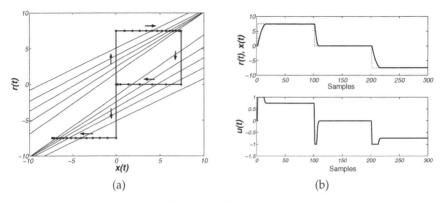

(a)   (b)

Fig. 11.2. Controller regions for first-order system and nonconstant reference

$$J = (\mathbf{Gu} + \mathbf{F}x(t) - \mathbf{1}_5 w(t))^T (\mathbf{Gu} + \mathbf{F}x(t) - \mathbf{1}_5 w(t)) + \lambda \mathbf{u}^T \mathbf{u}$$

where $\mathbf{1}_5 = [1\,1\,1\,1\,1]^T$. Defining $v(t) = [x(t)\ w(t)]^T$, $\mathbf{F}_v = [\mathbf{F} ; -\mathbf{1}_5]$

$$J = (\mathbf{Gu} + \mathbf{F}_v v(t))^T (\mathbf{Gu} + \mathbf{F}_v v(t)) + \lambda \mathbf{u}^T \mathbf{u}$$

The problem can be expressed as the mp-QP Problem (11.16) with:

$$\mathbf{H} = \begin{bmatrix} 8.8560\ 5.3956\ 3.9951\ 2.6390\ 1.3122 \\ 5.3956\ 7.9951\ 4.4390\ 2.9322\ 1.4580 \\ 3.9951\ 4.4390\ 6.9322\ 3.2580\ 1.6200 \\ 2.6390\ 2.9322\ 3.2580\ 5.6200\ 1.8000 \\ 1.3122\ 1.4580\ 1.6200\ 1.8000\ 4.0000 \end{bmatrix} \quad \mathbf{F}_v = \begin{bmatrix} 0.9000\ -1.0000 \\ 0.8100\ -1.0000 \\ 0.7290\ -1.0000 \\ 0.6561\ -1.0000 \\ 0.5905\ -1.0000 \end{bmatrix}$$

The constraint matrices of (11.16) are $\mathbf{R} = \begin{bmatrix} I_5 \\ -I_5 \end{bmatrix}$, $\mathbf{V} = \mathbf{0}$, $\mathbf{r}^b = \mathbf{1}_{10}$ and $\mathbf{S} = 2\mathbf{RH}^{-1}\mathbf{G}^T\mathbf{QF}_v$. The regions obtained when the mp-QP problem is solved can be seen in Figure 11.2-(a). Notice that the reference is shown in the vertical axis. The figure also shows a trajectory in the augmented state space (state plus reference) when the reference is changed according to the pattern shown in Figure 11.2-(b) which also shows the evolution of $x(t)$ in time. As can be seen the controller is able to follow the reference accurately.

11.3.4 The 1-norm and ∞-norm Cases

It has also been shown [18] that in the case of 1-norm and ∞-norm types of cost functions the resulting MPC is also a piecewise affine function of the state. The reason for this is that both problems can be expressed as a multi-parametric LP problem whose solution is an affine function of the parameters. Let us first consider the 1-norm case: the objective function is the sum of weighted reference absolute errors plus the sum of the weighted absolute values of the control signal.

The MPC problem can be stated as

$$\min_{\mathbf{u}} \ |\mathbf{w} - (\mathbf{Gu} + \mathbf{F}x(t))|^T \mathbf{q} + |\mathbf{u}|^T \mathbf{r} \tag{11.17}$$

$$\text{s. t.} \quad \mathbf{Ru} \leq \mathbf{V}x(t) + \mathbf{r}^b$$

where $\mathbf{q}$ and $\mathbf{r}$ are vectors penalizing the absolute errors of the output and control efforts. Problem (11.17) can be expressed as the following mp-LP problem

$$\min_{\mathbf{u},\alpha,\beta,\mu} \ \mu \tag{11.18}$$

$$\text{s. t.} \quad \mathbf{Ru} \leq \mathbf{V}x(t) + \mathbf{r}^b$$
$$-\alpha \leq \mathbf{w} - \mathbf{Gu} - \mathbf{F}x(t) \leq \alpha$$
$$-\beta \leq \mathbf{u} \leq \beta$$
$$-\mu \leq \alpha^T \mathbf{q} + \beta^T \mathbf{r} \leq \mu$$

$$\tag{11.19}$$

where α and β are vectors with nonnegative entries of appropriate dimensions and μ is a nonnegative scalar. Optimization Problem (11.18) can be expressed in a more compact form as

$$\min_{\mathbf{z}} \ \mathbf{c}^T \mathbf{z}$$
$$\text{s. t.} \quad \mathbf{R}_z \mathbf{z} \leq \mathbf{V}_x x(t) + \mathbf{r}^c \tag{11.20}$$
$$\mathbf{z} \geq 0$$

where $\mathbf{z}^T = [\mathbf{u}_+^T, \mathbf{u}_-^T, \alpha^T, \beta^T, \mu]$, $\mathbf{c}^T = [0 \cdots 0 \ 1]$ and

$$\mathbf{R}_z = \begin{bmatrix} \mathbf{R} & -\mathbf{R} & 0 & 0 & 0 \\ -\mathbf{G} & \mathbf{G} & -\mathbf{I} & 0 & 0 \\ \mathbf{G} & -\mathbf{G} & -\mathbf{I} & 0 & 0 \\ \mathbf{I} & -\mathbf{I} & 0 & -\mathbf{I} & 0 \\ -\mathbf{I} & \mathbf{I} & 0 & -\mathbf{I} & 0 \\ 0 & 0 & \mathbf{q}^T & \mathbf{r}^T & -1 \\ 0 & 0 & -\mathbf{q}^T & -\mathbf{r}^T & -1 \end{bmatrix} \quad \mathbf{V}_x = \begin{bmatrix} \mathbf{V} \\ \mathbf{F} \\ -\mathbf{F} \\ 0 \\ 0 \\ 0 \\ 0 \end{bmatrix} \quad \mathbf{r}^c = \begin{bmatrix} \mathbf{r}^b \\ -\mathbf{w} \\ \mathbf{w} \\ 0 \\ 0 \\ 0 \\ 0 \end{bmatrix} \tag{11.21}$$

Notice that by making $\mathbf{u} = \mathbf{u}_+ - \mathbf{u}_-$, the problem can be expressed as an LP in its standard form (nonnegative variables). Problem (11.20) can be considered as a multiparametric linear programming (mp-LP) problem where the components of $x(t)$ are the changing parameters of the problem. The solution of this problem results in a piecewise affine function of the parameters (in our case the parameters correspond to the process state variables).

Multiparametric linear programming ([71],[72]) is closely related to the concept of *critical region* which is defined as the set of points in the parameter space where a certain basis is optimal for problem (11.20). Let us consider the critical region Ω_i defined as the set of points in the parameter space where the optimal basis for problem (11.20) is B_i. Let $\mathbf{R}_z^i$ be the matrix formed by the rows of $\mathbf{R}_z$ corresponding to constraints which are active for basis B_i. Similarly let $\mathbf{V}_x^i$ and $\mathbf{r}^i$ be the corresponding rows and entries of matrix $\mathbf{R}_z$ and vector $\mathbf{r}^c$. That is, $\mathbf{R}_z^i \mathbf{z} = \mathbf{V}_x^i x(t) + \mathbf{r}^i$. The optimal solution for an LP problem is attained at one of the vertices of the feasible region, in this case the vertex is given by the solution of equation $\mathbf{R}_z^i \mathbf{z} = \mathbf{V}_x^i x(t) + \mathbf{r}^i$. Let us suppose that the number of rows of $\mathbf{R}_z^i$ is equal to $dim(\mathbf{z})$ and full rank. If this is not the case, just take a sufficient number of linearly independent rows of $\mathbf{R}_z^i$. Notice that this is always possible if no degeneracy occurs [20]. Then

$$\mathbf{z} = (\mathbf{R}_z^i)^{-1} \mathbf{V}_x^i x(t) + (\mathbf{R}_z^i)^{-1} \mathbf{r}^i \tag{11.22}$$

that is, the solution of the LP problem is an affine function of the state for all points in the critical region Ω_i.

To determine the conditions to be fulfilled by a point in the state space in order to belong to critical region Ω_i, let us call $\mathbf{R}_z^{ni}$ the matrix formed by the rows of $\mathbf{R}_z$ corresponding to constraints which are inactive for basis B_i. Similarly, let us call $\mathbf{V}_x^{ni}$ and $\mathbf{r}^{ni}$ the corresponding rows and entries of matrix $\mathbf{R}_z$ and vector $\mathbf{r}^c$, that is,

$$\mathbf{R}_z^{ni} \mathbf{z} < \mathbf{V}_x^{ni} x(t) + \mathbf{r}^{ni} \tag{11.23}$$

If (11.22) is substituted into Inequality (11.23) we get

$$\mathbf{R}_z^{ni} [(\mathbf{R}_z^i)^{-1} \mathbf{V}_x^i x(t) + (\mathbf{R}_z^i)^{-1} \mathbf{r}^i] < \mathbf{V}_x^{ni} x(t) + \mathbf{r}^{ni} \tag{11.24}$$

which can be reordered as

$$(\mathbf{R}_z^{ni} (\mathbf{R}_z^i)^{-1} \mathbf{V}_x^i - \mathbf{V}_x^{ni}) x(t) < \mathbf{r}^{ni} - \mathbf{R}_z^{ni} (\mathbf{R}_z^i)^{-1} \mathbf{r}^i \tag{11.25}$$

Inequalities (11.25) describe the open critical region. The closed critical region Ω_i is obtained by changing $<$ to $\leq$ in (11.25). This can be rewritten in a more compact way as

$$\mathbf{A}_i x(t) \leq \mathbf{a}_i \tag{11.26}$$

with $\mathbf{a}_i = \mathbf{R}_z^{ni} (\mathbf{R}_z^i)^{-1} \mathbf{V}_x^i - \mathbf{V}_x^{ni}$ and $\mathbf{a}_i = \mathbf{r}^{ni} - \mathbf{R}_z^{ni} (\mathbf{R}_z^i)^{-1} \mathbf{r}^i$.

Notice that it is also necessary for the vertex (of the constraint polytope in the $\mathbf{z}$ space) corresponding to this region to be optimal for the LP problem; therefore optimality has to be checked for each of the chosen bases. In principle this requires a point to be found in the interior of the region defined by (11.26), solving the LP problem and then verifying that the active constraints for the LP solution coincide with $\mathbf{R}_z^i$. The number of possibilities to be checked (and therefore the number of LP problems to be solved) is the maximum number of vertices of the region defined by $\mathbf{R}_z \mathbf{z} \leq \mathbf{V}_x x(t) + \mathbf{r}^c$, which depends of the number of constraints (m) and the dimension of $\mathbf{z}$ (n) as $\frac{m(m-1)\cdots(m-n)}{n!}$.

There are some algorithms proposed in literature for solving the multi-parametric LP problem explicitly (i.e. determining the critical regions). The first one, proposed in [72], is based on graphs. The algorithm associates a node of a connected graph to each of the optimal bases B_i. Two nodes are connected (neighbors) if their corresponding bases (B_i, B_j) are neighbors. That is, B_i and B_j are both optimal bases for a particular point in the state space and it is possible to obtain one of the bases from the other by just one pivoting step. The algorithm generates the critical region by constructing and exploring this graph.

Other algorithms have been proposed in literature. In [34], a geometric method that directly explores the parameter space is proposed. The idea of the method is to get a particular point in the state space (parameter space for the multiparametric problem) and then to solve the primal and dual LP problem. From the optimal solution, the corresponding optimal control law and critical region are defined using (11.22) and (11.26). The algorithm solves the cases where the solutions of the LP problem are affected by degeneracy (primal or dual). For the sake of simplicity all variables will be considered to be positive and the control move will be calculated as $\mathbf{u} = \mathbf{u}_+ - \mathbf{u}_-$ from the optimal solution $\mathbf{z}^T = [\mathbf{u}_+^T \ \mathbf{u}_-^T \ \alpha^T \ \beta^T \ \mu]$.

The ∞-norm Case

Let us turn our attention to the ∞-norm case. The MPC problem is expressed in this case as:

$$\min_{\mathbf{u}} \ \|\mathbf{w} - (\mathbf{G}\mathbf{u} + \mathbf{F}x(t))\|_\infty \tag{11.27}$$
$$\text{s. t.} \quad \mathbf{R}\mathbf{u} \leq \mathbf{V}x(t) + \mathbf{r}^b$$

Problem (11.27) can be expressed as

$$\min_{\mathbf{u},\mu} \ \mu \tag{11.28}$$
$$\text{s. t.} \quad \mathbf{R}\mathbf{u} \leq \mathbf{V}x(t) + \mathbf{r}^b$$
$$-\mu \leq \mathbf{w} - \mathbf{G}\mathbf{u} - \mathbf{F}x(t) \leq \mu$$
$$\tag{11.29}$$

where μ is a vector with nonnegative entries. Optimization Problem (11.28) can be expressed as the following mp-LP

$$\min_{\mathbf{z}} \ \mathbf{c}^T\mathbf{z} \tag{11.30}$$
$$\text{s. t.} \quad \mathbf{R}_z\mathbf{z} \leq \mathbf{V}_x x(t) + \mathbf{r}^c$$
$$\mathbf{z} \geq 0$$

where $\mathbf{z}^T = [\mathbf{u}_+^T \ \mathbf{u}_-^T \ \mu]$, $\mathbf{c} = [0 \cdots 0 \ 1]$ and

$$
\mathbf{R}_z = \begin{bmatrix} \mathbf{R} & -\mathbf{R} & \mathbf{0} \\ -\mathbf{G} & \mathbf{G} & -\mathbf{1} \\ \mathbf{G} & -\mathbf{G} & -\mathbf{1} \end{bmatrix}, \ \mathbf{V}_x = \begin{bmatrix} \mathbf{V} \\ \mathbf{F} \\ -\mathbf{F} \end{bmatrix}, \ \mathbf{r}^c = \begin{bmatrix} \mathbf{r}^b \\ -\mathbf{w} \\ \mathbf{w} \end{bmatrix}, \ \mathbf{u} = \mathbf{u}_+ - \mathbf{u}_-
$$

where $\mathbf{0}$ and $\mathbf{1}$ are vectors of appropriate dimensions whose entries are zeros and ones, respectively.

Notice that Problem (11.30) is a multiparametric LP problem whose solution is a piecewise affine function of the state.

Nonconstant References and Measurable Disturbances

As in the case of quadratic cost functions, the MPC results in a piecewise affine function of the reference trajectory and/or measurable disturbances. It is easy to see how mp-LP can be applied to this problem if the following augmented vector is considered

$$
\mathbf{v} = \begin{bmatrix} x(t) \\ \mathbf{w} \\ \mathbf{p} \end{bmatrix}
$$

where $\mathbf{w}$ is the vector of future references and $\mathbf{p}$ is the vector of predicted disturbances. The optimization Problems (11.20) and (11.30) can then be expressed as

$$
\min_{\mathbf{z}} \ \mathbf{c}^T \mathbf{z} \tag{11.31}
$$
$$
\text{s. t.} \ \ \mathbf{R}_z \mathbf{z} \leq \mathbf{V}_v \mathbf{v} + \mathbf{r}^c
$$

with constraints matrices defined accordingly (see Exercise 11.3). The solution to the multiparametric LP problem is again a piecewise linear function of the augmented state (process state plus future references and/or measurable disturbances).

11.4 Fast Implementation of MPC for Uncertain Systems

As was seen in Chapter 8, when uncertainties are considered, a very long computation time is required. The reason is that it is necessary to solve a min-max problem and at the same time to satisfy the constraints for any possible realization of the uncertainties. It can be seen that multiparametric programming can also be extended to this problem and that min-max MPC with ∞-norm (or 1-norm) results in a piecewise affine control law [19]. This fact can easily be deduced because ∞-norm (or 1-norm) min-max MPC can be expressed as an LP problem (as was shown in Chapter 7). As the solution of a multiparametric LP problem is a piecewise affine function of the parameters (i.e., the process state in this case) it follows that the resulting min-max controller is an affine function of the process state. This fact dramatically changes the way of implementing such control laws.

It has been proved in [173] that the min-max MPC control law with a quadratic objective function is also piecewise affine. This can be exploited to implement this type of control law to processes with fast dynamics.

Let us consider a min-max problem with bounded additive uncertainties

$$\min_{u} \max_{\theta \in \Theta} J(\theta, u, x(t))$$

where θ represents the sequence of future uncertainties and $x(t)$ is the process state. Notice that $x(t)$ will be used here like a parameter rather than a variable. When a linear prediction model is used, the set of j ahead optimal predictions for $j = 1, \ldots, N_2$ (where N_2 is the prediction horizon) can be written in condensed form as

$$\mathbf{y} = \mathbf{G}_u \mathbf{u} + \mathbf{G}_\theta \theta + \mathbf{F}_x x(t) \tag{11.32}$$

where $\mathbf{y} = [y(t+1) \cdots y(t+N_2)]^T$, $\mathbf{u} = [\Delta u(t) \cdots \Delta u(t+N_u-1)]^T$, $\theta = [\theta(t+1) \ldots \theta(t+N_2)]^T$, and $\mathbf{F}_x x(t)$ represents the free response, which depends linearly on the process state.

Without loss of generality, consider a constant setpoint of $w(t+j) = 0$ for $j = 1, \ldots, N_2$. Then function $J(\theta, \mathbf{u}, x(t))$ becomes

$$J(\theta, \mathbf{u}, x(t)) = \mathbf{u}^T \mathbf{M}_{uu} \mathbf{u} + \theta^T \mathbf{M}_{\theta\theta} \theta + 2\theta^T \mathbf{M}_{\theta u} \mathbf{u} + 2x(t)^T \mathbf{M}_{uf}^T \mathbf{u}$$
$$+ 2x(t)^T \mathbf{M}_{\theta f}^T \theta + x(t)^T \mathbf{F}_x^T \mathbf{F}_x x(t) \tag{11.33}$$

where $\mathbf{M}_{uu} = \mathbf{G}_u^T \mathbf{G}_u + \lambda I$, $\mathbf{M}_{\theta\theta} = \mathbf{G}_\theta^T \mathbf{G}_\theta$, $\mathbf{M}_{\theta u} = \mathbf{G}_\theta^T \mathbf{G}_u$, $\mathbf{M}_{uf} = \mathbf{G}_u^T \mathbf{F}_x$ and $\mathbf{M}_{\theta f} = \mathbf{G}_\theta^T \mathbf{F}_x$.

It is clear from (11.33) that function $J^*(u, x(t)) \triangleq \max_{\theta \in \Theta} J(\theta, u, x(t))$ can be expressed as the maximum of a quadratic function of θ for each value of $\mathbf{u}$ and $x(t)$

$$J^*(\mathbf{u}, x(t)) = \max_{\theta \in \Theta} \{\theta^T \mathbf{M}_{\theta\theta} \theta + \mathbf{M}_\theta'(\mathbf{u})\theta + \mathbf{M}'(\mathbf{u})\}$$

where $\mathbf{M}_\theta'(\mathbf{u}) = 2(x(t)^T \mathbf{M}_{\theta f}^T + \mathbf{u}^T \mathbf{M}_{\theta u}^T)$ and $\mathbf{M}'(\mathbf{u}) = \mathbf{u}^T \mathbf{M}_{uu} \mathbf{u} + 2x(t)^T \mathbf{M}_{uf}^T \mathbf{u} + x(t)^T \mathbf{F}_x^T \mathbf{F}_x x(t)$. Matrix $\mathbf{G}_\theta$ is a lower triangular matrix having all the elements of the main diagonal equal to one, thus $\mathbf{M}_{\theta\theta}$ is a positive definite matrix. This implies that function J^* is strictly convex (see [15], theorem 3.3.8) and the maximum of J will be reached at one of the vertices of the polytope Θ (see [15], theorem 3.4.6).

Function J^* is a piecewise quadratic function of $\mathbf{u}$. Therefore, the $\mathbf{u}$ domain (U) can be divided into different regions U_p so that $\mathbf{u} \in U_p$ if the maximum of J is attained for the polytope vertex θ_p. For the region U_p, J^* can be expressed as a function of $\mathbf{u}$ for a given $x(t)$

$$J^*(\mathbf{u}, x(t)) = \mathbf{u}^T \mathbf{M}_{uu} \mathbf{u} + \mathbf{M}_u^*(\theta_p)\mathbf{u} + \mathbf{M}^*(\theta_p) \tag{11.34}$$

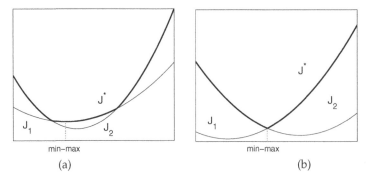

Fig. 11.3. Possible locations of the solution of min-max problem for two quadratic functions: (a) curve minima (b) intersection of curves

where $\mathbf{M}_u^*(\theta_p) = 2(\theta_p^T \mathbf{M}_{\theta u} + x(t)^T \mathbf{M}_{uf}^T)$ and $\mathbf{M}^*(\theta_p) = \theta_p^T \mathbf{M}_{\theta\theta}\theta_p + 2\theta_p^T \mathbf{M}_{\theta f} x(t)$ $+ x(t)^T \mathbf{F}_x^T \mathbf{F}_x x(t)$. The Hessian matrix of function $J^*(\mathbf{u})$ is $\mathbf{M}_{uu}$ which is positive definite for positive values of λ. This implies that the function is convex (see [15], theorem 3.3.8) and that there is only a unique minimizer, thus avoiding local minima problems (see [15], theorem 3.4.2).

Function J^* is a piecewise quadratic function of $\mathbf{u}$ for a given $x(t)$. Every region U_p is due to a different vertex θ_p of the polytope Θ. Also, each region can be seen as a region in which a different plant model is used to compute the worst case. The minima of this piecewise quadratic function is the solution of the min-max problem. This solution of the min-max problem has a graphical interpretation, as illustrated in the following.

Consider the simplest form of min-max MPC with bounded global uncertainties, e.g. $N_u = N_2 = 1$. In this case only two quadratic functions appear, J_1 and J_2, one for the maximum value of the uncertainty $\bar{\theta}$ and the other for the minimum value of the uncertainty $\underline{\theta}$. Thus the min-max problem consists of finding the minimum of the piecewise quadratic maximum curve J^*. The minimum will be on one of the minimizers of the quadratic functions J_1, J_2, or on the intersection point of J_1 and J_2. Both situations are depicted in Figure 11.3. This situation is generalized to arbitrary horizons in which more quadratic functions have to be taken into account. The solution of a min-max problem over a set of quadratic functions is in either a minimizer of one of them or in an intersection point of two or more of them.

Consider a linear prediction model and a process state $x(t)$ in which the min-max solution is in the minima of a quadratic function J_p related to the polytope vertex θ_p. Then the min-max solution is the same as that obtained by considering a linear plant whose model is the nominal one plus the contribution of the extreme uncertainty realization represented by vertex θ_p. The resulting model is also linear, and the control law in this case is affine on $x(t)$:

$$\mathbf{u}^* = -\mathbf{M}_{uu}^{-1}\mathbf{M}_{uf}x(t) - \mathbf{M}_{uu}^{-1}\mathbf{M}_{\theta u}^T \theta_p \tag{11.35}$$

In the case that the solution is attained at an intersection (i.e., $J_i = J_j$), the resulting problems is:

$$\min_{\mathbf{u}} J_i \text{ s. t.: } J_i = J_j$$

The equality equation $J_i = J_j$ results in a linear equation on $\mathbf{u}$ and the solution is therefore an affine function of the state.

It has been proven that the control law is piecewise affine and continuous in all the process state space where the problem is feasible for the min-max MPC unconstrained case [174] and the constrained case [175]. An algorithm to determine the min-max MPC controller explicitly (i.e., to determine the regions and the affine controller for each region) has been developed in [145].

However, the number of regions in which the state space has to be partitioned grows very rapidly with the prediction horizon. Thus, storage requirements and searching time for the appropriate region can be very high for practical values of the prediction and control horizons. Furthermore, if the process model changes, the computation of the regions has to be repeated. In [1] a method has been proposed to reduce the computation time required. The method is based on transforming the original min-max problem into a reduced min-max problem whose solution is much simpler. The idea is to consider only the active vertices (i.e., vertices of the uncertainty polytope that can be part of the min-max solution). Thus, for many processes in which time constants are measured in seconds or minutes, the reduced min-max problem can be solved online using standard numerical algorithms such as the ellipsoid method.

11.4.1 Example

Consider the system described by

$$x(t+1) = Ax(t) + Bu(t) + D\theta(t)$$

with

$$A = \begin{bmatrix} 1 & 1 \\ 0 & 1 \end{bmatrix}, \ B = \begin{bmatrix} 0 \\ 1 \end{bmatrix}, \ D = \begin{bmatrix} 1 & 0 \end{bmatrix}, \ -0.1 \le \theta(t) \le 0.1, \ -1 \le u(t) \le 1$$

The control objective consists of taking (and maintaining) the state vector as close to zero as possible by solving the following min-max problem:

$$\min_{\mathbf{u}\in[-1,1]^5} \max_{\theta\in[-0.1,0.1]^5} \sum_{j=1}^{5} x(t+j)^T x(t+j) + 10\,u(t+j-1)^2$$

Once the problem is solved[4], the resulting control regions can be seen in Figure 11.4 which also shows how the trajectory followed by the state from

---

[4] Note that the min-max problem to be solved is a min-max MPC with additive bounded uncertainties, constrained inputs and quadratic cost function.

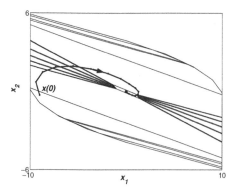

Fig. 11.4. Piecewise controller regions for a system with additive uncertainties

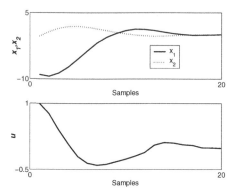

Fig. 11.5. State and manipulated variable evolution for a system with additive uncertainties

initial state $x(0) = [-6.5668 \ \ 0.5789]^T$ until it reaches a neighborhood around the origin. Notice that because of the perturbations it cannot be maintained at the origin exactly. The evolution of the state and control signal in time can be seen in Figure 11.5, which shows how the piecewise controller is able to take the state to a vicinity of the origin.

11.4.2 The Closed-Loop Min-max MPC

Recall from Chapter 8 that Dynamic Programming can be used to state the closed-loop min-max MPC. The problem can be expressed as the recursive problem

$$J_t^*(x(t)) \triangleq \min_{u(t)} J_t(x(t), u(t)) \tag{11.36}$$

s.t.

$$\left.\begin{array}{l} \mathbf{R}_x x(t) + \mathbf{R}_u u(t) \leq r \\ f(x(t), u(t), \theta(t)) \in \mathcal{X}(t+1) \end{array}\right\} \ \forall \theta(t) \in \Theta$$

$$J_t(x(t), u(t)) \triangleq \max_{\theta(t) \in \Theta} L(x(t), u(t)) + J_{t+1}^*(x(t+1)) \tag{11.37}$$

where $\mathcal{X}(t+1)$ is the region where function $J_{t+1}^*(x(t+1))$ is defined. In [21] it has been demonstrated that if the system is linear $x(t+1) = A(\theta(t))x(t) + B(\theta(t))u(t) + E(\theta(t))$, with $A(\theta(t)) = A_0 + \sum_{i=1}^{n_\theta(t)} A_i \theta^i(t)$, $B(\theta(t)) = B_0 + \sum_{i=1}^{n_\theta} B_i \theta^i(t)$, $E(\theta(t)) = E_0 + \sum_{i=1}^{n_\theta} E_i \theta^i(t)$, where $\theta^i(t)$ denotes the ith component of the uncertainty vector $\theta(t) \in \Theta$, and the stage cost of the objective function defined as: $L(x(t+j), u(t+j)) \triangleq \|Qx(t+j)\|_p + \|Ru(t+j)\|_p$ with the terminal cost defined as $J_{t+N}^*(x(t+N)) \triangleq \|Px(t+N)\|_p$ the solution is a piece affine function of the state.

The demonstration is based on the following results [21]:

1. If $J(x(t), \mathbf{u})$ is a convex piecewise affine function (i.e., $J(x(t), \mathbf{u}) = \max_{i=1,\cdots,s}\{L_i \mathbf{u} + H_i x(t) + k_i\}$), the problem

$$J^*(x(t)) \triangleq \min_{\mathbf{u}} J(x(t), \mathbf{u})$$

$$\text{s.t. } \mathbf{R}\mathbf{u} \leq r + \mathbf{S}x(t)$$

is equivalent to the following mp-LP problem:

$$\min_{\mathbf{u}, \epsilon} \ \epsilon$$

$$\text{s.t. } \mathbf{R}\mathbf{u} \leq r + \mathbf{S}x(t)$$

$$L_i \mathbf{u} + H_i x(t) + k_i \leq \epsilon \ \ i = 1, \dots, s$$

2. If $J(\mathbf{u}, x(t), \theta)$ and $g(\mathbf{u}, x(t), \theta)$ are convex functions in θ for all $(\mathbf{u}, x(t))$ with $\theta \in \Theta$, where Θ is a polyhedron with vertices θ_i ($i = 1, \cdots, N_\theta$). Then the problem

$$J^*(x(t)) \triangleq \min_{\mathbf{u}} \max_{\theta} J(x(t), \mathbf{u}, \theta) \tag{11.38}$$

$$\text{s.t. } g(\mathbf{u}, x(t), \theta) \leq 0 \ \forall \theta \in \Theta$$

is equivalent to the problem:

$$\min_{\mathbf{u}, \epsilon} \ \epsilon$$

$$\text{s.t. } \left.\begin{array}{l} J(x(t), \mathbf{u}, \theta_i) \leq \epsilon \\ g(\mathbf{u}, x(t), \theta_i) \leq 0 \end{array}\right\} i = 1, \cdots, N_\theta$$

3. If $J(\mathbf{u}, x(t), \theta)$ is convex and piecewise in $x(t)$ and $\mathbf{u}$ (i.e. $J(\mathbf{u}, x(t), \theta) = L_i(\theta)\mathbf{u} + H_i(\theta)x(t) + k_i(\theta)$) and $g(\mathbf{u}, \mathbf{x}(t), \theta)$ is affine in $x(t)$ and $\mathbf{u}$ (i.e. $g(\mathbf{u}, x(t), \theta) = L_g(\theta)\mathbf{u} + H_g(\theta)x(t) + k_g(\theta)$) with $L_i, H_i, k_i, L_g, H_g, k_g$ convex functions, then the min-max Problem (11.38) is equivalent to the problem:

$$\min_{\mathbf{u}, \epsilon} \epsilon$$

$$\text{s.t:} \quad \left. \begin{array}{l} L_j(\theta_i)\mathbf{u} + H_j(\theta_i)x(t) + k_j(\theta_i) \leq \epsilon \\ L_g(\theta_i)\mathbf{u} + H_g(\theta_i)x(t) + k_g(\theta_i) \leq 0 \end{array} \right\} \begin{array}{l} i = 1, \cdots, s \\ j = 1, \cdots, N_\theta \end{array}$$

Let us now consider the first step of the dynamic programming problem (11.36) with $L(x(t+j), u(t+j)) \triangleq \|Qx(t+j)\|_p + \|Ru(t+j)\|_p$ with the terminal cost defined as $J_{t+N}^*(x(t+N)) \triangleq \|P\mathbf{x}(t+N)\|_p$ and the linear system $x(t+1) = A(\theta(t))\mathbf{x}(t) + B(\theta(t))u(t) + E(\theta(t))$. The first step of the dynamic programming problem consists of solving the following problem

$$J_t^*(x(t)) \triangleq \min_{u(N-1)} J_{N-1}(x(N-1), u(N-1)) \tag{11.39}$$

s.t.

$$\left. \begin{array}{l} \mathbf{R}_x x(N-1) + \mathbf{R}_u u(N-1) \leq r \\ x(N) \in \mathcal{X}(N) \end{array} \right\} \forall \theta(N-1) \in \Theta$$

$$J_{N-1}(x(N-1), u(N-1)) \triangleq$$

$$\max_{\theta(N-1) \in \Theta} L(x(N-1), u(N-1)) + J_N^*(x(N)) \tag{11.40}$$

with:

$$L(x(N-1), u(N-1)) = \|Qx(N-1)\|_p + \|Ru(N-1)\|_p$$
$$J_N^*(x(N)) = \|Px(N)\|_p$$
$$x(N) = A(\theta(N-1))x(N-1) + B(\theta(N-1))u(N-1) + E(\theta(N-1))$$

The cost function (11.40) is piecewise affine convex with respect to the maximization variables $\theta(N-1)$. Furthermore, the constraints for the minimization Problem (11.39) are linear in $(x(N-1), u(N-1))$ for all $\theta(N-1)$. Applying the previous results, this is equivalent to the following mp-LP problem:

$$J_{N-1}^*(x(t)) \triangleq \min_{u(N-1), \mu} \mu \tag{11.41}$$

s.t.

$$\left. \begin{array}{l} \mathbf{R}_x x(N-1) + \mathbf{R}_u u(N-1) \leq r \\ x(N) = A(\theta_i)x(N-1) + B(\theta_i)u(N-1) + E(\theta_i) \\ x(N) \in \mathcal{X}(N) \\ \|Qx(N-1)\|_p + \|Ru(N-1)\|_p + \|Px(N)\|_p \leq \mu \end{array} \right\} i = 1, \cdots, n_\theta$$

$$\tag{11.42}$$

J_{N-1}^* is the solution of the mp-LP problem (11.41) which results in a piecewise affine function of the state $x(N-1)$. The corresponding control signal

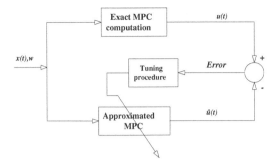

Fig. 11.6. Procedure to obtain an approximated MPC

$u^*(N-1)$ is also a continuous piecewise affine function of the state and the feasible set $\mathcal{X}(N-1)$ is a convex polyhedron. The same steps can be applied recursively for $j = N-2, N-3, \cdots 0$, and we reach the conclusion that $u^*(t)$ is a piecewise affine function of state $x(t)$. That is, that min-max MPC closed-loop control results in a piecewise linear control law. Notice that the argument does not hold when the objective function is quadratic as Function (11.42) would not be piecewise affine convex with respect to the maximization variables $\theta(N-1)$.

11.5 Approximated Implementation for MPC

As has been shown in this chapter, MPC results in a controller that, in principle, is quite easy to implement for small problems. However, the number of regions in which the state space has to be partitioned grows very rapidly with the prediction horizon. Thus, storage requirements and searching time for the appropriate region can be very high for practical values of the prediction and control horizons. A way to implement MPC in these cases is to use an approximated controller.

Consider an MPC controller defined by $u(t) = f_{\text{MPC}}(x(t), \mathbf{w})$. The idea is to choose a function $\hat{f}_{\text{MPC}}(x(t), \mathbf{w})$ such that the approximation error $e(x(t), \mathbf{w}) = f_{\text{MPC}}(x(t), \mathbf{w}) - \hat{f}_{\text{MPC}}(x(t), \mathbf{w})$ is small for the operating region (i.e., $x(t) \in \mathcal{X}$ and $\mathbf{w} \in \mathcal{W}$).

The approximation procedure starts by choosing a type of function that is able to approximate $f_{\text{MPC}}(x(t), \mathbf{w})$ by tuning some free parameters. These parameters are adjusted as indicated in Figure 11.6 to approximate the function. A set of representative points in $\mathcal{X} \times \mathcal{W}$ is chosen and for all these points, the MPC problem is solved as, indicated in Figure 11.6. Let us call $z_i = [x_i \ w_i]$ one of these points and u_i its corresponding solution. The approximating function is tuned to the set of points obtained. Normally this is done to minimize an error function such as $\sum_{i=1}^{M} \|f_{\text{MPC}}(x_i, \mathbf{w}_i) - \hat{f}_{\text{MPC}}(x_i, \mathbf{w}_i)\|_p$. Different types of functions have been used to approximate MPCs, such as neural networks and hinging hyperplanes.

Artificial Neural Network (ANN) based controllers exploit the possibilities of neural networks for *learning* nonlinear functions or the possibilities of neural networks to solve certain types of problems where massive parallel computation is required. The *learning* capability of ANN is used to make the controller learn a certain function, most of the time highly nonlinear, representing direct dynamics, inverse dynamics or any other characteristics of the process. This is usually done during a (normally long) training period when commissioning the controller in a supervised or unsupervised manner.

There are some applications of ANN to Model Predictive Control. In some cases, the ANNs are used to model the plant. For example, in [9], Neural Networks are used to model the free response of a solar plant in an MPC scheme; in [166], ANNs are used to implement nonlinear MPC. In other cases, ANNs are directly used to model the controller as indicated in this section. First, during a normally long training phase, the ANNs are adjusted to imitate the controller (see Figure 11.6) and then the controller is commissioned. Notice that the training phase of the ANN is analogous to the offline computation required to obtain the explicit controllers dealt with in this chapter. Once the NN is working, the amount of computation required is very small and can be compared to the online computation of the fast implementation methods. An example of how to implement an MPC for a nonlinear process (a mobile robot) using ANN is given in Chapter 12.

Hinging hyperplanes (HH) have also been used to model piecewise linear models. The HH technique is a nonlinear function approximation method that uses hinge functions, i.e., hyperplanes joined together. With this technique, piecewise linear functions, such as the resulting MPC controllers, can be described using a basis function expansion. The HH technique has been used in the MPC context in [51] and in [176] to implement a min-max MPC for a heat exchanger.

Notice that if the approximation is good enough (i.e., an approximation error below the conversion errors of the digital-analog conversion), the approximate implementation will not differ from the exact MPC at all from a practical viewpoint.

11.6 Fast Implementation of MPC and Dead Time Considerations

The fast methods for implementing MPC studied in this chapter consist of determining, in an offline manner, a function of the state[5] $f_{\text{MPC}}(x(t))$ that once computed in an online manner will give the optimal MPC control move

---

[5] Extended state in the case of nonconstant references or measurable disturbances as seen in Section 11.3.2.

$u^*(t)$. The complexity of the methods depends to a great extent on the state vector dimension which can be very high for processes with long dead times (as can be found frequently in industry). In the case of a dead time of d sampling instants, an augmented state vector $x_a(t) = [x(t)^t \quad u(t-1)^T \quad u(t-2)^T \dots \quad u(t-d)^T]$ is needed; that is, the online function has to be defined over the augmented state $(x_a(t))$ domain. The increase of the vector dimension due to the dead time could be very limiting for the applicability of the techniques described in this chapter.

A technique proposed in [43] can be used to overcome these problems. The idea consists of using the predicted state $\hat{x}(t+d \mid t)$ as indicated in Section 5.2.2 in order to compute $\hat{f}_{\text{MPC}}(\hat{x}(t+d \mid t))$ instead of $f_{\text{MPC}}(x_a(t))$. This can be done without problem using the information available at time t. Furthermore, the outcome is the same; i.e., $\hat{f}_{\text{MPC}}(\hat{x}(t+d \mid t)) = f_{\text{MPC}}(x_a(t))$.

To illustrate the gain that can be obtained, consider a process that can be modelled by the reaction curve method with a dead time equal to its time constant. If the sampling time chosen is one-tenth of the time constant, then $dim(x_a(t)) = 11$ while $dim(x(t)) = 1$ (see Exercise 11.6).

11.7 Exercises

11.1. Consider the double integrator MPC of the example in Section 11.3.1:

1. Write a computer program to simulate the controller given. Simulate the response with different initial states $x(t)$ such that $\|x(t)\|_\infty \leq 5$. Do all trajectories converge to the origin?
2. Find the attraction region for the controller.
3. Recompute the controller with $N = 3$ and repeat step 1. Is the attraction region enlarged by considering a higher control horizon?

11.2. Consider a linear system described with $dim(x(t)) = n_x$, $dim(y(t)) = n_y$, $dim(u(t)) = n_u$, $dim(p(t)) = n_p$ (measurable disturbances) and prediction and control horizon equal to N. The system is constrained by $\mathbf{R}u \leq \mathbf{V}x(t) + \mathbf{r}$ with $\mathbf{R}$ an $m \times (N \times n_u)$ matrix. Discuss the maximum number of regions, storage size required, and number of multiplications, additions and comparisons needed to implement the explicit MPC as a function of n_x, n_u, n_p, m and N.

11.3. Define constraint matrices and vectors $\mathbf{R}_z$, $\mathbf{V}_v$ and $\mathbf{r}^c$ of MPC problem (11.31).

11.4. Consider the system described by $x(t+1) = 0.9x(t)+u(t)+p(t)$ with the manipulated variable constrained by $-1 \leq u(t) \leq 1$ and an objective function $J = \sum_{j=1}^{N} \left(x(t+j)^2 + \lambda u(t+j-1)^2 \right)$. The signal $p(t)$ is a measurable disturbance:

1. Explain how to obtain an explicit MPC taking into account the measurable disturbances.
2. Obtain an explicit controller for $N = 2$, consider that the future disturbances are estimated as the present measured disturbance, i.e., $p(k + j) = p(k)$ for $j = 1, \ldots, N$ and that the control weight $\lambda = 1$.
3. Indicate what should be done if the future disturbances are computed as $p(k + j + 1) = 2p(k + j) - p(k + j - 1)$ for $j = 1, \ldots, N$.

11.5. For the problem described in Section 11.4.1 implement the closed-loop approach described in Section 11.4.2 with increasing values of N.

11.6. Consider the system described by

$$Y(s) = \frac{2e^{-10s}}{1 + 10s}U(s)$$

which is sampled with a sampling time of 1. The signals are bounded by $|u(t)| \leq 1$ and $|y(t)| \leq 1$:

1. Discuss how you would compute an explicit MPC with and without using a predictor for $y(t + d)$.
2. Formulate the explicit MPC using a predictor for $N = 3$.
3. Simulate the resulting controller with different values of N and λ and comment on the results.

12

Applications

This chapter is dedicated to presenting some MPC applications to the control of different real and simulated processes. The first application presented corresponds to a self-tuning and a gain scheduling GPC for a distributed collector field of a solar power plant. In order to illustrate how easily the control scheme shown in Chapter 5 can be used in any commercial control system, some applications concerning the control of typical variables such as flows, temperatures and levels of different processes of a pilot plant are presented. The description of two applications in the food industry (a sugar refinery and an olive oil mill) are included. Finally the application of an MPC to a highly nonlinear process (a mobile robot) is also described.

12.1 Solar Power Plant

This section presents an application of an adaptive long-range predictive controller to a solar power plant and shows how this type of controller, which normally requires a substantial amount of computation (in the adaptive case), can easily be implemented with few computation requirements. The results obtained when applying the controller to the plant are also shown.

The controlled process is the distributed collector field (Acurex) of the Solar Platform of Tabernas (Spain). The distributed collector field consists mainly of a pipeline through which oil is flowing and onto which the solar radiation is concentrated by means of parabolic mirrors, which follow the sun by rotating on one axis, in order to heat the oil. It consists of 480 modules arranged in 20 lines which form 10 parallel loops. A simplified diagram of the solar collector field is shown in Figure 12.1. The field is also provided with a sun-tracking mechanism which causes the mirrors to revolve around an axis parallel to that of the pipeline.

On passing through the field the oil is heated and then introduced into a storage tank to be used for the generation of electrical energy. The hot oil can also be used directly for feeding the heat exchanger of a desalination plant.

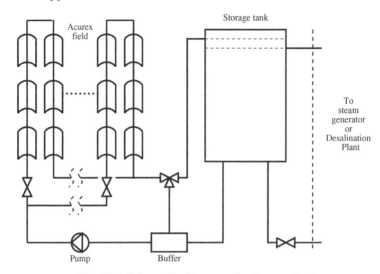

Fig. 12.1. Schematic diagram of collectors field

The cold inlet oil to the field is extracted from the bottom of the storage tank.

Each of the loops is formed by four twelve-module collectors, suitably connected in series. The loop is 172 metres long, the active part of the loop (exposed to concentrated radiation) measures 142 metres and the passive part 30 metres.

The system is provided with a three-way valve which allows the oil to be recycled in the field until its outlet temperature is adequate to enter the storage tank. A more detailed description of the field can be found in [101].

A fundamental feature of a solar power plant is that the primary energy source, whilst variable, cannot be manipulated. The intensity of the solar radiation from the sun, in addition to its seasonal and daily cyclic variations, is also dependent on atmospheric conditions such as cloud cover, humidity, and air transparency. It is important to be able to maintain a constant outlet temperature for the fluid as the solar conditions change, and the only means available for achieving this is via adjustment of the fluid flow.

The objective of the control system is to maintain the outlet oil temperature at a desired level in spite of disturbances such as changes in the solar irradiance level (caused by clouds), mirror reflectivity or inlet oil temperature. This is accomplished by varying the flow of the fluid through the field. The field exhibits a variable delay time that depends on the control variable (flow). The transfer function of the process varies with factors such as irradiance level, mirror reflectance and oil inlet temperature.

The distributed collector field is a nonlinear system which can be approximated by a linear system when considering small disturbances. The maintenance of a constant outlet temperature throughout the day as the solar conditions change requires a wide variation in the operational flow level. This leads to substantial variations in the general dynamic performance and in particular, from the control viewpoint, gives rise to a system time delay which varies significantly. The controller parameters need to be adjusted to suit the operating conditions, and self-tuning control offers one approach which can accommodate such a requirement.

Because of the changing dynamics and strong perturbations, this plant has been used to test different types of controllers [40],[41].

For self-tuning control purposes a simple, linear model is required which relates changes in fluid flow to changes in outlet temperature.

Observations of step responses obtained from the plant indicate that in the continuous time domain behaviour it can be approximated by a first-order transfer function with a time delay. Since the time delay τ_d is relatively small compared to the fundamental time constant τ, a suitable discrete model can be constructed by choosing the sample period T equal to the lowest value of the time delay τ_d. This corresponds to the operating condition where the flow level is highest. The discrete transfer function model then takes the form

$$g(z^{-1}) = z^{-k}\frac{bz^{-1}}{1 - az^{-1}}$$

and at the high flow level condition, k = 1.

12.1.1 Selftuning GPC Control Strategy

A particular feature of the system is the need to include a series feedforward term in the control loop [39]. The plant model upon which the self-tuning controller is based relates changes in outlet temperature to changes in fluid flow only. The outlet temperature of the plant, however, is also influenced by changes in system variables such as solar radiation and fluid inlet temperature. During estimation, if either of these variables changes it introduces a change in the system output unrelated to fluid flow which is the control input signal, and in terms of the model, it would result in unnecessary adjustments of the estimated system parameters.

Since both solar radiation and inlet temperature can be measured, this problem can be eased by introducing a feedforward term in series to the system, calculated from steady-state relationships, which makes an adjustment in the fluid flow input, aimed at eliminating the change in outlet temperature caused by the variations in solar radiation and inlet temperature. If the feedforward term perfectly countered the changes in solar radiation and inlet temperature, then the observed outlet temperature changes would only be caused by changes in the control input signal. Although exact elimination obviously cannot be achieved, a feedforward term based on steady-state considerations overcomes the major problems inherent in the single-input model and permits successful estimation of model parameters. The basic idea is to

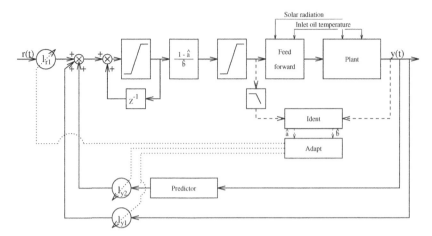

Fig. 12.2. Self-tuning control scheme

compute the necessary oil flow to maintain the desired outlet oil temperature given the inlet oil temperature and the solar radiation. The feedforward signal also provides control benefits when disturbances in solar radiation and fluid inlet temperature occur, but the basic reason for its inclusion is that of preserving the validity of the assumed system model in the self-tuning algorithm; for more details see [45].

In the control scheme, the feedforward term is considered as a part of the plant, using the setpoint temperature to the feedforward controller as the control signal.

In this section, the precalculated method described in Chapter 5 is used. As the dead time d is equal to 1, the control law is given by

$$\triangle u(t) = l_{y1}\hat{y}(t+1 \mid t) + l_{y2}y(t) + l_{r1}r(t) \qquad (12.1)$$

where the value $\hat{y}(t+1 \mid t)$ is obtained by use of the predictor:

$$\hat{y}(t+j+1 \mid t) = (1+a)\hat{y}(t+j \mid t) - a\hat{y}(t+j-1 \mid t) + b\triangle u(t+j-1) \quad (12.2)$$

The proposed control scheme is shown in Figure 12.2. The plant estimated parameters are used to compute the controller coefficients (l_{y1}, l_{y2}, l_{r1}) via the adaptation mechanism. Notice that in this scheme, the feedforward term is considered as a part of the plant (the control signal is the setpoint temperature for the feedforward controller instead of the oil flow). This signal is saturated and filtered before its use in the estimation algorithm. The controller also has a saturation to limit the increase of the error signal.

As suggested in Chapter 5, a set of GPC parameters were obtained for $\delta(i) = 1$, $\lambda(i) = 5$ and $N = 15$. The pole of the system has been changed with a 0.0005 step from 0.85 to 0.95, which are the values that guarantee the system stability if the parameter set estimation is not accurate enough. Notice that

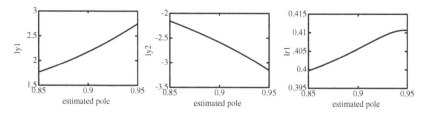

Fig. 12.3. Controller parameters for $\lambda = 5$

due to the fact that the closed-loop static gain must equal the value unity, the sum of the three parameters equals zero. The curves shown in Figure 12.3 correspond to the controller parameters l_{y1}, l_{y2}, l_{r1} for the values of the pole. The adjustment of analytical functions to the calculated values provide:

$$l_{y1} = 0.4338 - 0.6041\,\hat{a}\,/\,(1.11 - \hat{a})$$
$$l_{y2} = -0.4063 + 0.4386\,\hat{a}\,/\,(1.082 - \hat{a}) \qquad (12.3)$$
$$l_{r1} = -l_{y1} - l_{y2}$$

These expressions give a very good approximation to the true controller parameters and fit the set of computed data with a maximum error of less than 0.6 per % of the nominal values for the range of interest of the open loop-pole; for more details see [40].

The parameters of the system model in the control scheme are determined online via recursive least-squares estimation. The estimation algorithm incorporates a variable forgetting factor and only works when the input signal contains dynamic information. These considerations can be taken into account by checking the following conditions:

$$|\Delta u| \geq A$$

$$\sum_{k=-N}^{k=0} |\Delta u(k)| \geq B$$

If one of these conditions is true, the identifier is activated. Otherwise, the last set of estimated parameters is used. Typical values of A, B and N chosen from simulation studies are: $A = 9$, $7 \leq B \leq 9$ and $N = 5$. The covariance matrix $P(k)$ is also monitored by checking that its diagonal elements are kept within limits; otherwise $P(k)$ is reset to a diagonal matrix having the corresponding elements saturated to the violated limit.

With respect to the adaptation mechanism, it only works when the estimated parameters are contained within the ranges ($0.85 \leq \hat{a} \leq 0.95$ and $0.9 \leq \hat{k}_{est} \leq 1.2$, where $\hat{k}_{est}$ is the estimated static gain of the plant $\hat{b}/(1 - \hat{a})$) in order to avoid instability in cases of nonconvergence of the estimator. A backup controller is used in situations in which these conditions are not accomplished (for example, when daily operation starts).

In each sampling period the self-tuning regulator consists of the following steps:

1. Estimate the parameters of a linear model by measuring the inlet and outlet values of the process.
2. Adjust the parameters of the controller using Expressions (12.3).
3. Compute $\hat{y}(t + d \mid t)$ using the predictor in (12.2).
4. Calculate the control signal using (12.1).
5. Supervise the correct working of the control.

Plant Results

Figure 12.4 shows the outlet oil temperature and reference when the proposed self-tuning generalised predictive controller is applied to the plant. The value of the control weighting λ was made equal to 5 and, as can be seen, a fast response to changes in the setpoint is obtained (the rising time observed is approximately 7 minutes). When smaller overshoots are required, the control weighting factor has to be increased.

The evolution of the irradiation for the test can be seen in the same figure and it corresponds to a day with small scattered clouds. The oil flow changed from 4.5 l/s to 7 l/s, and the controller could maintain performance in spite of the changes in the dynamics of the process caused by flow changes.

12.1.2 Gain Scheduling Generalized Predictive Control

There are many situations in which it is known how the dynamics of a process change with the operating conditions. It is then possible to change the controller parameters taking into account the current operating point of the system. Gain scheduling is a control scheme with open-loop adaptation, which can be seen as a feedback control system in which the feedback gains are adjusted by a feedforward compensation. Gain scheduling control is a nonlinear feedback of a special type: it possesses a linear controller whose parameters are modified depending on the operating conditions in a pre-specified manner.

The working principle of this kind of controller is simple; it is based on the possibility of finding auxiliary variables which guarantee a good correlation with process changing dynamics. In this way, it is possible to reduce the effects of variations in the plant dynamics by adequately modifying the controller parameters as functions of auxiliary variables.

An essential problem is the determination of the auxiliary variables. In the case studied here, the behaviour and changes in the system dynamics mainly depend on the oil flow if very strong disturbances are not acting on the system (due to the existence of the feedforward controller in series with the plant). The oil flow has been the variable used to select the controller

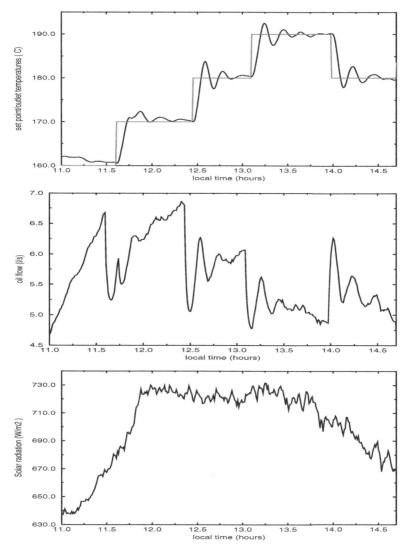

Fig. 12.4. Adaptive GPC: plant outlet oil temperature, flow and solar radiation

parameters (a low-pass filter is used to avoid the inclusion of additional dynamics due to sudden variations in the controller parameters).

Once the auxiliary variables have been determined, the controller parameters have to be calculated at a number of operating points, using an adequate controller design algorithm, which in this case is the GPC methodology. When coping with gain scheduling control schema, stability and performance of the controlled system are usually evaluated by simulation studies [142]. A crucial point here is the transition between different operating points. In those cases in which a nonsatisfactory behaviour is obtained, the number of inputs to the table of controller parameters must be augmented. As has been mentioned, it is important to point out that no feedback exists from the behaviour of the controlled system to the controller parameters. So this control scheme is not considered as an adaptive one, but rather as a special case of a nonlinear controller.

The main disadvantages of gain scheduling controllers are:

- It is an open-loop compenzation: there is no way to compensate for a wrong election of the controller parameters within the table.
- Another important inconvenience is that the design stage of the strategy often consumes too much time and effort. The controller parameters must be calculated for enough operating points, and the behaviour of the controlled system has to be checked under very different operating conditions.

Its main advantage is the ease of changing controller parameters in spite of changes in process dynamics. As classical examples of applications of this kind of controller, the following control fields can be mentioned: design of ship steering autopilots, pH control, combustion control, engine control, design of flight autopilots, etc. [12].

Plant Models and Fixed Parameter Controllers

The frequency response of the plant has been obtained by performing a PRBS PRBS test in different operating conditions, using both the plant and a nonlinear distributed parameter model[1] [25]. In this way, different linear models were obtained from input-output data in different working conditions. These models relate changes in the oil flow to those of the outlet oil temperature, and can take into account the antiresonance characteristics of the plant if they are adequately adjusted. The control structure proposed is shown in Figure 12.5. As can be seen, the output of the generalized predictive controller is the input (t_{rff}) of the series compensation controller, which also uses the solar radiation, inlet oil temperature and reflectivity to compute the value of the oil flow, which is sent to the pump controller.

The controller parameters were obtained from a linear model of the plant. From input-output data of the plant, the degrees of the polynomials A and B and the delay (of a CARIMA plant model) that minimizes Akaike's Information Theoretic Criterion (AIC) were found to be $n_a = 2$, $n_b = 8$ and $d = 0$. By a least squares estimation algorithm, the following polynomials were obtained using input-output data of one test with oil flow of around 6 l/s:

---

[1] The software simulation package for the solar distributed collector field with real data from the plant can be obtained by contacting the authors or by accessing http://www.esi.us.es/ eduardo.

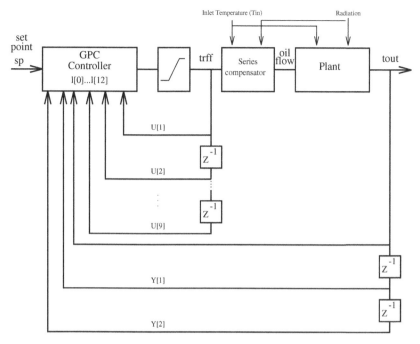

Fig. 12.5. Control scheme using high-order models

$$A(z^{-1}) = 1 - 1.5681z^{-1} + 0.5934z^{-2}$$
$$B(z^{-1}) = 0.0612 + 0.0018z^{-1} - 0.0171z^{-2} + 0.0046z^{-3} + 0.0005z^{-4}$$
$$+0.0101z^{-5} - 0.0064z^{-6} - 0.015z^{-7} - 0.0156z^{-8}$$

The most adequate value for the control horizon ($N = 15$) was calculated taking into account the values of the fundamental time constant and the sampling period used for control purposes. In this case, $N_1 = 1$ and $N_2 = 15$. The value of λ was determined by simulation studies using the nonlinear model and was found to be $\lambda = 6$ (fast) and $\lambda = 7$ (without overshoot). For smaller values of λ, faster and more oscillatory responses were obtained. Following the design procedure of the GPC methodology, the controller parameters corresponding to $\lambda = 7$ were obtained (Table 12.1).

Table 12.1. Fixed GPC controller coefficients

$l[0]$	$l[1]$	$l[2]$	$l[3]$	$l[4]$	$l[5]$	$l[6]$
−2.4483	6.8216	−4.7091	−0.0644	−0.0526	−0.0084	0.0629
$l[7]$	$l[8]$	$l[9]$	$l[10]$	$l[11]$	$l[12]$	
0.0161	0.0311	−0.0631	0.0231	1.0553	0.3358	

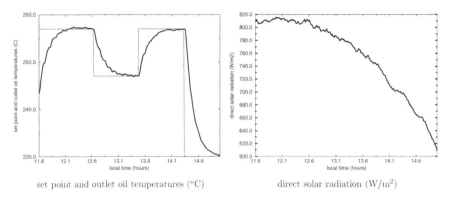

set point and outlet oil temperatures (°C)   direct solar radiation (W/m²)

Fig. 12.6. Test with the fixed GPC high-order controller

The control law can be written by

$$t_{rff} = l[2]t_{out} + l[1]y[1] + l[0]y[2] + l[6]u[6] + l[3]u[9] + l[4]u[8] + l[5]u[7]$$
$$+l[7]u[5] + l[8]u[4] + l[9]u[3] + l[10]u[2] + l[11]u[1] + l[12]sp \quad (12.4)$$

where

t_{rff}: reference temperature for the feedforward controller;
t_{out}: outlet temperature of the field;
sp: setpoint temperature;
$l[i]$: controller parameters;
$y[i]$: outlet temperature of the field at sampling time $(t - i)$and
$u[i]$: reference temperature for the feedforward controller at sampling time $(t - i)$.

With these values, the behaviour of this fixed parameter controller was analyzed in operation with the distributed solar collector field. The outlet oil temperature of the field evolution and corresponding setpoint can be seen in Figure 12.6. The evolution of the solar radiation during this test can also be seen in Figure 12.6. Although direct solar radiation goes from 810 W/m² to 610 W/m², the field was working in midflow conditions because the setpoint was also changed from 258°C to 230°C. When operating conditions in the field change, the dynamics of the plant also change and the controller should be redesigned to cope with the control objectives.

The dynamics of the field are mainly dictated by the oil flow, which depends on the general operating conditions of the field: solar radiation, reflectivity, oil inlet temperature, ambient temperature and outlet oil temperature setpoint. These changes in plant dynamics are illustrated in Figure 12.7, where the frequency response of the nonlinear distributed parameter dynamic model of the field can be seen. The curves shown in Figure 12.7 were obtained by a spectral analysis of the input-output signals of the model at different operating points (PRBS signals were used for the input).

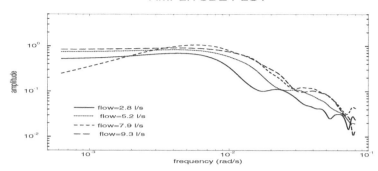

Fig. 12.7. Frequency response of the field under different operating conditions

As can be seen, the frequency response changes significantly for different operating conditions. The steady-state gain changes for different operating points, as does the location of the antiresonance modes.

Taking into account the frequency response of the plant and the different linear models obtained from it, it is clear that a self-tuning controller based on this type of model is very difficult to implement. The fundamental reason is the fact that estimation of the model parameters requires a lot of computation when the number of estimated parameters increases and the convergence of the estimation process is seldom accurate and fast enough.

Another way of coping with changing dynamics is to use a gain scheduling controller, making the controller parameters dependent on some variables which indicate the operating conditions.

With the input-output data used to obtain the frequency responses shown in Figure 12.7 and using the method and type of model previously described for the case of a high-order fixed parameter controller, process ($a[i]$ and $b[i]$) and controller ($l[i]$) parameters were obtained for several oil flow conditions ($q_1 = 2.8\,1/\mathrm{s}$, $q_2 = 5.2\,1/\mathrm{s}$, $q_3 = 7.9\,1/\mathrm{s}$ and $q_4 = 9.3\,1/\mathrm{s}$), using different values of the weighting factor λ. Tables 12.2 and 12.3 contain model and control parameters, respectively, for a weighting factor $\lambda = 6$. A value of $\lambda = 7$ has also been used to obtain responses without overshoot.

The controller parameters applied in real operation are obtained by using a linear interpolation with the data given in Table 12.3. It is important to point out that to avoid the injection of disturbances during the controller gain adjustment, it is necessary to use a smoothing mechanism of the transition surfaces of the controller gains. In this case, a linear interpolation in combination with a first-order filter has been used, given a modified flow $Q(t) = .95\,Q(t-1) + .05\,q(t)$ (where $q(t)$ is the value of the oil flow at instant t and $Q(t)$ is the filtered value used for controller parameter adjustment). The linear interpolation has also been successfully applied in [97]. Another kind of gain scheduling approach can be obtained by switching from one set of controller parameters to another depending on the flow condi-

Table 12.2. Coefficients of polynomials $A(z^{-1})$ and $B(z^{-1})$ for different flows

	q_1	q_2	q_3	q_4
$a[1]$	−1.7820	−1.438	−1.414	−1.524
$a[2]$	0.81090	0.5526	0.5074	0.7270
$b[0]$	0.00140	0.0313	0.0687	0.0820
$b[1]$	0.03990	0.0660	0.0767	0.0719
$b[2]$	−0.0182	−.0272	−.0392	−.0474
$b[3]$	−0.0083	0.0071	0.0127	0.0349
$b[4]$	0.00060	0.0118	0.0060	0.0098
$b[5]$	−.00001	0.0138	−.0133	−.0031
$b[6]$	0.00130	0.0098	−.0156	0.0111
$b[7]$	0.00160	0.0027	−.0073	0.0171
$b[8]$	0.00450	−.0054	0.0037	0.0200

Table 12.3. GPC controller coefficients in several operating points ($\lambda = 6$)

$l[0]$	−7.0481	−1.4224	−1.1840	−1.3603
$l[1]$	16.2223	3.84390	3.48440	3.02280
$l[2]$	−9.5455	−2.7794	−2.6527	−2.0142
$l[3]$	0.03910	−0.0139	0.00860	0.03740
$l[4]$	0.00980	0.00830	−0.0184	0.02730
$l[5]$	0.00560	0.02610	−0.0352	0.01080
$l[6]$	−0.0070	0.03390	−0.0239	−0.0197
$l[7]$	−0.0016	0.02480	0.02630	0.00460
$l[8]$	−0.0793	0.00880	0.03980	0.05070
$l[9]$	−0.1575	−0.0822	−0.0869	−0.1098
$l[10]$	0.36470	0.16410	0.19600	0.12480
$l[11]$	0.82620	0.83010	0.89360	0.87390
$l[12]$	0.37130	0.35800	0.35230	0.35170

tions, without interpolating between controller parameters. The set of controller parameters c can be obtained by choosing one of the sets c_i in Table 12.3, related to flow conditions q_i ($i = 1, 2, 3, 4$):

$$\text{if } \frac{q_{i-1} + q_i}{2} < q \le \frac{q_i + q_{i+1}}{2} \quad \text{then} \quad c = c_i, \ i = 2, 3$$

$$\text{if } q \le q_1 \ \text{ then } \ c = c_1$$

$$\text{if } q \ge q_4 \ \text{ then } \ c = c_4$$

The control structure is similar to the one obtained for the fixed controller previously studied. The optimal realization of the gain scheduling controller consists of calculating the controller parameters under a number of operating conditions and suppose that the values of the controller coefficients are

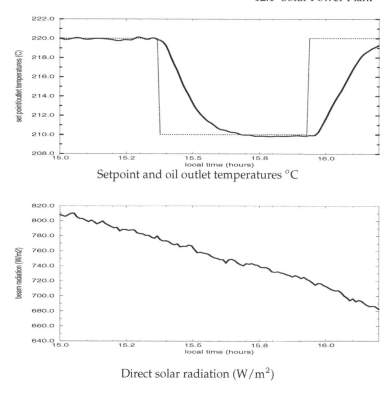

Setpoint and oil outlet temperatures ^{o}C

Direct solar radiation (W/m^{2})

Fig. 12.8. Test with the gain scheduling GPC controller, $\lambda = 7$

constant between different operating conditions, generating a control surface based on an optimization criterion which takes into account the tracking error and control effort. It is evident that if the procedure is applied at many working points, an optimum controller will be achieved for those operating conditions if there is a high correlation between the process dynamics and the auxiliary variable. The drawback to this solution is that the design process becomes tedious. This is one of the main reasons for including a linear interpolation between the controller parameters.

Plant Results

In the case of real tests, similar results were obtained and depending on the operating point, disturbances due to passing clouds, inlet oil temperature variations, etc., different performance was achieved.

Figure 12.8 shows the results of one of these tests with the gain scheduling GPC with $\lambda = 7$. The operating conditions correspond to a clear afternoon with the solar radiation changing from 800 W/m^{2} to 660 W/m^{2} and oil flow changing from 3.75 l/s to 2 l/s. As can be seen, the effect of the antireso-

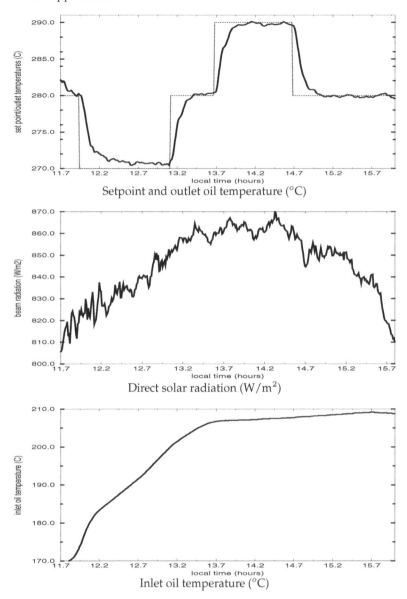

Setpoint and outlet oil temperature (^{o}C)

Direct solar radiation (W/m^2)

Inlet oil temperature (^{o}C)

Fig. 12.9. Test with the gain scheduling GPC controller, $\lambda = 7$

nance modes does not appear in the response, due to the use of an extended high order model which accounts for these system characteristics.

Figure 12.9 shows the result of a test with a weighting factor $\lambda = 7$ corresponding to a day of intermittent scattered clouds which produce large changes in the solar radiation level and the inlet oil temperature changing

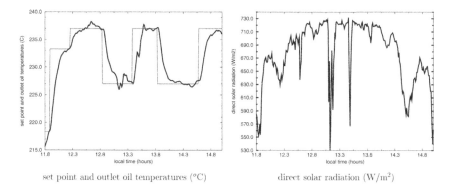

set point and outlet oil temperatures (°C)   direct solar radiation (W/m²)

Fig. 12.10. Test with the gain scheduling GPC controller, $\lambda = 7$

from 170°C to 207°C. As can be seen, the outlet oil temperature follows the setpoint in spite of changing operating conditions and the high level of noise in the radiation level produced by clouds.

The results of a test corresponding to a day with sudden changes in the solar radiation caused by clouds can be seen in Figure 12.10. As can be seen, the controller (also designed with $\lambda = 7$) is able to handle different operating conditions and the sudden perturbations caused by the clouds. After the tests presented using a weighting factor $\lambda = 7$, two new test campaigns were carried out to test the behaviour of the controller with a weighting factor $\lambda = 6$. In the first campaign, the evaluation of the controller performance was considered. In the second, the behaviour of the controller operating under extreme working conditions was studied.

Figure 12.11 shows the results obtained in the operation on a day with normal levels of solar radiation, but on which a wide range of operating conditions is covered (oil flow changing between 2 l/s and 8.8 l/s) by performing several setpoint changes. At the start of operation there is an overshoot of 6°C, due to the irregular conditions of the oil flowing through the pipes because the operation starts with a high temperature level at the bottom of the storage tank. After the initial transient, it can be observed that the controlled system quickly responds to setpoint changes under the whole range of operating conditions with a negligible overshoot. The rise time is about 6 minutes with a setpoint change of 15 degrees, as can be seen in Figure 12.11, with smooth changes in the control signal, constituting one of the best controllers implemented at the plant. It is important to note that the controller behaves well even with great setpoint changes.

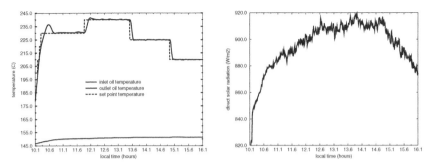

Fig. 12.11. Test with the gain scheduling GPC controller, $\lambda = 6$

12.2 Pilot Plant

In order to show how GPC can be implemented on an industrial SCADA, applications to the control of the most typical magnitudes found in the process industry (flow, level, temperatures) are introduced in this section.

The tests are carried out on a pilot plant existing in the *Departamento de Ingeniería de Sistemas y Automática* of the University of Seville. The pilot plant is provided with industrial instrumentation and is used as a testbed for new control strategies which can be implemented on an industrial SCADA connected to it. This plant is basically a system using water as the working fluid in which various thermodynamic processes with interchange of mass and energy can take place. It essentially consists of a tank with internal heating with a series of input-output pipes and recirculation circuit with a heat exchanger.

The design of the plant allows for various control strategies to be tested in a large number of loops. Depending on the configuration chosen, it is possible to control the types of magnitudes most frequently found in the process industry such as temperature, flow, pressure and level. For this, four actuators are available: three automatic valves and one electric heater that heats the interior of the tank. Later some of the possible loops are chosen (considered as being independent) for implanting the GPC controllers.

12.2.1 Plant Description

A diagram of the plant which shows its main elements as well as the localization of the various instruments is given in Figure 12.12.

The main elements are:

- feed circuit. The plant has two input pipes, a cold water one (at air temperature) and a hot water one (at about 70°C) with nominal flow and temperature conditions of 10 l/min and 2 bar for the cold water and 5 l/min and 1 bar for the hot. The temperatures and flows of the inputs are measured with thermocouples and orifice plates, respectively, with motorized valves for regulating the input flows.

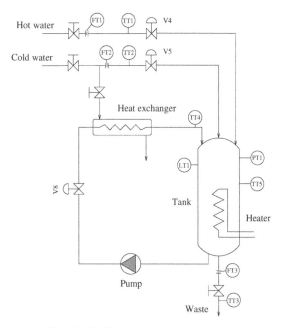

Fig. 12.12. Diagram of the pilot plant

- tank. It has a height of 1 m and an interior diameter of 20 cm, it is thermically insulated, and it has an approximate volume of 31 l. It can work pressurized (up to a limit of 4 bar) or at atmospheric pressure, depending on the position of the vent valve. In its interior there is a 15 kW electric resistance for heating, and an overflow, an output pipe and another pipe for recirculating the water through the exchanger.
- recirculation circuit. The hot water in the tank can be cooled by entering cold water through the cooling circuit. This circuit is composed of a centrifugal pump that circulates the hot water from the bottom of the tank through a tube bundle heat exchanger returning at a lower temperature at its top.

12.2.2 Plant Control

To control the installation there is an ORSI Integral Cube distributed control system, composed of a controller and a supervisor connected by a local data highway. The former is in charge of carrying out the digital control and analogous routines whilst the latter acts as a programming and communication platform with the operator. On this distributed control system the GPC algorithms seen before will be implemented. This control system constitutes a typical example of an industrial controller, having the most normal characteristics of medium-size systems to be found in the market today. As in most control computers the calculation facilities are limited and there is little time available for carrying out the control algorithm because of the attention called for by other operations. It is thus an excellent platform for implanting precalculated GPC in industrial fields.

From all the possible loops that could be controlled the results obtained in certain situations will be shown. These are: control of the cold water flow FT_2 with valve V_5, control of the output temperature of the heat exchanger TT_4 with valve V_8, control of the tank level LT_1 with the cold water flow by valve V_5 and control of the tank temperature TT_5 with the resistance.

12.2.3 Flow Control

The control of the cold water flow has been chosen as an example of regulating a simple loop. Because all the water supplied to the plant comes from only one pressure group, the variations affecting the hot water flow or the cold water flow of the heat exchanger will affect the cold water flow as disturbances. Regulating the cold water flow is not only important as a control loop but it may be necessary as an auxiliary variable to control the temperature or level in the tank.

The dynamics of this loop are mainly governed by the regulation valve. This is a motorized valve with a positioner with a total open time of 110 seconds, thus causing slow dynamics in the flow variation. The flow behaviour will approximate that of a first-order system with delay.

First the parameters identifying the process are obtained using the reaction curve, and then the coefficients of the GPC are found using the method described in Chapter 5. In order to do this, working with the valve 70% open (flow of 3.98 l/min), a step of up to 80% is produced, obtaining after the transition a stationary value of 6.33 l/min. From the data obtained it can be calculated that

$$K = 0.25 \qquad \tau = 10.5 \text{ seconds} \qquad \tau_d = 10 \text{ seconds}$$

when a sampling time $T = 2$ seconds is used, the parameters for the corresponding discrete model are:

$$a = 0.8265 \qquad b = 0.043 \qquad d = 5$$

The control signal can easily be computed using the expression

$$u(t) = u(t-1) + (l_{y1}\hat{y}(t+5) + l_{y2}\hat{y}(t+4) + l_{r1}r(t))/K$$

where $u(t)$ is the position of the valve V_5 and $y(t)$ is the value of the flow FT_2. Using the approximation formulas (5.10) with $\lambda = 0.8$, the controller gains result as:

$$l_{y1} = -2.931$$
$$l_{y2} = 1.864$$
$$l_{r1} = 1.067$$

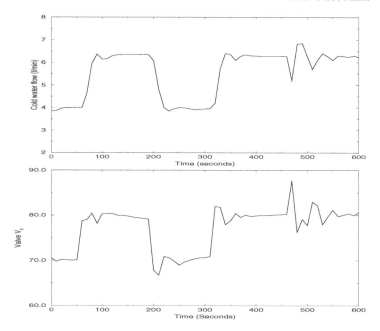

Fig. 12.13. Flow control

The behaviour of the closed loop when setpoint changes are produced can be seen in Figure 12.13.

The disturbances appearing in the flow are sometimes produced by changes in other flows of the installation and at other times by electrical noise produced by equipment (mainly robots) located nearby. The setpoint was changed from 4 to 6.3 litres per minute. The measured flow follows the step changes quite rapidly (taking into account the slow valve dynamics) with an overshoot of 13%. The manual valve at the cold water inlet was partially closed in order to introduce an external disturbance. As can be seen, the controller rejected the external perturbation quite rapidly.

12.2.4 Temperature Control at the Exchanger Output

The heat exchanger can be considered to be an independent process within the plant. The exchanger reduces the temperature of the recirculation water, driven by the pump, using a constant flow of cold water for this. The way of controlling the output temperature is by varying the flow of the recirculation water with the motorized valve V_8; thus the desired temperature is obtained by variations in the flow. In brief, the heat exchanger is nothing more than a tube bundle with hot water inside that exchanges heat with the exterior cold water. It can thus be considered as being formed of a large number of first-order elements that together act as a first-order system with pure dead time

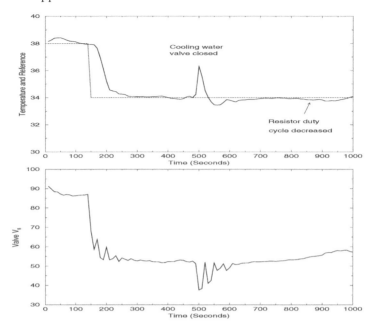

Fig. 12.14. Behaviour of the heat exchanger

(see Chapter 5). Thus the TT_4-V_8 system will be approached by a transfer function of this type.

Following the procedure used until now the system parameters which will be used for the control law are calculated. Some of the results obtained are shown in Figure 12.14. It should be born in mind that as the exchanger is not independent from the rest of the plant, its output temperature affects, through the tank, that of the input, producing changes in the operating conditions. In spite of this its behaviour is reasonably good.

The setpoint was changed from 38°C to 34°C. As can be seen in figure 12.14, the heat exchanger outlet temperature evolved to the new setpoint quite smoothly without exhibiting oscillations. Two different types of external disturbances were introduced. First the manual valve of the refrigerating cold water was closed for a few seconds. As was expected, the outlet temperature of the heat exchanger increased very rapidly because of this strong external perturbation and then it was taken back to the desired value by the GPC. The second perturbation is caused by decreasing the duty cycle of the resistor in the tank, thus decreasing the inlet hot water temperature and changing the heat exchanger operating point. As can be seen, the GPC rejects almost completely this perturbation, caused by a change in its dynamics.

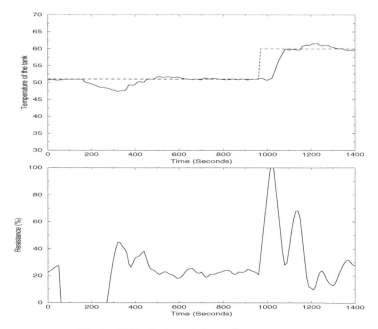

Fig. 12.15. Evolution of the tank temperature

12.2.5 Temperature Control in the Tank

The next example chosen is also that of a very typical case in the process industry: the temperature of the liquid in a tank. The manipulated variable in this case is the duty cycle of the heating resistor.

The process has integral effect and was identified around the nominal operating conditions (50°C). The following model was obtained:

$$G(s) = \frac{0.41}{s(1 + 50s)} e^{-50s}$$

The GPC was applied with a sampling time $T = 10$ seconds, $\lambda = 1.2$ and $N = 15$. As in the previous case, the controller parameters were computed by the formulas given in Chapter 5 for integrating processes. The results obtained are shown in Figure 12.15. A perturbation (simulating a major failure of the actuator) was introduced. As can be seen, after the initial drop in the temperature of the tank, caused by the lack of actuation, the control system is able to take the tank temperature to the desired value with a very smooth transient. A change in the setpoint from 50°C to 60°C was then introduced. The temperature of the tank evolves between both setpoints without big oscillations.

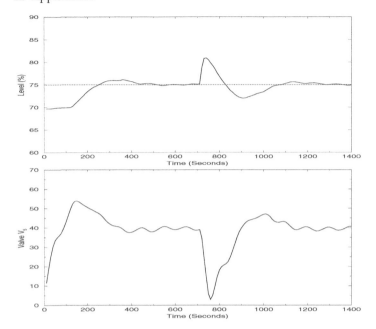

Fig. 12.16. Evolution of the level

12.2.6 Level Control

Level is another of the most common variables in the process industry. In the pilot plant, the level of the tank can be controlled by the input flows (cold or hot water). In this example, the valve V_5 will be used. The system was identified around the nominal operating point (70%) by the reaction curve method. The model transfer function is:

$$G(s) = \frac{1.12}{1 + 87s} e^{-45s}$$

For a sampling time $T = 10$ seconds, the dead time results to be non-integer, thus the controller parameters must be calculated as shown in Section 5.2. The results obtained when working with a weighting factor of 1 and a prediction horizon of 15 are shown in Figure 12.16. The setpoint was changed from 70 to 75%. As can be seen, the level of the tank moves smoothly between both setpoints. An external perturbation was introduced to test the disturbance rejection capabilities of the GPC. The perturbation consisted of opening the hot water inlet and thus increasing the level of the tank. As can be seen, the perturbation is rejected after a well-damped transient.

12.2.7 Remarks

The main objective of the control examples presented in this section was to show how easily GPC can be implemented on a commercial distributed control system using the implementation technique presented in Chapter 5. The GPCs were implemented without difficulty using the programming language (ITER) of the Integral Cube distributed control system.

Although comparing the results obtained by GPC with those obtained using other control techniques was not one of the objectives, GPC has been shown to produce better results than the traditional PID on the examples treated. In all the processes, the results obtained by PID, tuned by the Ziegler-Nichols open-loop tuning rules, were very oscillatory, better results were obtained [203] after a long commissioning period where *optimal* PID parameters were found. The commissioning of the GPC controllers was done in virtually no time; they worked from the word go. The results obtained by GPC were superior in all cases to the ones obtained by the PID controllers as reported in [203].

12.3 Model Predictive Control in a Sugar Refinery

This section shows an application of Precomputed GPC to a process in a sugar refinery. The implementation was carried out by the authors in collaboration with the firm PROCISA. The refinery is located in Peñafiel (Valladolid, Spain) and belongs to *Ebro Agricolas*. The controller runs in a ORSI Integral Cube control system, where the GPC has been included as a library routine which can be incorporated in a control system as easily as the built-in PID routine. The factory produces sugar from sugar beet by means of a series of processes such as precipitation, crystallization, etc. The process to be controlled is the temperature control of the descummed juice in the diffusion.

In order to extract the sugar from the beet it is necessary to dilute the saccharose contained in the tuber tissue in water to form a juice from which sugar for consumption is obtained.

The juice is obtained in a process known as diffusion. Once the beet has been cut into pieces (called chunks) to increase the interchangeable surface, it enters the macerator (which revolves at a velocity of 1 rpm) where it is mixed with part of the juice coming from the diffusion process (see Figure 12.17). Part of the juice inside the macerator is recirculated to be heated by means of steam and in this way it maintains the appropriate temperature for maceration. The juice from the maceration process passes into the diffusor (a slowly revolving pipe 25 m long with a diameter of 6 m) where it is mixed with water and all the available sugar content is extracted, leaving the pulp as a subproduct. The juice coming out of the diffusor is recirculated to the macerator, from which the juice already prepared is extracted for the next process.

For the diffusor to work correctly it is necessary to supply thermal energy to the juice during maceration. To obtain this objective, part of the juice from the macerator ($150 \text{ m}^3/\text{h}$) is made to recirculate through a battery of exchangers; within these the steam proceeding from the general services of the

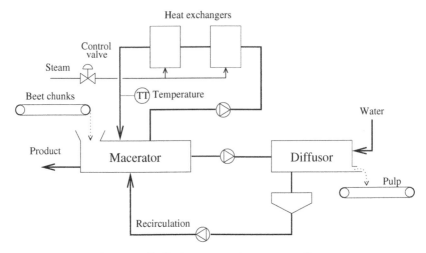

Fig. 12.17. Diffusion process in a sugar refinery

factory provides the heat needed to obtain optimum maceration. Therefore the controller must adjust the steam valve (u) to achieve a determined return temperature to the macerator (y).

The system response is seriously disturbed by changes in the steam pressure, which are frequent because the steam used in the exchangers has to be shared with other processes which can function in a noncontinuous manner.

The process is basically a thermal exchange between the steam and the juice in the pipes of the exchanger, with overdamped behaviour and delay associated to the transportation time of the juice through pipes about 200 metres long. These considerations, together with the observation of the development of the system in certain situations, justify the use of a first-order model with delay.

A model was identified by its step response. Starting from the conditions of 82.42 °C and the valve at 57 %, the valve was closed to 37 % in order to observe the evolution; the new stationary state is obtained at 78.61 °C. The values of gain, time constant and delay can easily be obtained from the response (as seen in previous examples):

$$K = \frac{82.42 - 78.61}{57 - 37} = 0.1905\,\frac{{}^0C}{\%} \qquad \tau = 5 \text{ min} \qquad \tau_d = 1 \text{ min } 45 \text{ s}$$

However, it is seen that the system reacts differently when heated to when cooled, the delay being much greater in the first case. A similar test changing the valve to 57 % again provides values of

$$K = 0.15 \qquad \tau = 5 \text{ min } 20 \text{ s} \qquad \tau_d = 4 \text{ min } 50 \text{ s}$$

Although an adaptive strategy could be used (with the consequent computational cost), a fixed parameter controller was employed, showing, at the

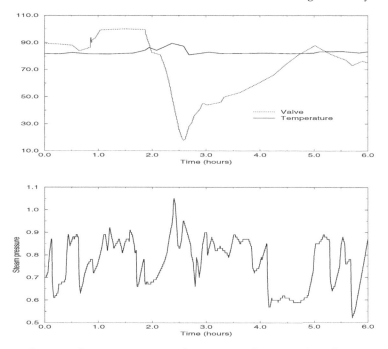

Fig. 12.18. System response in the presence of external disturbances

same time, the robustness of the method when using the T-polynomial in the presence of modelling errors. The error in the delay, which is the most dangerous, appears in this case. The following values of the model were chosen for this

$$K = 0.18 \qquad \tau = 300 \text{ s} \qquad \tau_d = 190 \text{ s}$$

and sampling time of T = 60 s.

It should be noticed that there are great variations in the delay (that produced on heating is about three times greater than that on cooling), due to which it is necessary to introduce a filter $T(z^{-1})$ as suggested in [210] to increase the robustness.

The following figures show various moments in operating the temperature control. The behaviour of the controller rejecting the disturbances (sudden variations in the steam pressure and macerator load) can be seen in Figure 12.18. On the other hand, Figure 12.19 shows the response to a setpoint change in the juice temperature.

It should be emphasized that this controller worked satisfactorily and without interruption until the end of the year's campaign, being handled without difficulty by the plant operators who continuously changed the values of the model throughout the operation time. Following many operational days the operators themselves concluded that a satisfactory model was given

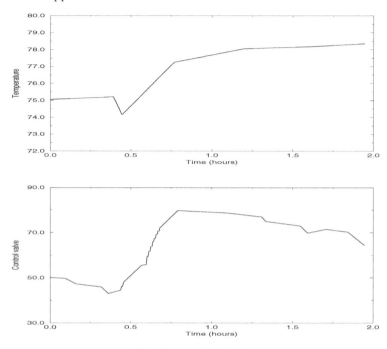

Fig. 12.19. Setpoint change

by:

$$K = 0.25 \qquad \tau = 250 \text{ s} \qquad \tau_d = 220 \text{ s} \qquad \lambda = 0.1$$

The sampling time was set to 50 seconds and the following robust filter was used

$$T(z^{-1}) = A(z^{-1})(1 - 0.8az^{-1})$$

with a equal to the discrete process pole.

12.4 Olive Oil Mill

This example describes the application of a predictive controller that deals with measurable disturbances in the extraction process in an olive oil mill. The application focuses on the thermal part of the process, where the raw material is prepared for the mechanical separation. The system under consideration can be viewed as composed of several changing level stirred tanks. The example shows the development of the controller based upon a model obtained from first principles combined with experimental results and validated with real data. Strong disturbances and large time delays appear in the process, so predictive control strategies have been tested under simulation and have been implemented on the real plant. A study about the consideration of different models for the estimation of measurable disturbances along

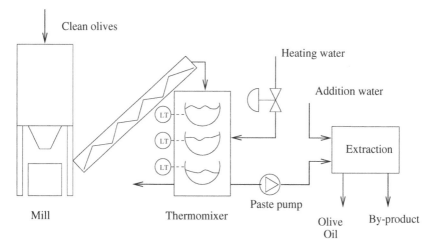

Fig. 12.20. Process

the prediction horizon shows that good performance can be obtained by use of an appropriate model. A new idea that can improve periodic disturbance rejection in Model Based Predictive controllers is also presented.

The process is composed of several operations: recollection and reception of raw material (olives), washing, preparation, extraction, and storage of the produced oil. Figure 12.20 shows the most important phases of the process: preparation and extraction.

The preparation phase is crucial for the whole process; it consists of two subprocesses. The first is olive crushing by a special mill, whose objective is to destroy the olive cells where oil is stored. The second aims at homogenising the paste by stirring it while its temperature is kept constant at a specific value (around 35 ° C). This is performed by a machine called a *thermomixer*, which homogenises the three phases of the paste (oil, water and a by-product) while it exchanges energy with surrounding pipes of hot water. This is done to facilitate oil extraction in the following process: mechanical separation in the *decanter*. This case study is focused on thermomixer control since homogenisation is crucial in the entire process. Bad operational conditions in the thermomixer can dramatically reduce the quality and quantity of the final product.

There are three main obstacles that appear when trying to maintain the optimal operating conditions in the thermomixer. The first is the existence of large delays (around one and a half hours) because of the thermal nature of the process. The predictive controller treats the delays in a convenient way. The second obstacle is caused by the on-off mechanism feeding the paste, so the inlet paste flow does not reach a constant value. These changes introduce periodic variations in level and therefore temperature changes since the quantity of product inside the machine varies. As the level can be easily measured, it can be considered as a measurable disturbance and hence can be taken into account by the predictive algorithm as a feedforward action. The third difficulty usually takes place at the beginning or the end of the campaign. The process is frequently interrupted because of the heterogeneity and low quantity (and often, low quality) of raw material. When the process is stopped and restarted, the temperature inside the thermomixer increases rapidly.

12.4.1 Plant Description

The system considered corresponds to a thermomixer, whose main objective is to homogenise the three phases of the paste (oil, water and the by-product) and keep it at a desired temperature to facilitate oil extraction. Heating the paste is achieved by means of hot water circulating through a jacket. The machine is divided into different (usually two, three or four) tanks or bodies, each with revolving blades to facilitate homogenising. The bodies are composed of semicylinders about 3 metres long with a diameter of 1 metre. Paste is dropped over one side of the first body and pushed by the revolving blades which make the paste fall down to the second body through the overflow and so on. The existence of several bodies allows a gradual temperature increment along the thermomixer, since abrupt changes in paste temperature would affect the quality of the end product. Each body has its own water jacket. The water circuit is connected in parallel, so the same quantity of water flows for each jacket. We can only set the total water flow (which is shared equally amongst all the bodies).

The paste is heated to facilitate mixing since the paste turns more liquid as the temperature rises. However, there exists a maximum temperature at which olive oil loses quality (flavour, fragrance, etc.) due to the oxidation process and the loss of volatile components. The heating water comes from a boiler that supplies hot water to several processes in the factory, so it is affected by load changes and is therefore another disturbance.

Therefore, the outlet temperature presents oscillations at the frequency of level variations with changes produced by heating water variation. The controller must be able to reduce the effect of these disturbances as much as possible. Level and water temperature can easily be measured and future level evolution can be predicted as shown later. Figure 12.21 shows level evolution in the last body and the random variations of inlet water temperature. It is clear that a more efficient performance could be obtained by a good design in the feeding system that keeps the level constant; this is the kind of level control system that is installed in most olive oil mills for cost reasons. Therefore, to address the current problem, this disturbance has been considered as something external to the proposed control solution.

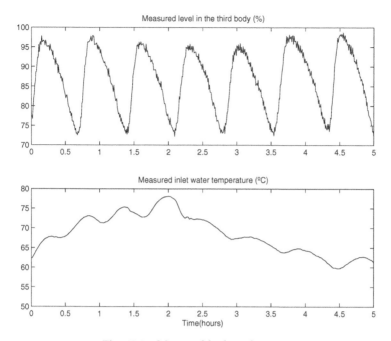

Fig. 12.21. Measurable disturbances

12.4.2 Process Modelling and Validation

The attainment of the process model is described in [32], where thermody-namical equations together with input-output data allows construction of a nonlinear simulator that can be used to test the controller. The model was validated using real data obtained from an olive oil mill located in Málaga (Spain). These data were used to estimate many of the parameters that appear in the model which are not perfectly known, since they depend on several circumstances: type and moisture of olives, soil in the heating circuit, etc.

A linear model must be developed to design a simple linear predictive controller that will run on a Programmable Logic Controller (PLC) with low computational capabilities. The linear model is obtained by performing tests in the nonlinear model, where all manipulated variables can be changed independently to see their influence on temperature behaviour. The models needed for control must predict correctly the final paste temperature (the paste that leaves the last body of the thermomixer) as a function of hot water flow, level, and temperature of the heating water.

For the CARIMA model, numbers that give temperature as a function of the hot water flow are

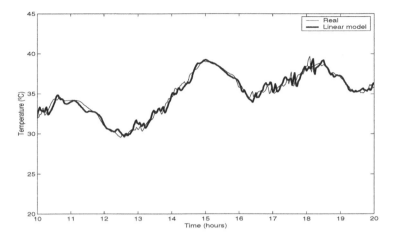

Fig. 12.22. Validation linear prediction model

$$B(z^{-1}) = 4.315 \cdot 10^{-5} z^{-5}$$
$$A(z^{-1}) = 1 - 0.92 z^{-1}$$

temperature with respect to level:

$$B(z^{-1}) = -0.0028 z^{-3}$$
$$A(z^{-1}) = 1 - 0.8825 z^{-1}$$

and temperature with respect to water temperature

$$B(z^{-1}) = 0.0093 z^{-3}$$
$$A(z^{-1}) = 1 - 0.926 z^{-1}$$

The model used in simulation has a sample time of 100 seconds. Figure 12.22 shows the predictions of temperature using a linear model with a prediction horizon of seven, which is the one that will be used by the predictive controller, compared with real data. This corresponds to a new set of data not used to build the model.

12.4.3 Controller Synthesis

The control objective is to maintain the operating conditions in the thermomixer, that is equivalent to keep the temperature of the last body as constant as possible in spite of disturbances (level and variations in hot water temperature). The manipulated variable is the hot water flow.

The process is characterized by big dead times in temperature dynamics. Moreover, the effect of disturbances, mainly level, on the controlled variable shows fast dynamics (mainly at high production rates). This fact makes disturbance rejection more difficult and implies the necessity of a control strategy that eliminates the effect of level disturbance, at least in the nominal case.

MPC can be an interesting candidate to control this system and disturbance rejection capabilities can be improved by the estimation of the future disturbance values in the controlled variable prediction model. That is, the current value of the disturbance is known at the sampling time, but its future evolution along the horizon can also be estimated since it affects the process predicted output (temperature). This is a slight improvement with respect to standard MPC algorithms, which consider that disturbances are kept constant (and equal to their current value) in the future. The information that provides this future evolution is very important in this case, allowing the controller to anticipate its influence on the process output.

In this application, as the main disturbance acting on the output (level) exhibits a predictable behaviour, the control law is calculated considering an Auto-Regressive (AR) model of disturbance. Periodic disturbance rejection is also treated in [29]. Therefore the AR model is not used to estimate the actual values of the disturbances, but utilized as an instrument to improve the prediction of future values of the disturbances. So the prediction of the controlled variable must also be enhanced. The identified AR polynomial is:

$$A(z^{-1}) = 1 - 1.589z^{-1} + 1.696z^{-2} - 1.178z^{-3} + 0.697z^{-4}$$

When a disturbance model is included in the prediction, the control law that minimizes the general cost function

$$J = \sum_{i=1}^{p} [\hat{y}(t+i) - w(t+i)]^2 + \lambda \sum_{i=0}^{m-1} [\Delta u(t+i)]^2 \qquad (12.5)$$

is given by (see [31])

$$\mathbf{u} = (\mathbf{G}^T\mathbf{G} + \lambda\mathbf{I})^{-1}\mathbf{G}^T \left[\mathbf{w} - (\mathbf{f}_u + \hat{\mathbf{f}}_d)\right] \qquad (12.6)$$

where

- $\mathbf{u}$ is the vector of future control action increments;
- $\mathbf{f}_u$ is the calculated free response without disturbances;
- $\mathbf{w}$ is the reference trajectory and
- $\hat{\mathbf{f}}_d$ is the estimated value of the free response due to measurable disturbances and can be calculated using the following equation

$$\hat{f}_d(t+i) = \sum_{k=0}^{i-1} g_k \Delta\hat{d}(t+i-k) + \sum_{k=i}^{N} g_k \Delta d(t+i-k) \qquad (12.7)$$

where g_k are the samples of the truncated step response and $\Delta d(t)$ is the increment of the disturbance signal at time t and the AR prediction is given by

$$\hat{d}(t) = \sum_{i=1}^{N} a_i d(t - i) \tag{12.8}$$

The terms that depend on future values of the disturbances are separated from the directly measured ones. The first part in the sum is assumed to be zero in standard MPC formulations. This means that the polynomial AR is equal to $1 - z^{-1}$, which is equivalent to the assumption that future values of disturbance equal its current value.

Therefore, the control algorithm (MPC with Auto-Regressive model for the measurable disturbances or MPC-AR) is reduced to the following:

1. Compute the free response as in a basic MPC algorithm.
2. Add the term due to measurable AR disturbance as shown in (12.7 - 12.8).
3. Calculate the control law with Equation (12.6).

The controller presented here has been devised to be implemented in low-cost control equipment, as is the case of a PLC. This can be done since only part of the algorithm is calculated in real time. The most demanding part of MPC (the optimization that corresponds to Equation (12.6)) is done beforehand, since the model is known and the only part to be calculated at every sampling time is the free response of the system. Should the model change with time, the control law parameters will be updated by a higher-level routine running on the SCADA computer.

12.4.4 Experimental Results

Several tests have been done on the real plant. The experiments have been carried out in different situations since it is very difficult to obtain the same operational conditions in different days because of the continuously changing raw material. Several tunings of the controller (using different weighting factors and different disturbance models) have been tested. The controller runs on a PLC, where the control law can be switched to an existing PID controller for comparison. The PID was tuned manually by trained personnel: see [32].

Table 12.4 shows the root mean square error of the temperature, comparing a PID with feedforward to the proposed MPC. Results with changes in temperature setpoint are also presented.

Table 12.4. Fits obtained with PID and proposed MPC in real plant

Error ($^\circ$ C)	PID	MPC-AR
Regulation	2.1	0.5
Tracking	2.9	0.9

Figure 12.23 shows the controlled temperature in the last body when the setpoint changes in real tests. MPC-AR is able to reach the set point faster

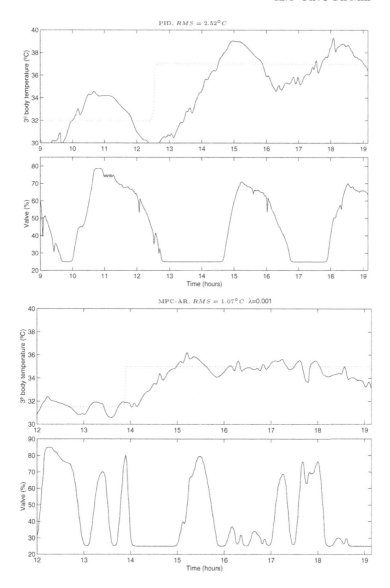

Fig. 12.23. Changes in the setpoint. PID control (two upper graphs) and MPC-AR control (two lower graphs).

than PID, which ever continuously oscillates around the setpoint. The control action of PID is obviously slower, and it cannot be accelerated much more without losing stability.

Figure 12.24 presents two different tests of the designed MPC in the real plant, performed during the intermediate dates of the campaign, in which

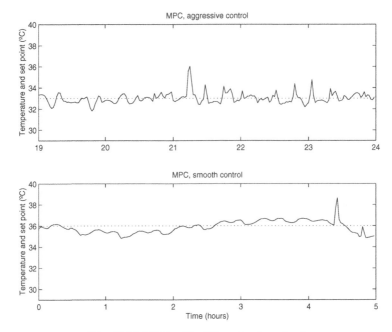

Fig. 12.24. MPC with different values for λ

the process is often stabilized. The first corresponds to a controller with a small value for the control weighting factor (λ) and no measurable disturbances considered. The second shows the same controller with a bigger value of λ, giving a slower control. The λ parameter defines the *aggressiveness* of the controller. If λ is set to a very small value, the closed-loop system behaves faster, but it loses robustness. In this application, the fits achieved in both cases are similar, because a faster control induces more high-frequency noise. Although the fit in the first case is slightly better, it is preferable to have a more robust and smoother control.

The biggest benefits of the inclusion of AR model for measurable disturbances can be obtained at the end of the olive season, when the operational conditions are not stationary and effect level evolution. In the first case (Figure 12.25, top graph) the behaviour of the temperature is not good, with considerable oscillations around the setpoint, although it is the same controller that has shown good performance during the intermediate dates of the campaign. The performance is clearly improved with the AR model (Figure 12.25, bottom graph), under the same conditions (similar evolution of level and water temperature), showing that the proposed algorithm is a viable solution to the problem.

Plant results have shown that an MPC considering the prediction of future level variations can be a recommendable solution for the problem that exists in the real plant.

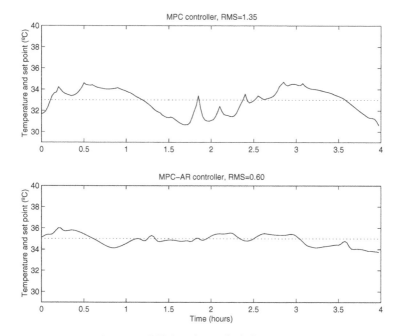

Fig. 12.25. MPC at the end of olive season

12.5 Mobile Robot

This section shows the application of NMPC to mobile robot navigation. This implementation tries to solve one of the main problems in the development of autonomous mobile robots: the problem of path tracking in an environment with unexpected obstacles. MPC is a suitable technique to apply to this problem because the objective is to drive future outputs (robot position and heading) close to the desired values in some way, bearing in mind the control activity required to do so.

If the control signal is constrained and the system model is nonlinear the MPC results in a very complex and time-consuming problem. A neural network is used to solve the problem, as described in Section 9.4.

12.5.1 Problem Definition

The problem of driving a mobile robot to follow a previously calculated desired path has been addressed in [77] where an NN is used to solve the NMPC problem online with the following objective function

$$J_1(N_1, N_2, N_u) = \sum_{i=N_1}^{N_2} [\hat{Y}(t+i|t) - Y_d(t+i)]^2$$

$$+ \sum_{i=1}^{N_u} \lambda_1([\Delta\omega_r(t+i-1)]^2 + [\Delta\omega_l(t+i-1)]^2) +$$

$$+ \sum_{i=1}^{N_u} \lambda_2[\omega_r(t+i-1) - \omega_l(t+i-1)]^2$$

where $\hat{Y}(t+i|t) = \{\hat{x}(t+i|t), \hat{y}(t+i|t)\}$ is an i step prediction of the robot position made at instant t; ω_r and ω_l are the right and left angular velocities of the two driving wheels, which are the control variables; and λ_1, λ_2 and ψ are constant weighting factors. The first term in J penalizes the position error; the second term penalizes the acceleration and the third penalizes the robot angular velocity. These last two terms ensure smooth robot guidance.

When unexpected static obstacles are taken into account, a new term must be added to the cost function in order to penalize the proximity between the robot and the obstacles, which are detected with an ultrasound proximity system placed on board the mobile robot. Therefore, the function to be minimized now is:

$$J(N_1, N_2, N_u) = J_1(N_1, N_2, N_u) + \sum_{j=1}^{NFO} \left(\sum_{i=N_1}^{N_2} \frac{\psi}{[\text{dist}_f(\hat{Y}(t+i|t), FO_j)]^2} \right)$$

The new term is a potential function term, where $\text{dist}_f(\cdot)$ is a measurement in $t+i$ of the distance between the robot and a fixed obstacle FO_j, which is considered to have a polygonal geometry in the plane. A more precise description of this function is presented here. A block diagram of the system is shown in Figure 12.26. Notice that the consideration of unexpected obstacles makes the objective function not quadratic and thus the computational burden is increased.

In the following N_1 and N_2 will be considered to be $N_1 = d + 1$ and $N_2 = N$, and N_u will be given a value of $N_2 - d$. So the controller has only one free parameter N. The predictive problem, formulated under these circumstances, has to be solved with numerical optimization methods, which are not acceptable for real-time control. The controller is implemented using a neural network scheme, which allows real time.

12.5.2 Prediction Model

For an MBPC formulation, a model of the mobile platform is needed to predict the future positions and headings of the robot. As a testbed for the experiments, a TRC LABMATE mobile robot has been used.

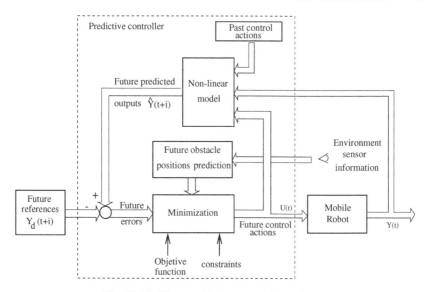

Fig. 12.26. The predictive controller scheme

A model of the LABMATE mobile robot, which takes into account the dead time produced by communication with the host processor, was obtained by using kinematic equations and identification tests. The following kinematic model (which corresponds to a differential-drive vehicle) is used for computing the predictions

$$\theta(k+1) = \theta(k) + \mathcal{A}T$$

$$x(k+1) = x(k) + \frac{V}{\mathcal{A}}(\sin(\theta(k) + \mathcal{A}T) - \sin(\theta(k)))$$

$$y(k+1) = y(k) - \frac{V}{\mathcal{A}}(\cos(\theta(k) + \mathcal{A}T) - \cos(\theta(k)))$$

$$\mathcal{A} = R\frac{\omega_r(k-1) - \omega_l(k-1)}{2W}$$

$$V = R\frac{\omega_r(k-1) + \omega_l(k-1)}{2}$$

where x, y and θ are the position and heading of the robot in a fixed reference frame (see Figure 12.27), T is the sample interval and W is the half-distance between wheels, which value has been estimated to be 168 mm (Figure 12.27). V is the linear velocity of the mobile robot, $\mathcal{A}$ is the steering speed, and $\omega_r(k-1)$, $\omega_l(k-1)$ and R are the right and left wheel angular velocities (which are considered to be constant for each sample interval) and the wheel radius, respectively. In the case of a linear trajectory ($\mathcal{A} = 0$), the equations of motion are given by:

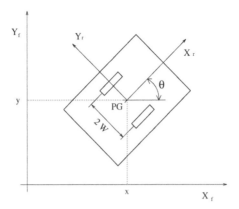

Fig. 12.27. Reference frame

$$\theta(k+1) = \theta(k)$$
$$x(k+1) = x(k) + VT \cos \theta(k)$$
$$y(k+1) = y(k) + VT \, sen \, \theta(k)$$

Using the maximum acceleration value, the velocities of both wheels have been considered to be constant for each sample period.

12.5.3 Parametrization of the Desired Path

The reference path is given to the MPC controller as a set of straight lines and circular arcs. The MPC approach needs the desired positions and headings of the mobile platform at the next N time instants. So, given the current position and heading of the robot, it is necessary to parameterize the desired path for the next N periods of time to calculate the N future path points desired. As is shown in Figure 12.28, the desired point for the current instant $(X_d(k), Y_d(k))$ is obtained first. It is located at the intersection between the desired path and its perpendicular, traced from the actual robot position $(X_r(k), Y_r(k))$. The next N points are spaced equally on the path, with a separation between them of ΔS, which is a design parameter.

12.5.4 Potential Function for Considering Fixed Obstacles

As stated earlier, the fixed obstacles are considered to have polygonal geometry in the plane, and the surfaces of the obstacles are considered to be perpendicular to the moving plane of the mobile robot.

Two different potential functions have been used for the polygonal geometry case: one for the convex polygon and the other for the concave polygon:

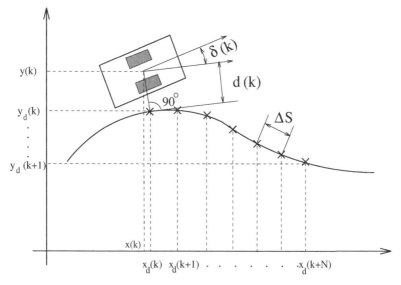

Fig. 12.28. Parametrization of the desired path

- potential function for a convex polygon [94]. A convex region will be described by a set of inequalities

$$g(x) \leq 0, \quad g \in L^m, \ x \in R^n$$

where L is the set of linear functions and n is the space dimension (in this case $n = 2$). The function

$$f(x) = \sum_{i=1}^{NSF} g_i(x) + |g_i(x)|$$

is zero inside the convex region and increases linearly out of it, as the distance to the frontier is augmented. NSF is the number of segments of the obstacle frontier. The following potential function is used

$$p_{cvx}(x) = [\delta + f(x)]^{-1} = \cfrac{1}{\delta + \sum\limits_{i=1}^{NSF} (g_i(x) + |g_i(x)|)}$$

where δ is a small constant that limits the value of $p_{cvx}(x)$ inside the convex region; $p_{cvx}(x)$ reaches its maximum value δ^{-1} inside the region occupied by the obstacle, and decreases with the distance between the robot and the obstacle. A graphic example of this function is shown in Figure 12.29, where two rectangular static obstacles are present in the proximity of the robot.

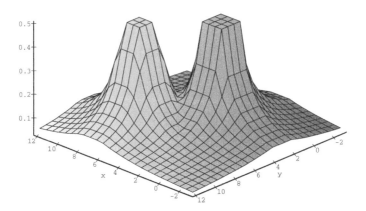

Fig. 12.29. Convex regions potential function

- potential function for a concave polygon. A concave region will be described by a set of inequalities

$$v(x) \geq 0, \quad v \in L^m, \, x \in R^n$$

The potential function used in this case is

$$p_{ccv}(x) = \frac{1}{\delta + g_{ccv}(x)}$$

where δ is a small constant and $g_{ccv}(x)$ is the minimum of the distances between the robot position and every straight line that defines the obstacle frontier. This function has the same characteristics as $p_{ccx}(x)$.

12.5.5 The Neural Network Approach

As was mentioned before, the minimization of the cost function J has to be carried out by a numerical optimization method which requires too much computation to be used in real time. A neural network solution is proposed, which guarantees real time for the robot control.

  The modules of the control scheme used in this work (see Figure 12.30) are:

- **artificial neural network controller**. The NN architecture chosen here is a Multilayer Perceptron, with one hidden layer (see Figure 12.31).
  The input layer consists of twelve neurons. The first two inputs correspond to the previous linear and angular velocities of the robot. The next

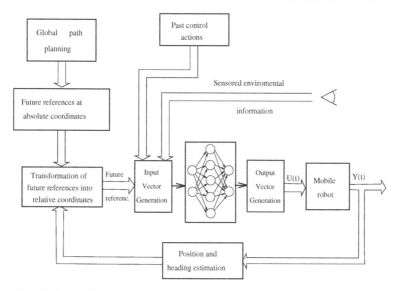

Fig. 12.30. Predictive neural network scheme for mobile robot navigation

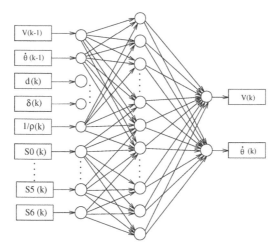

Fig. 12.31. Neural network scheme

three inputs are associated to the parameterization of the desired trajectory over the prediction horizon. To reduce the number of inputs, the parameters given to the network are the distance d from the robot guide point to the path, the angle δ between the robot heading and the path orientation and an average of the inverse of the curvature of the future desired points $(1/\rho)$ (see Figure 12.28). The last seven inputs correspond to the distances measured directly by the ring of sonar sensors. This fact avoids the high-level process that usually has to be carried out with sen-

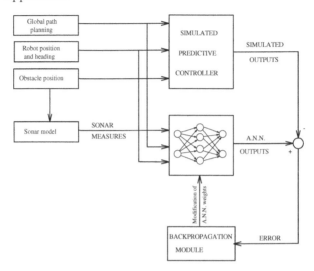

Fig. 12.32. ANN supervised training scheme

sor data to provide useful environment information, which is a difficult and high computation time-consuming phase. The output layer consists of two nodes which correspond to the control commands (the linear and angular robot velocities).

- **input vector generation module**. A symmetry analysis is made here to reduce the number of training patterns needed to provide good performance of the neural network controller [78]. Also a normalization is made which leads to better performance at the NN training stage.
- **output vector generation module**. This performs the inverse transformation of that made at the symmetry input module when required.
- **reference path coordinates transformation module**. The desired path coordinates are transformed from a global reference system to a local reference system, attached to the mobile robot. This avoids the use of additional NN inputs for the robot position and heading, which are implicitly given to the NN in the reference path.
- **past control actions**. These are needed for NN to consider the delay time of the robot system.
- **sonar range measurements**. Their measurements are directly used as inputs for NN. ANN learns from the input patterns set, where different situations of static obstacles are present. Thus, there is no need for a high-level sonar measurement preprocess, which guarantees real time.

12.5.6 Training Phase

The controller has been trained using a classic supervised training scheme as the backpropagation algorithm (see Figure 12.32). The training patterns set have been obtained from an offline simulation system. For the minimization phase a Powell iterative algorithm has been used, where constraints on the control variables are considered. Also, the sonar system measurements have been simulated using a sonar model where the objects in the environment are described as a set of geometric primitives such as planes, cylinders, edges and corners.

12.5.7 Results

The proposed control structure has been tested with the LABMATE mobile robot.

The sampling interval T was given a value of 2 s. The value of N chosen for the MPC was made equal to 7, thus N_1, N_2 and N_u were given the values 2, 7 and 6, respectively, and the weighting factors were given the following values: $\lambda_1 = 35$, $\lambda_2 = 5$ and $\psi = 0.5$. The maximum and minimum linear and angular velocities were given the following values, respectively: 0 m/s, 0.4 m/s, -20 °/s and 20 °/s. For Δs, a value of 0.15 m was chosen, which leads to an average linear robot velocity of 0.25 m/s.

Figure 12.33 shows some of the experimental results obtained in the laboratory when applying the proposed algorithm to the LABMATE mobile robot. Although it is drawn as straight lines in the figures, the real environment includes laboratory objects such as chairs, tables, etc. This shows the robustness of the neural controller as it has been trained with only a set of geometric primitives such as planes, corners, etc.

Figure 12.33(a) shows the desired trajectory and the real trajectory through the laboratory where unexpected static obstacles have been positioned. It is important to notice how the mobile robot returns to the reference path after an unexpected obstacle has been avoided.

Figure 12.33(b) shows an experiment where an unexpected obstacle is situated in the path that the robot must follow to avoid a previous unexpected obstacle. Again, the controller performance is quite good. Finally, Figure 12.33(c) shows another test for a path where small curvature radii are specified. The tracking error observed in Figure 12.33(c) is due to saturation in the angular velocity and to the penalization chosen for the control actions.

More details about the implementation of the controller can be found in [78], [79].

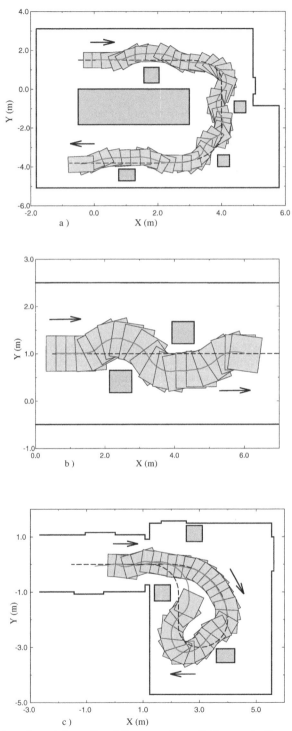

Fig. 12.33. Experimental results of the neural predictive controller for mobile robot navigation

A

Revision of the Simplex Method

The Simplex method [129] is the algorithm most used for solving linear programming problems, such as the ones appearing when using a 1-norm MPC. The Simplex algorithm finds successive and better feasible solutions at the extreme points of the feasible region until it stops, in a finite number of steps, either at the optimum or finds that the optimal solution is not bound by the feasible region. This appendix is dedicated to revising the basic ideas behind the Simplex method.

A.1 Equality Constraints

The problem of minimizing a linear function subject to equality constraints will be considered first

$$\text{Minimize } c^t x$$
$$\text{subject to } Ax = b \tag{A.1}$$
$$x \geq 0$$

where A is a $q \times p$ real matrix with $q < p$ and full rank.

If the equality constraint equation is multiplied by a matrix T and the columns of A and corresponding components of x are interchanged in such a way that $T A = [I_{q \times q} \ N]$ and $T b = \bar{b}$, the point $x = [x_b^t \ x_n]^t = [\bar{b}^t \ 0]$ is a basic feasible solution. The components x_b are called *basic variables* whereas the remaining components (corresponding to N) are called *nonbasic variables*. Note that this can be done by applying elementary row transformations to matrix A and interchanging the columns of A (and the corresponding x variables) to take matrix A to the form $[I \ N]$. If the same transformations are applied to I and b, matrix T and vector $\bar{b}$ are obtained.

The objective function takes the value $z_0 = c_b^t x_b + c_n^t x_n = c_b^t x_b$. The basic variables can be expressed as a function of the nonbasic variables from the transformed constraint equation:

$$x_b = \bar{b} - N x_n$$

by substituting in the cost function

$$z = c^t x = c_b^t(\bar{b} - N x_n) + c_n^t x_n = z_0 + (c_n^t - c_b^t N) x_n$$

As $x_n \geq 0$, the objective function decreases if any component of $(c_n^t - c_b^t N)_i$ is negative and the corresponding nonbasic variable x_{n_i} increases. This gives an indication of how to obtain a more feasible solution and is the basic idea behind the algorithm. The problem is determining which of the nonbasic variables should be increased (become basic) and which of the basic variables should leave the basis.

This is done as follows:

1. Find an initial basic solution.
2. Form the following tableau:

	x_b^t	x_n^t	
x_b	I	N	$\bar{b}$
J	0	$c_n - c_b^t N$	$c_b^t \bar{b}$

3. If $c_n - c_b^t N \geq 0$ then STOP, the actual basic solution is optimal.
4. Choose one of the negative elements (say, the jth) of row $c_n - c_b^t N$ (the most negative is usually chosen).
5. Choose i such that the ratio $\bar{b}_i / N_{ij}$ is the minimum for all $N_{ij} > 0$. If there are no nonnegative elements in that column of the tableau, the problem is not bounded.
6. Make x_{n_j} a basic variable and x_{b_i} a non basic variable by *pivoting*:
   a) Divide the ith row of the tableau by N_{ij}.
   b) Make zero the remaining elements of the jth column of the nonbasic variable block by multiplying the resulting the ith row by N_{kj} and subtracting it from the kth row.
7. Go to step 2.

A.2 Finding an Initial Solution

The Simplex method starts from an initial feasible extreme point. An initial point can be found by applying elementary row transformations to matrix A and vector b and interchanging the columns of A (and the corresponding x variables) to take matrix A to the form $[I\ N]$.

A solution can be obtained by using the Simplex algorithms in different ways. One way, known as the two-phase method, consists of solving the following augmented system:

$$\begin{aligned}
& \text{minimize } 1^t x_a \\
& \text{subject to } x_a + Ax = b \\
& \qquad\qquad x \geq 0, \quad x_a \geq 0
\end{aligned} \tag{A.2}$$

Note that the constraint matrix is now $[I A]$ and that the obvious solution $x = 0$ and $x_a = b$ is an extreme point of the augmented problem. The variables x_a are called *artificial* variables and are all in the basis for the initial solution. If the algorithm does not find a solution with $x_a = 0$, the problem is not feasible. Otherwise, the solution constitutes an initial solution to the original problem and the second phase of the algorithm can be started for the original problem with this solution.

Another way of dealing with initial conditions, known as the big-M method [15], solves the whole problem in only one phase. Artificial variables are also introduced as in the two-phase method but a term is added to the objective function penalizing the artificial variables with high weighting factors in order to force artificial variables out of the basis. The problem is now stated as:

$$\text{minimize } c^t x + m^t x_a$$
$$\text{subject to } x_a + Ax = b \qquad \qquad \text{(A.3)}$$
$$x \geq 0, \quad x_a \geq 0$$

If all the artificial variables are out of the basis at termination, a solution to the optimization problem has been found. Otherwise, if the variable entering the basis is the one with the most positive cost coefficient, we can conclude that the problem has no feasible solution.

A.3 Inequality Constraints

Consider the problem:

$$\text{minimize } c^t x$$
$$\text{subject to } Ax \leq b \qquad \qquad \text{(A.4)}$$
$$x \geq 0$$

The Simplex method can be used to solve inequality constraint problems by transforming the problem into the standard format by introducing a vector of variables, $x_s \geq 0$, called *slack* variables, such that the inequality constraint $Ax \leq b$ is transformed into the equality constraint $Ax + x_s = b$. The problem can now be stated in the standard form as:

$$\text{minimize } c^t x$$
$$\text{subject to } [A \ I] \begin{bmatrix} x \\ x_s \end{bmatrix} = b \qquad \qquad \text{(A.5)}$$
$$\begin{bmatrix} x \\ x_s \end{bmatrix} \geq 0$$

The number of variables is now $q + p$ and the number of constraints is q. Notice that the form of the equality constraint matrix is $[A \ I]$ and the point $[x^t \ x_s^t] = [0 \ b^t]$ is an initial basic solution to this problem.

Duality

The number of *pivoting* operations needed to solve a standard linear programming problem is on the order of q [129], while the number of floating-point operations needed for each pivoting operation is on the order of $q \times p$.

The number of floating-point operations needed to solve a standard LP problem is therefore on the order of $(q^2 + q \times p)$. For the inequality constraint problem, the number of variables is increased by the number of slack variables which is equal to the number of inequality constraints. In the linear programming problems resulting from robust MPC, all of the constraints are inequality constraints, so the number of operations needed is on the order of $q(q \times (q + p)) = q^3 + q^2 p$. That is, it is linear in the number of variables and cubic in the number of inequality constraints. The problem can be transformed into an LP problem with a different and more convenient structure using duality.

Given the problem (*primal*)

$$\begin{aligned}
& \text{minimize } c^t x \\
& \text{subject to } Ax \leq b \\
& \qquad\qquad x \geq 0
\end{aligned} \tag{A.6}$$

the *dual* problem is defined as [129]:

$$\begin{aligned}
& \text{minimize } -b^t \lambda \\
& \text{subject to } -A^t \lambda \leq -c \\
& \qquad\qquad \lambda \geq 0
\end{aligned} \tag{A.7}$$

The number of operations required by the *dual* problem is on the order of $p(p \times (p + q)) = p^3 + p^2 q$. That is, it is cubic in the number of variables and linear in the number of inequality constraints. So, for problems with more inequality constraints than variables, solving the *dual* problem will require less computation than the primal.

The solutions to both problems are obviously related (Bazaraa and Shetty [15], theorem 6.6.1.). If x_o and λ_o are the optimal of the *primal* and *dual* problem then, $c^t x_o = b^t \lambda_o$ (the cost is identical) and $(c^t - \lambda_o^t A)x_o = 0$. If the primal problem has a feasible solution and has an unbounded objective value the *dual* problem is unfeasible. If the *dual* problem has a feasible solution and an unbounded objective value the *primal* problem is unfeasible.

B

Dynamic Programming and Linear Quadratic Optimal Control

Model Predictive Control is closely related to Linear Quadratic (LQ) Optimal Control. This appendix shows the main characteristics of LQ and its relation to MPC.

Dynamic Programming can be used to find the solution of the LQ problem, since it provides an efficient means for sequential decision making. It is based on Bellman's principle of optimality [16], which states that an optimal policy has the property that whatever the initial state and the initial decision are, the remaining decisions must constitute an optimal policy with regard to the state resulting from the first decision. Notice that a *decision* is the control action at a particular time instant while *policy* is equivalent to the control sequence.

If the goal is to move the process from an initial state to a final state with minimum cost, it is clear that the optimal solution can be obtained calculating the cost associated to every possible route and choosing the route with minimum cost. This implies an evaluation of all possible alternatives. However, the problem can be solved by defining the cost associated to a particular state as the sum of two terms: the part attributable to the current decision and the part representing the minimum value of all future costs, starting with the state which results from the first decision.

The principle of optimality replaces a choice among all alternatives by a sequence of decisions among fewer alternatives. Dynamic Programming allows one to concentrate on a sequence of current decisions rather than being concerned about all decisions simultaneously.

B.1 Linear Quadratic Problem

When the cost is quadratic and the system is linear, the problem can be solved analytically and the controller results in a linear state feedback.

The process is modelled by

$$x(t + 1) = Ax(t) + Bu(t) \tag{B.1}$$

with **x**(0) known, and the objective is to find the control sequence $u(0)$, $u(1)$, ..., $u(N - 1)$ that drives the process from the initial to the final state minimizing the cost given by

$$J = x(N)^T Q_N x(N) + \sum_{k=0}^{N-1} x(k)^T Q_k x(k) + u(k) R_k u(k)$$

where Q_k are symmetric positive semidefinite matrices and $R_k > 0$.

The procedure to obtain the control sequence is based on solving the problem in reverse sense, that is, start computing $u(N - 1)$ and finishing with $u(0)$. Let us define I_1^* as the optimal cost of the last stage (from state $x(N - 1)$ to the end), which is expressed as

$$I_1^*(x(N - 1)) = \min_{u(N-1)} x(N)^T Q_N x(N) + u(N - 1) R_{N-1} u(N - 1)$$

and can be computed analytically deriving with respect to $u(N - 1)$, giving:

$$u(N - 1) = -(B^T Q_N B + R)^{-1} B^T Q_N A x(N - 1) = K_{N-1} x(N - 1) \tag{B.2}$$

Therefore, the control action is a linear feedback of the state vector. The last stage cost is then given by:

$$I_1^* = (Ax + BK_{N-1}x)^T Q_N (Ax + BK_{N-1}x) + x^T K_{N-1}^T R_{N-1} K_{N-1} x$$

Defining

$$P_{N-1} = (A + BK_{N-1})^T Q_N (A + BK_{N-1}) + K_{N-1}^T R_{N-1} K_{N-1}$$

this cost can be written as a quadratic form of the state:

$$I_1^* = x(N - 1)^T P_{N-1} x(N - 1)$$

At this point Bellman's optimality principle is used to calculate the next cost:

$$I_2^* = \min_{u(N-2)} x(N - 1)^T Q_{N-1} x(N - 1) + u(N - 2) R_{N-2} u(N - 2) + I_1^*$$

$$= \min_{u(N-2)} x(N - 1)^T (Q_{N-1} + P_{N-1}) x(N - 1) + u(N - 2) R_{N-2} u(N - 2)$$

As $x(N - 1)$ can be expressed as a function of $x(N - 2)$ and $u(N - 2)$ using (B.1), I_2^* depends on these last values and the optimal control action can be obtained analytically in the same fashion as (B.2); that is,

$$u(N - 2) = K_{N-2} x(N - 2)$$

This procedure can be extended to all the states leading to the general expression of the control law

$$u(k) = K_k x(k) \text{ with } K_k = -(B^T P_{k+1} B + R)^{-1} B^T P_{k+1} A \qquad \text{(B.3)}$$

and the symmetric semidefinite matrix P_k is given by:

$$P_k = (A + BK_k)^T P_{k+1}(A + BK_k) + K_k^T R_k K_k + Q_k$$

which after a few manipulations is transformed into:

$$P_k = A^T P_{k+1} A + A^T P_{k+1} B K_k + Q_k \qquad \text{(B.4)}$$

This is called the *discrete-time Riccati equation* and can be used to compute the value of P_k recursively starting with $P_N = Q_N$.

The problem is solved backwards, starting at time N and calculating $u(k)$ using (B.3) and matrix P from (B.4).

As the controller is a linear feedback of the state, the use of a state estimator or observer is required to compute the control action. If the observer is a Kalman filter, then it gives rise to the well-known control strategy called Linear Quadratic Gaussian (LQG). A detailed study of the relationship between MPC (especially GPC) and LQG can be found in [27].

B.2 Infinite Horizon

In some situations it is justifiable to assume that the terminal time is infinitely far in the future. This so-called infinite horizon case leads to a constant feedback gain matrix, which can be calculated from (B.4) considering that the weighting matrices are constant. Then $P_k \rightarrow P_\infty \geq 0$ and is calculated using the *algebraic Riccati equation*:

$$P_\infty = A^T P_\infty A + A^T P_\infty B K_\infty + Q$$

Now the control action becomes the constant state feedback law:

$$u(k) = K_\infty x(k) \text{ with } K_\infty = -(B^T P_\infty B + R)^{-1} B^T P_\infty A$$

It can be proved that this control law is stabilizing by defining the Lyapunov function as $V(x(k)) = x(k)^T P_\infty x(k)$.

The close relationship between LQ and MPC can be used to derive stability properties of MPC based on the well-known LQ properties, as shown in [27]. However, there are some differences between the two methods, the main one being that LQ does not take constraints into account. Also the cost function, although similar, is not exactly the same, since MPC uses increments in the control actions and LQ weights the control actions. The predictive control problem can be put in the standard LQ framework using the incremental state space model of Equation (2.8) with $\bar{x}(t) = [x(t) \; u(t-1)]^T$. In this case, the new error weighting matrix is:

$$\overline{Q} = \begin{bmatrix} Q & 0 \\ 0 & 0 \end{bmatrix}$$

Notice that the control weight R_k remains unchanged but now it has the meaning of weights on the control increments. The concept of control horizon does not exist in LQ but can easily be added setting $R_k = \infty$ for $k \geq N_u$.

References

1. T. Alamo, D.R. Ramirez, and E.F. Camacho. Efficient Implementation of Min-Max Model Predictive Control with Bounded Uncertainties. In *Preprints IEEE CCA'02. Glasgow. UK.*, 2002.
2. P. Albertos and R. Ortega. On Generalized Predictive Control: Two Alternative Formulations. *Automatica*, 25(5):753–755, 1989.
3. F. Allgöwer, T.A. Badgwell, J.S. Qin, J.B. Rawlings, and S.J. Wright. *Advances in Control (Highlights of ECC 99)*, chapter Nonlinear Predictive Control and Moving Horizon Estimation - An Introductory Overview. Springer, 1999.
4. A.Y. Allidina and F.M. Hughes. Generalised Self-tuning Controller with Pole Assignment. *Proceedings IEE, Part D*, 127:13–18, 1980.
5. J.C. Allwright. *Advances in Model-Based Predictive Control*, chapter On min-max Model-Based Predictive Control. Oxford University Press, 1994.
6. T. Alvarez and C. Prada. Handling Infeasibility in Predictive Control. *Computers and Chemical Engineering*, 21:577–582, 1997.
7. T. Alvarez, M. Sanzo, and C. Prada. Identification and Constrained Multivariable Predictive Control of Chemical Reactors. In 4th *IEEE Conference on Control Applications, Albany*, pages 663–664, 1995.
8. P. Ansay and V. Wertz. Model Uncertainties in GPC: A Systematic Two-step Design. In *Proceedings of the 3rd European Control Conference, Brussels*, 1997.
9. M.R. Arahal, M. Berenguel, and E.F. Camacho. Neural Identification Applied to Predictive Control of a Solar Plant. *Control Engineering Practice*, 6:333–344, 1998.
10. L.V.R. Arruda, R. Lüders, W.C. Amaral, and F.A.C Gomide. An Object-oriented Environment for Control Systems in Oil Industry. In *Proceedings of the 3rd Conference on Control Applications, Glasgow, UK*, pages 1353–1358, 1994.
11. K.J. Aström and B. Wittenmark. *Computer Controlled Systems. Theory and Design*. Prentice-Hall. Englewood Cliffs, NJ, 1984.
12. K.J. Aström and B. Wittenmark. *Adaptive Control*. Addison-Wesley, 1989.
13. R. Babuska, J. Oosterhoff, A. Oudshoorn, and P.M. Brujin. Fuzzy Self-tuning PI Control of pH in Fermentation. *Engineering Applications of Artificial Intelligence*, 15:3–15, 2002.
14. T.A. Badgwell and S.J. Qin. *Nonlinear Predictive Control*, chapter Review of Nonlinear Model Predictive Control Applications. IEE Control Engineering series, 2001.
15. M.S. Bazaraa and C.M. Shetty. *Nonlinear Programming*. Wiley, 1979.

16. R. Bellman and S.E. Dreyfus. *Applied Dynamic Programming*. Princeton University Press, Princeton, NJ, 1962.

17. A. Bemporad. Reducing Conservativeness in Predictive Control of Constrained Systems with Disturbances. In *IEEE Conference on Decision and Control*, 1998.

18. A. Bemporad, F. Borrelli, and M. Morari. Explicit Solution of LP-based Model Predictive Control. In 39th *IEEE Conference on Decision and Control, Sydney, Australia*, 2000.

19. A. Bemporad, F. Borrelli, and M. Morari. Robust Model Predictive Control: Piecewise Linear Explicit Solution. In *Proc. European Control Conference, ECC'01*, August 31 - September 3 2001.

20. A. Bemporad, F. Borrelli, and M. Morari. Model Predictive Control Based on Linear ProgrammingThe Explicit Solution. *IEEE Trans. on Automatic Control*, 47(12):1974–1985, 2002.

21. A. Bemporad, F. Borrelli, and M. Morari. Min-max Control of Constrained Uncertain Discrete-time Linear Systems. *IEEE Tran. on Automatic Control*, 48(9):1600 – 1606, 2003.

22. A. Bemporad and M. Morari. Control of Systems Integrating Logic, Dynamics and Constraints. *Automatica*, 35(3):407–427, 1999.

23. A. Bemporad, M. Morari, V. Dua, and E.N. Pistikopoulos. The Explicit Linear Quadratic Regulator for Constrained Systems. *Automatica*, 38(1):3–20, 2002.

24. M. Berenguel, M.R. Arahal, and E.F. Camacho. Modelling Free Response of a Solar Plant for Predictive Control. *Control Engineering Practice*, 6:1257–1266, 1998.

25. M. Berenguel, E.F. Camacho, and F.R. Rubio. *Simulation Software Package for the Acurex Field*. Departamento de Ingeniería de Sistemas y Automática, ESI Sevilla (Spain), Internal Report, 1994.

26. L. T. Biegler. *Nonlinear Model Predictive Control*, chapter Efficient Solution of Dynamic Optimization and NMPC Problems. Birkhäuser, 2000.

27. R.R. Bitmead, M. Gevers, and V. Wertz. *Adaptive Optimal Control. The Thinking Man's GPC*. Prentice-Hall, 1990.

28. H.H.J. Bloemen, T.J.J. van den Boom, and H.B. Verbruggen. Model-based Predictive Control for Hammerstein systems. In *Proceedings of the 39th IEEE Conference on Decision and Control. Sydney, Australia*, 2000.

29. M. Bodson and S.C. Douglas. Adaptive Algorithms for the Rejection of Sinusoidal Disturbances with Unknown Frequency. *Automatica*, 33(12):2213–2221, 1997.

30. C. Bordons. *Control Predictivo Generalizado de Procesos Industriales: Formulaciones Aproximadas*. PhD thesis, Universidad de Sevilla, 1994.

31. C. Bordons and J.R. Cueli. Modelling and Predictive Control of an Olive Oil Mill. In *Proceedings European Control Conference, Porto*, September 2001.

32. C. Bordons and J.R. Cueli. Predictive Controller with Estimation of Measurable Disturbances. Application to an Olive Oil Mill. *Journal of Process Control*, 14(3):305–315, 2004.

33. C. Bordons and F. Dorado. Non-linear Models for a Gypsum Kiln. A Comparative Analysis. In *Proceedings of the IFAC Triennal World Congress*, Barcelona, Spain, 2002.

34. F. Borrelli, A. Bemporad, and M. Morari. A Geometric Algorithm for Multi-Parametric Linear Programming. *Journal of Optimization Theory and Applications*, 118(3):515–540, 2003.

35. S. Boyd, L. El Ghaouni, E. Feron, and V. Balakrishnan. *Linear Matrix Inequalities in Systems and Control Theory*. SIAM Books, 1994.

36. E.H. Bristol. On a New Measure of Interaction for Multivariable Process Control. *IEEE Trans. on Automatic Control*, 11(1):133–4, 1966.
37. A.E. Bryson. *Control of Spacecraft and Aircraft*. Princeton University Press, NJ, 1994.
38. E.F. Camacho. Constrained Generalized Predictive Control. *IEEE Trans. on Automatic Control*, 38(2):327–332, 1993.
39. E.F. Camacho and M. Berenguel. *Advances in Model-Based Predictive Control*, chapter Application of Generalized Predictive Control to a Solar Power Plant. Oxford University Press, 1994.
40. E.F. Camacho, M. Berenguel, and C. Bordons. Adaptive Generalized Predictive Control of a Distributed Collector Field. *IEEE Trans. on Control Systems Technology*, 2:462–468, 1994.
41. E.F. Camacho, M. Berenguel, and F.R. Rubio. Application of a Gain Scheduling Generalized Predictive Controller to a Solar Power Plant. *Control Engineering Practice*, 2(2):227–238, 1994.
42. E.F. Camacho, M. Berenguel, and F.R. Rubio. *Advanced Control of Solar Power Plants*. Springer-Verlag, London, 1997.
43. E.F. Camacho and C. Bordons. Implementation of Self Tuning Generalized Predictive Controllers for the Process Industry. *Int. Journal of Adaptive Control and Signal Processing*, 7:63–73, 1993.
44. E.F. Camacho and C. Bordons. *Model Predictive Control in the Process Industry*. Springer-Verlag, 1995.
45. E.F. Camacho, F.R. Rubio, and F.M. Hughes. Self-tuning Control of a Solar Power Plant with a Distributed Collector Field. *IEEE Control Systems Magazine*, 2(2):72–78, 1992.
46. R.G. Cameron. The Design of Multivarible Systems. In *V Curso de Automática en la Industria*, La Rábida, Huelva, Spain, 1985.
47. P.J. Campo and M. Morari. Robust Model Predictive Control. In *American Control Conference, Minneapolis, Minnesota*, 1987.
48. H. Chen and F. Allgöwer. A Quasi-Infinite Horizon Nonlinear Model Predictive Control Scheme for Constrained Nonlinear Systems. In *Proceedings 16^{t}h Chinese Control Conference*, Qindao, 1996.
49. H. Chen and F. Allgöwer. A Quasi-infinite Horizon Nonlinear Predictive Control Scheme with Guaranteed Stability. *Automatica*, 34(10):1205–1218, 1998.
50. T.L. Chia and C.B. Brosilow. Modular Multivariable Control of a Fractionator. *Hydrocarbon Processing*, pages 61–66, 1991.
51. Y. Chikkula, J.H. Lee, and B. Ogunnaike. Dynamic Scheduled Model Predictive Control Using Hinging Hyperplane Models. *AIChE Journal*, 44:1691–1724, 1998.
52. L. Chisci, J. A. Rossiter, and G. Zappa. Systems with persistent disturbances: Predictive control with restricted constraints. *Automatica*, 37:1019–1028, 2001.
53. C.M. Chow and D.W. Clarke. *Advances in Model-Based Predictive Control*, chapter Actuator nonlinearities in predictive control. Oxford University Press, 1994.
54. D.W. Clarke. Application of Generalized Predictive Control to Industrial Processes. *IEEE Control Systems Magazine*, 122:49–55, 1988.
55. D.W. Clarke and P.J. Gawthrop. Self-tuning Controller. *Proceedings IEE*, 122:929–934, 1975.
56. D.W. Clarke and P.J. Gawthrop. Self-tuning Control. *Proceedings IEEE*, 123:633–640, 1979.
57. D.W. Clarke and C. Mohtadi. Properties of Generalized Predictive Control. *Automatica*, 25(6):859–875, 1989.

58. D.W. Clarke, C. Mohtadi, and P.S. Tuffs. Generalized Predictive Control. Part I. The Basic Algorithm. *Automatica*, 23(2):137–148, 1987.
59. D.W. Clarke, C. Mohtadi, and P.S. Tuffs. Generalized Predictive Control. Part II. Extensions and Interpretations. *Automatica*, 23(2):149–160, 1987.
60. D.W. Clarke, E. Mosca, and R. Scattolini. Robustness of an Adaptive Predictive Controller. In *Proceedings of the 30th Conference on Decision and Control*, pages 979–984, Brighton, England, 1991.
61. D.W. Clarke and R. Scattolini. Constrained Receding-horizon Predictive Control. *Proceedings IEE*, 138(4):347–354, july 1991.
62. C.R. Cutler and B.C. Ramaker. Dynamic Matrix Control- A Computer Control Algorithm. In *Automatic Control Conference, San Francisco*, 1980.
63. P.B. Deshpande and R.H. Ash. *Elements of Computer Process Control*. ISA, 1981.
64. F.J. Doyle, R.K. Pearson, and B.A. Ogunnaike. *Identification and Control Using Volterra Models*. Springer, 2001.
65. J.C. Doyle and G. Stein. Multivariable Feedback Design: Concepts for a Classical/Modern Synthesis. *IEEE Trans. on Automatic Control*, 36(1):4–16, 1981.
66. G. Ferretti, C. Manffezzoni, and R. Scattolini. Recursive Estimation of Time Delay in Sampled Systems. *Automatica*, 27(4):653–661, 1991.
67. R. Findeisen, L. Imsland, F. Allgöwer, and B.A. Foss. State and Output Nonlinear Model Predictive Control: An Overview. *European Journal of Control*, 9:190–206, 2003.
68. C.A. Floudas. *Non-Linear and Mixed Integer Optimization*. Oxford Academic Press, 1995.
69. Y.K. Foo and Y.C. Soh. Robust Stability Bounds for Systems with Structured and Unstructured Perturbations. *IEEE Trans. on Automatic Control*, 38(7), 1993.
70. J.B. Froisy and T. Matsko. IDCOM-M Application to the Shell Fundamental Control Problem. In *AIChE Annual Meeting*, 1990.
71. T. Gal. *Postoptimal Analyses, Parametric Programming and Related Topics*. McGraw-Hill, 1979.
72. T. Gal and J. Nedona. Multi-parametric Linear Programming. *Management Science*, 18:406–422, 1972.
73. C.E. García, D.M. Prett, and M. Morari. Model Predictive Control: Theory and Practice-a Survey. *Automatica*, 25(3):335–348, 1989.
74. W. García-Gabín. *Control Predictivo Multivariable para Sistemas de Ceros de Transmisión en el Semiplano Derecho*. PhD thesis, Universidad de Sevilla, 2002.
75. W. García-Gabín and E.F. Camacho. Application of Multivariable GPC to a Four Tank Process with Unstable Transmission Zeros. In *Conference on Control Applications*, Glasgow, Scotland, Sep 2002.
76. W. García-Gabín, E.F. Camacho, and D. Zambrano. Multivariable Model Predictive Control of Process with Unstable Transmission Zeros. In *American Control Conference*, Anchorage, AK, May 2002.
77. J. Gómez Ortega and E.F. Camacho. Neural Network MBPC for Mobile Robots Path Tracking. *Robotics and Computer Integrated Manufacturing Journal*, 11(4):271–278, December 1994.
78. J. Gómez Ortega and E.F. Camacho. Mobile Robot Navigation in a Partially Structured Environment using Neural Predictive Control. *Control Engineering Practice*, 4:1669–1679, 1996.
79. J. Gómez Ortega and E.F. Camacho. Neural Predictive Control for Mobile Robot Navigation in a Partially Structured Static Environment. In *Proceedings of the 13th IFAC World Congress, San Francisco, CA*, june 1996.

80. G. Goodwin and K. Sin. *Adaptive Filtering, Predicition and Control*. Prentice-Hall, 1984.
81. F. Gordillo and F.R. Rubio. Self-tuning Controller with LQG/LTR Structure. In *Proceedings 1st European Control Conference, Grenoble*, pages 2159–2163, july 1991.
82. J.R. Gossner, B. Kouvaritakis, and J.A. Rossiter. Stable Generalized Predictive Control with Constraints and Bounded Disturbances. *Automatica*, 33:551–568, 1997.
83. C. Greco, G. Menga, E. Mosca, and G. Zappa. Performance Improvement of Self Tuning Controllers by Multistep Horizons: The MUSMAR Approach. *Automatica*, 20:681–700, 1984.
84. M.J. Grimble. Generalized Predictive Optimal Control: an Introduction to the Advantages and Limitations. *International Journal of Systems Science*, 23(1):85–98, 1992.
85. P. Grosdidier, J.B. Froisy, and M. Hamman. *IFAC Workshop on Model Based Process Control*, chapter The IDCOM-M Controller. Pergamon Press, Oxford, 1988.
86. R. Haber, R. Bars, and O. Lengvel. Long-range Predictive Control of the Parametric Hammerstein Model. In *Proceedings of the IFAC NOLCOS'98*, pages 434–439, Enschede, The Netherlands, 1998.
87. W.P.M.H. Heemels, B. De Schutter, and A. Bemporad. Equivalence of Hybrid Dynamical Models. *Automatica*, 37(7):1085–1091, 2001.
88. M. Henson. Nonlinear Model Predictive Control: Current Status and Future Directions. *Computers and Chemical Engineering*, pages 187–202, 1998.
89. M. Henson and D.E. Seborg. Adaptive Nonlinear Control of a pH Neutralization Process. *IEEE Transactions on Control Systems Technology*, 2(3):169–182, 1994.
90. E. Hernandez and Y. Arkun. A Nonlinear DMC Controller: Some Modeling and Robustness Considerations. In *Proceedings of the American Control Conference*, Boston, MA, 1991.
91. F.J. Hill and G.R. Peterson. *Introduction to switching theory and logical design*. John Wiley and Sons, Inc., 1968.
92. K. Hornik, M. Stinchombe, and H. White. Multilayer Feedforward Networks are Universal Approximators. *Neural networks*, pages 359–366, 1989.
93. B. Hu and A. Linnemann. Towards Infinite-horizon Optimality in Nonlinear Model Predictive Control. *IEEE Transactions on Automatic Control*, 47(4):679–682, 2002.
94. Y. K. Hwang and N. Ahuja. A Potential Field Approach to Path Planning. *IEEE Transactions on Robotics and Automation*, 8(1):23–32, February 1992.
95. R. Isermann. *Digital Control Systems*. Springer-Verlag, 1981.
96. A. Jadbabaie, J. Yu, and J. Hauser. Unconstrained Receding-Horizon Control of Nonlinear Systems. *IEEE Transactions on Automatic Control*, 46(5):776–783, 2001.
97. J. Jiang. Optimal Gain Scheduling Controller for a Diesel Engine. *IEEE Control Systems Magazine*, pages 42–48, 1994.
98. Y. Jin, X. Sun, and C. Fang. Adaptive Control of Bilinear Systems with Bounded Disturbances, and its Application. *Control Engineering Practice*, pages 815–822, 1996.
99. T.A. Johansen, J.T. Evans, and B.A. Foss. Identification of Nonlinear System Structure and Parameters Using Regime Decomposition. *Automatica*, pages 321–326, 1995.
100. K.H. Johansson. The Quadruple-Tank Process: A Multivariable Laboratory Process with an Adjustable Zero. *IEEE Transaction on Control Systems Technology*, 8(3):456–465, May 2000.

101. A. Kalt. Distributed Collector System Plant Construction Report. *IEA/SSPS Operating Agent DFVLR, Cologne*, 1982.
102. M.R. Katebi and M.A. Johnson. Predictive Control Design for Large Scale Systems. In *IFAC Conference on Integrated System Engineering*, pages 17–22, Baden-Baden, Germany, 1994.
103. S.S. Keerthi and E.G. Gilbert. Optimal Infinite-horizon Feedback Laws for a General Class of Constrained Discrete-time Systems: Stability and Moving-horizon Approximations. *J. Optim. Theory Appl.*, 57(2):265–293, 1988.
104. E. Kerrigan. *Robust Constraint Satisfaction: Invariant Sets and Predictive Control*. PhD thesis, University of Cambridge, 2000.
105. R.M.C. De Keyser. Basic Principles of Model Based Predictive Control. In 1st *European Control Conference, Grenoble*, pages 1753–1758, july 1991.
106. R.M.C. De Keyser. A Gentle Introduction to Model Based Predictive Control. In *PADI2 International Conference on Control Engineering and Signal Processing, Piura, Peru*, 1998.
107. R.M.C. De Keyser and A.R. Van Cuawenberghe. Extended Prediction Self-adaptive Control. In *IFAC Symposium on Identification and System Parameter Estimation, York,UK*, pages 1317–1322, 1985.
108. R.M.C. De Keyser, Ph.G.A. Van de Velde, and F.G.A. Dumortier. A Comparative Study of Self-adaptive Long-range Predictive Control Methods. *Automatica*, 24(2):149–163, 1988.
109. M.V. Kothare, V. Balakrishnan, and M. Morari. Robust Constrained Predictive Control using Linear Matrix Inequalities . *Automatica*, 32:1361–1379, 1996.
110. B. Kouvaritakis, M. Cannon, and J.A. Rossiter. Stability, Feasibility, Optimality and the Number of Degrees of Freedom in Constrained Predictive Control. In *Symposium on Non-linear Predictive Control*. Ascona, Switzerland, 1998.
111. B. Kouvaritakis, M. Cannon, and J.A. Rossiter. Non-linear Model Based Predictive Control. *International Journal of Control*, 72(10):919–928, 1999.
112. B. Kouvaritakis, J.A. Rossiter, and A.O.T Chang. Stable Generalized Predictive Control: An Algorithm with Guaranteed Stability. *Proceedings IEE, Part D*, 139(4):349–362, 1992.
113. K. Krämer and H. Ubehauen. Predictive Adaptive Control. Comparison of Main Algorithms. In *Proceedings 1st European Control Conference, Grenoble*, pages 327–332, julio 1991.
114. A.G. Kutnetsov and D.W. Clarke. *Advances in Model-Based Predictive Control*, chapter Application of Constrained GPC for Improving Performance of Controlled Plants. Oxford University Press, 1994.
115. W.H. Kwon and A.E. Pearson. On Feedback Stabilization of Time-Varying Discrete Linear Systems. *IEEE Trans. on Automatic Control*, 23:479–481, 1979.
116. J.H. Lee, M. Morari, and C.E. García. State-space Interpretation of Model Predictive Control. *Automatica*, 30(4):707–717, 1994.
117. J.H. Lee and Z. Yu. Worst-case Formulations of Model Predictive Control for Systems with Bounded Parameters. *Automatica*, 33(5):763–781, 1997.
118. M.A. Lelic and P.E. Wellstead. Generalized Pole Placement Self Tuning Controller. Part 1. Basic Algorithm. *International J. of Control*, 46(2):547–568, 1987.
119. M.A. Lelic and M.B. Zarrop. Generalized Pole Placement Self Tuning Controller. Part 2. Application to Robot Manipulator Control. *International J. of Control*, 46(2):569–601, 1987.
120. C.E. Lemke. *Mathematics of the Decision Sciences*, chapter On Complementary Pivot Theory. G.B. Dantzig and A.F. Veinott (Eds.), 1968.

121. J.M. Lemos and E. Mosca. A Multipredictor-based LQ Self-tuning Controller. In *IFAC Symp. on Identification and System Parameter Estimation, York, UK*, pages 137–141, 1985.
122. W. S. Levine. *The Control Handbook*, chapter Volterra and Fliess Series Expansions for Nonlinear Systems. CRC/IEEE Press, 1996.
123. W.S. Levine. *The Control Handbook*. IEEE Press, Boca de Ratón, 1996.
124. D. Limon, T. Alamo, and E.F. Camacho. Enlarging the domain of attraction of MPC controller using invariant sets. In *Proceedings of the IFAC World Congress*, 2002.
125. D. Limon, T. Alamo, and E.F. Camacho. Input-to-state Stable MPC for Constrained Discrete-time Nonlinear Systems with Bounded Additive Uncertainties. In *IEEE Conference on Decision and Control*, 2002.
126. D. Limon, T. Alamo, and E.F. Camacho. Stable Constrained MPC without Terminal Constraint. In *Proceedings of the American Control Conference*, 2003.
127. D.A. Linkers and M. Mahfonf. *Advances in Model-Based Predictive Control*, chapter Generalized Predictive Control in Clinical Anaesthesia. Oxford University Press, 1994.
128. L. Ljung. *System Identification. Theory for the user*. Prentice-Hall, 1987.
129. D.E. Luenberger. *Linear and Nonlinear Programming*. Addison-Wesley, 1984.
130. J. Lunze. *Robust Multivariable Feedback Control*. Prentice-Hall, 1988.
131. J.M. Maciejowski. *Predictive Control with Constraints*. Prentice Hall, Harlow, 2001.
132. L. Magni, G. De Nicolao, L. Magnani, and R. Scattolini. A Stabilizing Model-based Predictive Control Algorithm for Nonlinear Systems. *Automatica*, 37:1351–1362, 2001.
133. B.R. Maner, J.C. Doyle, B.A. Ogunnaike, and R.K. Pearson. Nonlinear Model Predictive Control of a Multivariable Polymerization Reactor unsing Second Order Volterra Models. *Automatica*, 32:1285–1302, 1996.
134. J.M. Martin-Sanchez and J. Rodellar. *Adaptive Predictive Control. From the concepts to plant optimization*. Prentice -Hall International (UK), 1996.
135. D.Q. Mayne and H. Michalska. Receding Horizon Control of Nonlinear Systems. *IEEE Trans. on Automatic Control*, 35:814–824, 1990.
136. D.Q. Mayne and H. Michalska. Robust Receding Horizon Control of Constrained Nonlinear Systems. *IEEE Trans. on Automatic Control*, 38:1623–1633, 1993.
137. D.Q. Mayne, J.B. Rawlings, C.V. Rao, and P.O.M. Scokaert. Constrained Model Predictive Control: Stability and Optimality. *Automatica*, 36:789–814, 2000.
138. H. Michalska and D.Q. Mayne. Robust receding horizon control of constrained nonlinear systems. *IEEE Trans. on Automatic Control*, 38(11):1623–1633, 1993.
139. C. Mohtadi, S.L. Shah, and D.G. Fisher. Frequency Response Characteristics of MIMO GPC. In *Proceedings 1st European Control Conference, Grenoble*, pages 1845–1850, july 1991.
140. M. Morari. *Advances in Model-Based Predictive Control*, chapter Model Predictive Control: Multivariable Control Technique of Choice in the 1990s? Oxford University Press, 1994.
141. M. Morari and E. Zafiriou. *Robust Process Control*. Prentice-Hall, 1989.
142. E. Mosca. *Optimal, Predictive and Adaptive Control*. Prentice Hall, 1995.
143. E. Mosca, J.M. Lemos, and J. Zhang. Stabilizing I/O Receding Horizon Control. In *IEEE Conference on Decision and Control*, 1990.
144. E. Mosca and J. Zhang. Stable Redesign of Predictive Control. *Automatica*, 28:1229–1233, 1992.

145. D. Muñoz, T. Alamo, and E.F. Camacho. Explicit Min-max Model Predictive Control. *Internal Report GAR 2003/03. University of Seville*, 2003.

146. K.R. Muske, E.S. Meadows, and J.B. Rawlings. The Stability of Constrained Receding Horizon Control with State Estimation. In *Proceedings of the American Control Conference*, pages 2837–2841, Baltimore, MD, 1994.

147. K.R. Muske and J. Rawlings. Model Predictive Control with Linear Models. *AIChE Journal*, 39:262–287, 1993.

148. O. Nelles. *Nonlinear System Identification*. Springer, 2001.

149. R.B. Newell and P.L. Lee. *Applied Process Control. A Case Study*. Prentice-Hall, 1989.

150. G. De Nicolao, L. Magni, and R. Scattolini. *Nonlinear Model Predictive Control*, chapter Nonlinear Receding Horizon Control of Internal Combustion Engines. Birkhäuser, 2000.

151. G. De Nicolao and R. Scattolini. *Advances in Model-Based Predictive Control*, chapter Stability and Output Terminal Constraints in Predictive Control. Oxford University Press, 1994.

152. M. Norgaard, O. Ravn, N.K. Poulsen, and L.K. Hansen. *Neural Networks for Modelling and Control of Dynamic Systems*. Springer, London, 2000.

153. J.E. Normey, E.F. Camacho, and C. Bordons. Robustness Analysis of Generalized Predictive Controllers for Industrial Processes. In *Proceedings of the 2nd Portuguese Conference on Automatic Control*, pages 309–314, Porto, Portugal, 1996.

154. J.E. Normey-Rico, C. Bordons, and E.F. Camacho. Improving the Robustness of Dead-Time Compensating PI Controllers. *Control Engineering Practice*, 5(6):801–810, 1997.

155. J.E. Normey-Rico and E.F. Camacho. A Smith Predictor Based Generalized Predictive Controller. *Internal Report GAR 1996/02. University of Sevilla*, 1996.

156. J.E. Normey-Rico and E.F. Camacho. Robustness Effect of a Prefilter in Generalized Predictive Control. *IEE Proc. on Control Theory and Applications*, 146:179–185, 1999.

157. J.E. Normey-Rico, J. Gómez-Ortega, and E.F. Camacho. A Smith Predictor Based Generalized Predictive Controller for Mobile Robot Path-Tracking. In 3rd *IFAC Symposium on Intelligent Autonomous Vehicles*, pages 471–476, Madrid, Spain, 1998.

158. S. Norquay, A. Palazoglu, and J.A. Romagnoli. Application of Wiener Model Predictive Control (WMPC) to a pH Neutralization Experiment. *IEEE Transactions on Control Systems Technology*, 7(4):437–445, 1999.

159. M. Ohshima, I. Hshimoto, T. Takamatsu, and H. Ohno. Robust Stability of Model Predictive Control. *International Chemical Engineering*, 31(1), 1991.

160. A.W. Ordys and D.W. Clarke. A State-space Description for GPC Controllers. *International Journal of System Science*, 24(9):1727–1744, 1993.

161. A.W. Ordys and M.J. Grimble. *Advances in Model-Based Predictive Control*, chapter Evaluation of Stochastic Characteristics for a Constrained GPC Algorithm. Oxford University Press, 1994.

162. G.C. Papavasilicu and J.C. Allwright. A Descendent Algorithm for a Min-Max Problem in Model Predictive Control. In *Proceedings of the 30th Conference on Decision and Control*, Brighton, UK, 1991.

163. T. Parisini and R. Zoppoli. A Receding-Horizon Regulator for Nonlinear Systems and a Neural Approximation. *Automatica*, 31(10):1443–1451, 1995.

164. D. Peña, E.F. Camacho, and S. Pinón. Hybrid Systems for Solving Model Predictive Control of Piecewise Affine System. In *Proceedings IFAC Conference on Analysis and Design of Hybrid Systems ADHS'03. Saint-Malo (France)*, pages 76–81, 2003.

165. V. Peterka. Predictor-based Self-tuning Control. *Automatica*, 20(1):39–50, 1984.

166. S. Piche, B. Sayyar-Rodsari, D. Johnson, and M. Gerules. Nonlinear Model Predictive Control Using Neural Networks. *IEEE Control Systems Magazine*, 20(3):53–62, 2000.

167. D.M. Prett and M. Morari. *Shell Process Control Workshop*. Butterworths, 1987.

168. D.M. Prett and R.D. Morari. Optimization and Constrained Multivariable Control of a Catalytic Cracking Unit. In *Proceedings of the Joint Automatic Control Conference.*, 1980.

169. A.I. Propoi. Use of LP Methods for Synthesizing Sampled-data Automatic Systems. *Automatic Remote Control*, 24, 1963.

170. S.J. Qin and T.A. Badgwell. An Overview of Industrial Model Predictive Control Technology. In Chemical Process Control: Assessment and New Directions for Research. In *AIChE Symposium Series 316, 93. Jeffrey C. Kantor, Carlos E. Garcia and Brice Carnahan Eds. 232-256*, 1997.

171. S.J. Qin and T.A. Badgwell. An Overview of Nonlinear Model Predictive Control Applications. In *IFAC Workshop on Nonlinear Model Predictive Control. Assessment and Future Directions. Ascona (Switzerland)*, 1998.

172. J.M. Quero and E.F. Camacho. Neural Generalized Predictive Controllers. In *Proc. IEEE International Conference on System Engineering, Pittsburg, PA*, 1990.

173. D.R. Ramírez and E.F. Camacho. On the piecewise linear nature of Min-Max Model Predictive Control with bounded uncertainties. In *Proc. 40th Conference on Decision and Control, CDC'2001*, December, 4-7 2001.

174. D.R. Ramirez and E.F. Camacho. On the Piecewise Linear Nature of Min-max Model Predictive Control with Bounded Global Uncertainties. In *Proc. 40th IEEE Conference on Decision and Control CDC'01, Orlando FL*, 2001.

175. D.R. Ramirez and E.F. Camacho. On the Piecewise Linear Nature of Constrained Min-max Model Predictive Control with Bounded Global Uncertainties. In *Proc. of the American Control Conference ACC'03, Denver, CO*, 2003.

176. D.R. Ramírez, E.F. Camacho, and M. R. Arahal. *Proc. of the IFAC World Congress, B'02, Editors: E.F. Camacho, L. Basanez and J.A. de la Puente*, chapter Implementation of Min-Max MPC Using Hinging Hyperplanes. Application to a Heat Exchanger. Elsevier Science, 2002.

177. J. Rawlings and K. Muske. The Stability of Constrained Receding Horizon Control. *IEEE Trans. on Automatic Control*, 38:1512–1516, 1993.

178. J. Richalet. *Practique de la commande predictive*. Hermes, 1992.

179. J. Richalet. Industrial Applications of Model Based Predictive Control. *Automatica*, 29(5):1251–1274, 1993.

180. J. Richalet, S. Abu el Ata-Doss, C. Arber, H.B. Kuntze, A. Jacubash, and W. Schill. Predictive Functional Control. Application to Fast and Accurate Robots. In *Proc. 10th IFAC Congress, Munich*, 1987.

181. J. Richalet, A. Rault, J.L. Testud, and J. Papon. Algorithmic Control of Industrial Processes. In 4th IFAC *Symposium on Identification and System Parameter Estimation. Tbilisi USSR*, 1976.

182. J. Richalet, A. Rault, J.L. Testud, and J. Papon. Model Predictive Heuristic Control: Application to Industrial Processes. *Automatica*, 14(2):413–428, 1978.

183. B.D. Robinson and D.W. Clarke. Robustness Effects of a Prefilter in Generalized Predictive Control. *Proceedings IEE, Part D*, 138:2–8, 1991.

184. A. Rossiter, J.R. Gossner, and B. Kouvaritakis. Infinite Horizon Stable Predictive Control. *IEEE Trans. on Automatic Control*, 41(10), 1996.

185. J.A. Rossiter and B. Kouvaritakis. Constrained Stable Generalized Predictive Control. *Proceedings IEE, Part D*, 140(4), 1993.

186. J.A. Rossiter and B. Kouvaritakis. *Advances in Model-Based Predictive Control*, chapter Advances in Generalized and Constrained Predictive Control. Oxford University Press, 1994.

187. R. Rouhani and R.K. Mehra. Model Algorithmic Control: Basic Theoretical Properties. *Automatica*, 18(4):401–414, 1982.

188. P.O.M. Scokaert and D.W. Clarke. *Advances in Model-Based Predictive Control*, chapter Stability and Feasibility in Coinstrained Predictive Control. Oxford University Press, 1994.

189. P.O.M. Scokaert and D.Q. Mayne. Min-max feedback model predictive control for constrained linear systems. *IEEE Transactions on Automatic Control*, 43(8):1136–1142, 1998.

190. P.O.M. Scokaert, D.Q. Mayne, and J.B. Rawlings. Suboptimal model predictive control (feasibility implies stability). *IEEE Transactions on Automatic Control*, 44(3):648–654, 1999.

191. S.L. Shah, C. Mohtadi, and D.W. Clarke. Multivariable Adaptive Control without a Prior Knowledge of the Delay Matrix. *Systems and Control Letters*, 9:295–306, 1987.

192. I. Skrjanc and D. Matko. *Advances in Model-Based Predictive Control*, chapter Fuzzy Predictive Controller with Adaptive Gain. Oxford University Press, 1994.

193. O.J.M. Smith. Close Control of Loops with Deadtime. *Chemical Engineering Progress*, 53(5):217, 1957.

194. R. Söeterboek. *Predictive Control. A unified approach*. Prentice-Hall, 1992.

195. E.D. Sontag. Nonlinear Regulation: The Piecewise Linear Approach. *IEEE Trans. on Automatic Control*, 26(2):346–358, 1981.

196. E. Srinivasa and M. Chidambaram. Robust Control of a Distillation Column by the Method of Inequalities. *Journal of Process Control*, 1(3):171–176, 1993.

197. G.W. Stewart. *Introduction to Matrix Computations*. Academis Press, Inc., 1973.

198. Y. Tan and R. De Keyser. *Advances in Model-Based Predictive Control*, chapter Neural Network Based Predictive Control. Oxford University Press, 1994.

199. P. Tøndel and T.A. Johansen. *Proc. of the IFAC World Congress, B'02, Editors: E.F. Camacho, L. Basanez and J.A. de la Puente*, chapter Complexity Reduction in Explicit Linear Model Predictive Control. Elsevier Science, 2002.

200. S. Townsend and G.W. Irwin. *Nonlinear Predictive Control*, chapter Nonlinear Model Based Predictive Control Using Multiple Local Models. IEE Control Engineering series, 2001.

201. C.A. Tsiligiannis and S.A Svoronos. Multivariable Self-tuning Control via the Right Interactor Matrix. *IEE Trans. Aut. Control*, 31:987–989, 1986.

202. M.L. Tyler and M. Morari. Propositional Logic in Control and Monitoring Problems. *Technical Report AUT96-15, Institut fur Automatik, ETH- Swiss Federal Institute of Technology, Zurich, Switzerland*, 1996.

203. E.R. Velasco. *Control y Supervisión de Planta Piloto mediante Sistema de Control Distribuido*. P.F.C. Univesidad de Sevilla, 1994.

204. R.A.J. De Vries and H.B. Verbruggen. *Advances in Model-Based Predictive Control*, chapter Multivariable Unified Predictive Control. Oxford University Press, 1994.

205. P.E. Wellstead, D. Prager, and P. Zanker. A Pole Assignment Self Tuning Regulator. *Proceedings IEE, Part D*, 126:781–787, 1978.

206. T.H. Yang and E. Polak. Moving Horizon Control of Nonlinear Systems with Input Saturations, Disturbances and Plant Uncertainties. *Int. Journal of Control*, pages 875–903, 1993.

207. B.E. Ydstie. Extended Horizon Adaptive Control. In *Proc. 9th IFAC World Congress, Budapest, Hungary*, 1984.

208. T.W. Yoon and D.W. Clarke. Prefiltering in Receding-Horizon Predictive Control. *Internal Report 1995/93, University of Oxford, Department of Engineering Science*, 1993.

209. T.W. Yoon and D.W. Clarke. *Advances in Model-Based Predictive Control*, chapter Towards Robust Adaptive Predictive Control, pages 402–414. Oxford University Press, 1994.

210. T.W. Yoon and D.W. Clarke. Observer Design in Receding-Horizon Control. *International Journal of Control*, 2:151–171, 1995.

211. L.A. Zadeh and B.H. Whalen. On Optimal Control and Linear Programming. *IRE Trans. Automatic Control*, 7(4), 1962.

212. E. Zafiriou. Robust Model Predictive Control of Processes with Hard Constraints. *Computers and Chemical Engineering*, 14(4/5):359–371, 1990.

213. A. Zheng and M. Morari. Stability of Model Predictive Control with Soft Constraints. *Internal Report. California Institute of Technology*, 1994.

214. A. Zheng and W. Zhang. *Nonlinear Predictive Control*, chapter Computationally Efficient Nonlinear Model Predictive Control Algorithm for Control of Constrained Nonlinear Systems. IEE Control Engineering series, 2001.

215. Y. Zhu and T. Backx. *Identification of Multivariable Industrial Processes*. Springer-Verlag, 1993.

Index